Japanese-Russian Relations, 1907–2007

This book provides a comprehensive survey of Japanese-Russian relations from the end of the Russo-Japanese War until the present. Based on extensive original research in both Japanese and Russian sources, it traces the development of relations from the tumultuous pre-war period, through the Second World War, Cold War and post-Cold War periods. Considering the wider international situation, domestic influences and ideological factors throughout, it shows how the hopeful period of the late 1990s – when Japanese-Russian relations briefly ceased to be acrimonious, and it seemed that normal relations might be established – was not unique. Joseph P. Ferguson argues there have been several previous occasions when *rapprochement* seemed possible, which in the end proved elusive: *rapprochement* frequently becoming the victim of domestic factors which often worked against and took precedence over good relations. The book concludes with an assessment of the present situation and of how relations are likely to develop in the immediate future.

Joseph P. Ferguson is Vice President of the National Council for Eurasian and East European Research and Adjunct Professor at the University of Washington. He is also Associate Editor of the journal *Problems of Post-Communism*. He received his Ph.D. in International Relations from the John Hopkins University Nitze School of Advanced International Studies (SAIS).

Routledge Contemporary Japan Series

1 **A Japanese Company in Crisis**
Ideology, strategy, and narrative
Fiona Graham

2 **Japan's Foreign Aid**
Old continuities and new directions
Edited by David Arase

3 **Japanese Apologies for World War II**
A rhetorical study
Jane W. Yamazaki

4 **Linguistic Stereotyping and Minority Groups in Japan**
Nanette Gottlieb

5 **Shinkansen**
From bullet train to symbol of modern Japan
Christopher P. Hood

6 **Small Firms and Innovation Policy in Japan**
Edited by Cornelia Storz

7 **Cities, Autonomy and Decentralization in Japan**
Edited by Carola Hein and Philippe Pelletier

8 **The Changing Japanese Family**
Edited by Marcus Rebick and Ayumi Takenaka

9 **Adoption in Japan**
Comparing policies for children in need
Peter Hayes and Toshie Habu

10 **The Ethics of Aesthetics in Japanese Cinema and Literature**
Polygraphic desire
Nina Cornyetz

11 **Institutional and Technological Change in Japan's Economy**
Past and present
Edited by Janet Hunter and Cornelia Storz

12 **Political Reform in Japan**
Leadership looming large
Alisa Gaunder

13 **Civil Society and the Internet in Japan**
Isa Ducke

14 Japan's Contested War Memories
The 'memory rifts' in historical consciousness of World War II
Philip A. Seaton

15 Japanese Love Hotels
A cultural history
Sarah Chaplin

16 Population Decline and Ageing in Japan – The Social Consequences
Florian Coulmas

17 *Zainichi* Korean Identity and Ethnicity
David Chapman

18 A Japanese Joint Venture in the Pacific
Foreign bodies in tinned tuna
Kate Barclay

19 Japanese-Russian Relations, 1907–2007
Joseph P. Ferguson

20 War Memory, Nationalism and Education in Post-War Japan, 1945–2007
The Japanese history textbook controversy and Ienaga Saburo's court challenges
Yoshiko Nozaki

Japanese-Russian Relations, 1907–2007

Joseph P. Ferguson

LONDON AND NEW YORK

First published 2008 by Routledge
2 Park Square, Milton Park, Abingdon, Oxon, OX14 4RN

Simultaneously published in the USA and Canada
by Routledge
270 Madison Avenue, New York, NY 10016

Routledge is an imprint of the Taylor & Francis Group, an informa business

Typeset in Times by
Pindar New Zealand
Printed and bound in Great Britain by Biddles Ltd, King's Lynn

British Library Cataloguing in Publication Data
A catalogue record for this book is available from the British Library

Library of Congress Cataloging in Publication Data
Ferguson, Joseph P.
Japanese-Russian relations: 1907–2007/Joseph P. Ferguson. –1st ed.
p. cm.–(Routledge contemporary Japan series)
Includes bibliographical references.
1. Russia–Foreign relations–Japan. 2. Japan–Foreign relations–Russia.
3. Soviet Union–Foreign relations–Japan. 4. Japan–Foreign relations–Soviet Union. 5. Russia (Federation)–Foreign relations–Japan.
6. Japan–Foreign relations–Russia (Federation) I. Title
DK68.7.J3F47 2008
327.5204709'04–dc22
2007038500

ISBN10: 0-415-45314-3 (hbk)
ISBN10: 0-203-92920-9 (ebk)

ISBN13: 978-0-415-45314-1 (hbk)
ISBN13: 978-0-203-92920-9 (ebk)

Contents

Abstract

The period 1996–2007 marked a time of tremendous anticipation in bilateral relations between Japan and Russia. For the first time in over two hundred years neither country viewed the other as a direct security threat. Bilateral political contacts of the highest level proliferated, highlighted by warm personal contacts between the leaders of both nations. Relations between the defense establishments of both nations sprouted seemingly overnight, perhaps even surpassing bilateral political relations in terms of atmospherics and good will. And by the end of the 1990s Japanese-Russian and international energy consortia in the Russian Far East actually began producing and exporting energy. For the first time since the end of the Second World War there was a palpable mood in both countries that the long-standing enmity could be put aside and that the relationship could finally be normalized. By 2007, however, it was clear the two nations were still far from settling the territorial dispute that had prevented a normalization of relations since the end of the Second World War. This study explores why the two nations have been unable to normalize relations. Japanese-Russian bilateral interaction is examined in the context of international structural forces, domestic politics, and ideational factors. The two nations have been brought together in brief periods of *rapprochement* throughout the twentieth century, due primarily to international structural factors, but domestic political and ideational factors have often hindered the relationship from advancing beyond a mere temporary dalliance of convenience. This study also explores how states react to change wrought by structural transformation, and how this affects their diplomacy. This study contributes to the literature on Japanese and Russian foreign policy making, and on the international relations of Northeast Asia in the post-Cold War era, by examining the different influences and factors behind the decision-making process of the dominant political elite in both countries, based on a review of Japanese and Russian language sources and interviews with a wide range of decision makers and experts in both countries.

Keywords

Japan, Russia, Foreign Policy, Diplomatic Relations, Northeast Asia

Acknowledgments

Much of the research that I have compiled is based on my experiences in Tokyo as a *Monbusho* Fellow at Aoyama Gakuin University in 1996–97, and in Moscow as a Fulbright Fellow at the Institute of World Economy and International Relations (Russian Academy of Sciences) from 1999 to 2001. In addition to the lengthy stays in each country I traveled back to Japan and Russia each year through to 2007, conducting research on each occasion. I was able to conduct interviews with experts in the fields, and with politicians and diplomats involved in the decision-making process of both countries. For the period prior to the Second World War I chose to do little archival research, as I was able to find relevant material dealing with this era in English, as well as in Japanese and Russian. For the post-Second World War and post-Cold War eras there is an abundance of written material (books, monographs, and magazines) available in all three languages.

In both Japan and Russia I traveled widely and was also able to meet with experts and officials from various regions that are affected by Japanese-Russian relations. In this way, I tried to avoid falling into the trap of relying on the mindset of the center, or the seats of government. Additionally, interviewing non-governmental experts, such as academics and journalists, provided me with additional insights into the policymaking realm. My travels and personal experiences in Japan and Russia have provided me with a unique opportunity to better understand the mentality and the psyche of the people of these two wonderful nations that are undergoing difficult transformations.

Thanks are due to innumerous people, and in this space it would perhaps be impossible to acknowledge them all, although I will try. Perhaps the two people I should acknowledge most of all are my two grandfathers, Harry Weidel Ferguson and Henry Hamilton Dewar. I owe them thanks first and foremost for having not only put me through college, but also through graduate school. Two self-made men that rose up through the hard days of the Depression and the Second World War, they managed to support and succor three generations of my family, both immediate members, and in some cases more distant relatives. They were intelligent, attentive, generous, and loving. Both men had a great interest in international affairs and traveled widely throughout their lifetimes. My maternal grandfather, Hal Dewar, had a particular interest in Russian culture and history. He traveled to the Soviet Union on several occasions, and amassed quite a collection of books on Russia.

He even had the opportunity to meet Nikita Khrushchev one evening in New York City in 1959. He wrote an interesting and humorous account of the meeting and of the speech Khrushchev delivered at a gathering of American business executives over dinner. I was fortunate to have received most of his collection (including the Khrushchev letter) when he passed away more than thirty years ago. He recognized my interest in history early on, and encouraged me to read and explore. He also frequently took me on excursions to the Alamo in San Antonio, and to the other Spanish missions of that fabulous city, where he was a long-time resident. My interest in Russia, no doubt, partially came through him. Both of my grandfathers would have been fascinated with my travels and experiences, and amazed at the changes that have transpired in the world that have allowed me to undertake the path that I have. No doubt they are watching with satisfied smiles.

Thanks need to be given to people in three different countries. First, the United States. My thanks go to my mentors, teachers, and advisors at the Paul H. Nitze School of Advanced International Studies (SAIS) of the Johns Hopkins University in Washington, DC. Thanks to Nathaniel Thayer and Michael Green for all they taught me about Japan and its political intricacies. Thanks are also due to both of them for encouraging me to apply for the Ph.D. program at SAIS, and to Nat for accepting me. Both Nat and Mike also provided me with excellent introductions across Japan that opened many hidden doors. George Packard served both as my Dean and my Professor of Japanese History, before he left SAIS to go to the US-Japan Foundation. Lyman Miller (now at the Hoover Institution of Stanford University) introduced me to the fascinating world of Chinese history and diplomacy in East Asia. Bruce Parrott taught me much about Russia and the Soviet Union, and was by far the most generous with his time. He was the leading light for me as I trudged through the process of completing this study. Eliot Cohen was perhaps the most inspiring teacher I have ever had, and I appreciate the time he took not only with this dissertation, but also the many papers that I wrote for him in the other fascinating classes that he taught us about international strategy and policy. He also took the time to take camping trips with his students to Civil War battlefields, and to visit modern defense facilities around the country. Karl Jackson also spent quality time with me in the early stages of my dissertation work, and read through it and gave me good advice and insight. I could not have done the research I did without the tremendous training provided to me by my Japanese teacher at SAIS, Meiko Inoue, and my Russian teachers there, Lyudmila Guslistova and Natasha Simes. After studying with them, getting by in Japanese and Russian almost seemed like a breeze. Gilbert Rozman of Princeton University was also on my dissertation committee, and without his knowledge of the subject matter, this work would have fallen far short of its intended aim. Gil provided me with great insight, not only through his writings over the years, but also through the many correspondences and meetings we had over the past several years. Gil also invited me to spend a year with him at Princeton as a post-doctoral scholar in 2004–05. At Princeton, I helped Gil edit two books on Japanese and Russian strategic thinking in Asia. All of these individuals have become not only great role models and mentors, but also good friends. Lastly, thanks to the Fulbright Commission for seeing to it that I was awarded a fellowship to spend almost two years in Russia doing research. To put it simply, it was a life-changing experience for me. The Fulbright representative in Moscow, Joe McCormick, had much to do with my being selected from among a wide field of candidates.

In Japan thanks foremost to the great people of Kochi Prefecture on Shikoku, where I spent my first extended time in that country (and in Asia for that matter) teaching English. Their hospitality, generosity, and friendship ensured that I would be spending a great deal of my future dealing with Japan in some capacity. My thanks to the Japanese Ministry of Education for paying my teaching salary and then, several years later, for paying me a stipend which allowed me to do research on Japanese-Russian relations for almost a year in 1996–97. During that time I was in residence at Aoyama Gakuin University in Tokyo.

At Aoyama Gakuin I learned a great deal under two professors, Shigeki Hakamada and the late Tomohisa Sakanaka. These two generous men introduced me to a number of influential people in Japan that were connected with Russia, including the late Ichirô Suetsugu, and numerous officials from the Foreign Ministry's Russia School. At the Foreign Ministry I wish to thank especially Minoru Tamba and Kazuhiko Tōgō, two men with important duties, but who nevertheless gave of their time to conduct interviews with me on more than one occasion. I also wish to thank Kuninori Matsuda, also of the Russia School, newly promoted to head the Russia Desk in Tokyo, and with whom I spent quality time in both Washington and Moscow. Thanks also in the Foreign Ministry to Akio Kawata, Kyōji Komachi, Mutsuyoshi Nishimura, and Kenji Shinoda, among others. Hiroshi Kimura of Kyoto was one of the greatest sources of knowledge on Russia for me in Japan, and he faithfully read everything that I sent to him. He never failed to give me tremendous feedback. He has also become a good friend. Nobuo Shimotomai of Tokyo has also provided me with great insight on Japanese-Russian relations, and is one of the most conscientious scholars I have encountered anywhere. His former colleague at Hosei University and long-time resident of Japan, Konstantin Sarkisov, was instrumental under Gorbachev in shaping new views toward Japan, and continues to provide tremendous insight into the policymaking processes of both countries.

As for the private sector in Japan, I owe heartfelt thanks to Kōji Hitachi, who spent many hours in discussions with me in Tokyo and Moscow on numerous occasions. His knowledge of not only Russian current events but also Russian culture and history are something to be aspired to. In Washington I was able to form a great friendship with Yoichi Nishimura of the *Asahi Shimbun*. Not only he is a wonderful journalist with a tremendous future, but he also happens to have one of the keenest insights on the intricacies of policymaking in both Japan and Russia, and is one of Japan's most knowledgeable experts on Russia. He is a great source, and a great friend. Lastly, thanks to the late Ichirō Suetsugu for his immense knowledge of Japanese-Russian relations, for his willingness to sit down on numerous occasions to share this knowledge with me, and for his tireless efforts to ameliorate Japanese-Russian relations over the past several decades. It was through Suetsugu that I had the fortune of making many acquaintances in Russia. It was also due to his efforts that I was given an office in the Institute for World Economy and International Relations of the Russian Academy of Sciences (IMEMO) in Moscow from 1999 to 2001.

Which leads me finally to Russia. There is no person to whom I owe a bigger debt of gratitude than to Georgii Kunadze, former Deputy Foreign Minister, Ambassador to the Republic of Korea, and most recently Deputy Director of IMEMO (Institute of World Economy and International Relations). His knowledge of Japan and the Japanese language is extraordinary, and his having imparted a small portion of this expertise to me was above and beyond his call of duty. Kunadze also provided me with invaluable introductions throughout Moscow. In fact, practically everyone that I met and interviewed in Moscow was through his introduction. Without his kind urging, I would not have been able to meet with a former member of the Politburo of the Soviet Union, with current Duma members, with Japan experts at the Foreign Ministry, or with all of the top Japan experts throughout Moscow. Additionally, Kunadze repeatedly read portions of this dissertation and gave me valuable critiques, and not to mention tremendous insight into the workings of Soviet and Russian foreign policy. I only hope that I can repay a portion of his kindness. Also at IMEMO, I must thank Valerii Zaitsev, a fountain of information not just about Russian policy in Japan, but also across Northeast Asia. Thanks also to another Japan expert, Vadim Ramzes, who – like myself – not only has a fascination for Japan, but also a love for jazz music. The two Directors of IMEMO who overlapped my time there were Vladen Martynov and Nodari Simonia. Their allowing me to base myself at the Institute was a tremendous gesture on their parts. Oleg Kirioushin, also a Deputy Director of IMEMO, kindly helped out in every way administratively at the Institute, and even let me keep my computer in a safe there when I traveled. The late Sergei Blagovolin also was a source of great information, and I had the fortune of meeting him first in Japan. In fact, he was the first Japan expert from Russia that I met. He also shared his time with me in Moscow.

Aleksandr Yakovlev, the architect behind *perestroika* and *glasnost'* and Mikhail Gorbachev's right hand man for some time, took time out from his busy, yet vital task of accumulating the lists of those killed in the Soviet Union during the Great Purges of the 1930s. Yakovlev and I had something in common – we both were recipients of fellowships of that wonderful program, the Fulbright Scholar program. Sergei Troush of the Institute for Canada and the United States has been a dear friend. Not only did he provide me with wonderful information about Russian-Chinese relations, but he also helped me settle into Moscow, and in fact met me when I first arrived at the airport. I got to know Sergei when he was a Visiting Fellow at the Brookings Institution in Washington, DC. In the Russian Far East, Sergei Verolainen of Vladivostok, who I got to know when he was an IREX scholar at SAIS, provided me with the proper introductions in the region, and he even put me up in his apartment on more than one occasion when I traveled out there. Lyudmila Troshina of Moscow provided me with a unique insight into the Russian mind and psyche. In fact, I became so dependent on her that I decided that I could not return to the United States without her, so I married her in Moscow and brought her back with me. We now live in Seattle and have three children, Alex, Paulina, and Phillip. Thanks to them for their patience while I grappled with this study.

As is *de rigueur*, I accept all responsibility for any inaccuracies or misinterpretations present in this study.

Seattle, Washington

Introduction

Continuing patterns

As the twenty-first century begins, intellectuals have been debating over the past several years what sort of period the global community is entering. 'Post-Cold War' is the term generally settled upon, but various other designations, such as 'the New World Order,' 'the End of History,' and 'the Clash of Civilizations,' have been used to describe the world system as it enters the third millennium AD Some people feel that the Cold War has not truly ended, and that the collapse of the Soviet Union served as a mere respite from an East-West conflict, in which the eastern half has merely shifted some 4,000 miles farther to the east from Moscow to Beijing. Whichever argument is most valid, the world community is clearly in a transitional period marking not just a new millennium but also a new political order. And yet two countries, Japan and Russia, are mired in an acrimonious territorial dispute that dates from the Second World War. The two nations still have not signed a peace treaty, more than six decades after the Pacific War ended. Why have the two nations failed to normalize relations, and move forward into the twenty-first century to meet the new challenges confronting each one at home and abroad? Why, incredibly, are the two nations still unable to sign a peace treaty ending the Second World War, despite the domestic changes in each country and the global changes that have transpired since the end of the Cold War?

This anomaly of 'Post-Cold War' politics is important to understand, because it is not as if two small nations, on the far periphery from great power intersection, are arguing over an issue or geographical area of minimal importance. Japan and Russia are two major powers. The area of contention itself lies within a region that many predict could be the tinderbox of the next great power conflict – Northeast Asia. Furthermore, by exploring this particular bilateral relationship in an era of international change we can better understand the nature of bilateral relationships in periods of international systemic change. The history of this particular bilateral relationship can tell us something about how changes in the international system and in domestic politics (in this case domestic change associated with the broader systemic shifts) can affect bilateral relationships. How do nations react to major international change, particularly when this change has such a great influence on domestic processes? There is a broader analytical value that should be recognized. International systemic change and domestic political change affect not only nation-states themselves, but also the nature of discourse and relations with other countries. In the case of Japan and Russia, this it is especially pertinent given the fact that the larger systemic changes associated with the end of the Cold War have been closely linked to domestic institutional changes in each country.

The late 1990s and the early 2000s marked a period of tremendous anticipation in bilateral relations between Japan and Russia, particularly the years between 1996 and 2007. For the first time in over two hundred years neither country viewed the other as a direct security threat. Bilateral political contacts of the highest level proliferated, highlighted by warm personal contacts between the leaders of both nations. Relations between the defense establishments of both nations sprouted seemingly overnight, perhaps even surpassing bilateral political relations in terms of atmospherics and goodwill. And although economic relations were somewhat stagnant, by the end of the 1990s Japanese-Russian and international energy consortia in the Russian Far East actually began producing energy after years of mere talk and planning. Oil produced off of Sakhalin Island in the Russian Far East began being shipped to Japan. At the grass roots level, citizens in the border areas of Hokkaido and the southern Kuril Islands were permitted visa-free visits between the two regions. For the first time since the end of the Second World War there was a palpable mood in both countries that the long-standing enmity could be put aside and that the relationship could finally be normalized. As the 1990s came to a close, however, it was clear the two nations were still far from settling the territorial dispute that had prevented a normalization of relations since the end of the Second World War.[1] The optimism that pervaded the relationship during the latter half of the 1990s had seemingly evaporated as the self-imposed deadline for a peace treaty expired at the end of the year 2000.[2] And by the middle of the decade after 2006–07 the awkward – if not acrimonious – dialogue that marked Japanese-Russian relations during the Cold War also seemed to reappear at times.

The *rapprochement* of the late 1990s and the early 2000s was not the first instance over the past century that Japanese-Russian relations have undergone a sudden improvement, marked by a dramatic increase in political, diplomatic, and/or military contacts. Each time the two nations seemed to be on the brink of a genuine political breakthrough, however, constraining factors have emerged, thus leaving relations between Moscow and Tokyo to their prior cold state. This is a pattern that has repeated itself on at least half a dozen occasions since the end of the Russo-Japanese War in 1905. This pattern demonstrates that the roots of bilateral tension preceded and transcended the Cold War, and even the Second World War, and suggests that the failure to normalize relations is probably not due solely to the territorial dispute. Attempts to explain the impasse by pointing solely to the territorial dispute miss the mark. The mistrust has existed longer than just six decades.

This work frequently utilizes the French term '*rapprochement*' to describe any warming in Japanese-Russian relations that includes a step-up in good faith diplomatic efforts to resolve not only the territorial dispute but other issues of economic and political importance (such as fisheries agreements). Such efforts are often accompanied by an increase in high-level political interaction, and by attempts to bolster trade and investment, to shape public understanding, and to increase military-to-military contacts. During periods of Japanese-Russian *rapprochement* there are numerous references in Japanese writings to *sekkin* and in Russian writings to *sblizhenie*. These are the nearest translations to *rapprochement* in the two languages. Therefore, when one detects a noticeable increase in the usage of these terms over a certain period of time, particularly in official statements or writings, it is a good indication that the two governments are making sincere efforts to improve relations. In this work the term 'normalization' of Japanese-Russian relations refers to the signing of a peace treaty ending the Second World War and acknowledgements from both sides that the territorial dispute has been resolved to mutual satisfaction.[3] Normalization is used more loosely in reference to events prior to the Second World War, when territorial disputes also from time to time dogged the relationship, and at times threatened war. Nevertheless, diplomatic relations prior to the Second World War were never for long held hostage to these disputes. The often-bloody confrontations from the early part of the twentieth century and the resentment that accompanied them, however, have remained to seethe in the respective national psyches. There they remain today.

For the purposes of this work the term *rapprochement* does not necessarily denote 'normalization.' In fact, one could argue that since the end of the Second World War, Japan and Russia have not had 'normal' relations, because they have not actually signed a peace treaty ending the war. Prior to the Second World War, Japan and Russia/the Soviet Union had 'normal' diplomatic relations (except in times of actual war, of which there were four during that period) in that they recognized one another's government and one another's territorial integrity. After the Second World War, the two nations went more than ten years without diplomatic recognition and still do not have a peace treaty. Therefore, the question of normalization is the focus of the late 1990s and the early 2000s. Why are relations not 'normal,' and why is there no peace treaty? What has prevented this?

Although this work examines earlier periods of Japanese-Russian *rapprochement* during the twentieth century, its focus remains on the period 1996–2007. Therefore, attempts to describe the earlier periods should be seen as summaries, not comprehensive histories. Accordingly, bibliographic references to the earlier periods are also not meant to be comprehensive; they are meant to give a general overview. It should also be pointed out that the main focus of the research in this work is on the upper level of each nation's leadership. The relationship between Japan and Russia throughout the twentieth century up until the 1990s was dominated by exclusive groups of top leaders in each nation. This changed somewhat in the 1990s given the birth of democracy in Russia and its further development in Japan. As such, there is little reference to public opinion polls before the 1990s. Where appropriate, references to such opinion polls appear in the chapters dealing with the 1990s and 2000s. Unless noted (as in Chapter 6), when reference is made to national 'attitudes,' it is being made with dominant elite attitudes in mind. It is the elite that have, for the most part, dominated policymaking toward one another.

The reference to "three dimensions" denotes the three areas that frame this historical analysis of bilateral relations: international structural forces, domestic politics, and ideational factors. Each of the periods of rapprochement in the twentieth century that are the focus of this study will be examined against the background of international events, domestic politics, and the general ideas and ideologies that influenced each society during the specific periods.

Historical patterns of Japanese-Russian relations

Periods of cooperation and *rapprochement* between Japan and Russia during the twentieth century were numerous, however short-lived each one may have been. Within two years of the end of the Russo-Japanese War of 1904–05[4] (hence the beginning date of this study – 1907) Japan and Russia had begun a political accommodation that culminated in a military alliance in 1916 during the First World War. In 1925, after the withdrawal of Imperial Japanese forces from the Russian Far East and Northern Sakhalin Island following the Russian Civil War, Japan and the Soviet Union established diplomatic relations and engaged in political and economic cooperation (which some historians think was aimed at opposing the United States).[5] The Soviet-Japanese neutrality pact that lasted from 1941 to 1945 was another instance of bilateral cooperation. In 1956, although no peace treaty was signed, the two nations agreed to recognize one another and commence economic cooperation. Again in 1972–73, Japan and the Soviet Union pledged to increase economic and political cooperation. The accession of Mikhail Gorbachev as Secretary-General in the Soviet Union in 1985 was welcomed enthusiastically in Japan. It was hoped in Japan that under Gorbachev bilateral relations would finally be fully normalized. The same type of euphoria in Japan came about with the accession of Russian President Boris Yeltsin just a few years later in 1991–92.

Each brief period of political *rapprochement* ended in mutual suspicion at best, and in conflict in three cases. The 1917 Russian Revolution ended the Japanese-Russian alliance, and Japanese troops soon occupied large parts of the Russian Far East. Even after the Japanese government recognized the Soviet Union in 1925, the growth of conservative power in Japan and Soviet mistrust of Japanese intentions in China and Northeast Asia killed off any hope of a 'normal' relationship. Conflict ensued in the late 1930s. The Soviet Union invaded Manchuria and seized both Sakhalin and the Kuril Islands in August 1945, violating the neutrality treaty that had governed relations since 1941. After mutual recognition in 1956, bilateral political cooperation and all hopes for a resolution of the territorial dispute died with the controversial renewing of the US-Japan Security Treaty in 1960 and with the ending of the first period of US-Soviet détente. The visit of Japanese Prime Minister Kakuei Tanaka to Moscow in 1973 marked a supposedly 'new era' in Soviet-Japanese relations. Promises to expand Japanese investment in the Russian Far Eastern energy complex seemed logical given the worldwide energy crisis of that year. But by the mid-1970s realities and obstacles brought to an end the brief warming of relations. The accession of Mikhail Gorbachev to Soviet leadership also raised hopes in Japan. However, by the year of his visit to Tokyo (1991), it was clear no major concessions would be forthcoming from either side. The excitement surrounding the birth of a Russian democratic government in late 1991 was soon dampened after Russian President Boris Yeltsin cancelled a scheduled visit to Japan in the fall of 1992. As was the case in each of the previous eras of Japanese-

Russian/Soviet reconciliation, the palpable hope surrounding the *rapprochement* of 1996–2007 also seemed to taper off by the middle of the first decade of the new millennium.

Was the political *rapprochement* of the late 1990s and early 2000s similar to the previous eras of *rapprochement*? Can similarities be discerned in the respective domestic political contexts and in international relations during these periods of political amicability? Can similarities be found in the re-emergence of confrontational relations that disrupted each of these periods? Answering these questions must be done against the background of the continuing pattern in bilateral relations between Moscow and Tokyo during the twentieth century.

New areas for examination in Japanese-Russian relations

Japanese and Russian journalists, scholars and politicians have filled the scholarly journals and newspapers of both nations over the past few years with articles on the bilateral relationship. In Japan,[6] Shigeki Hakamada, Hiroshi Kimura, Nobuo Shimotomai, Kazuhiko Tōgō and others have closely followed and described unfolding events in the pages of publications such as *Chuuō Kōron*, *Foresight Magazine*, *Sekai*, the *Asahi Shimbun*, the *Yomiuri Shimbun*, and the *Sankei Shimbun*. In Russia, Vasilii Golovnin, Georgii Kunadze, Aleksandr Panov, Konstantin Sarkisov, and others have done the same in the pages of *Izvestia*, *Mezhdunarodnaya Zhizhn'*, *MEiMO*, *Novoe Vremya*, and other publications. The accounts are descriptive, sometimes prescriptive, and oftentimes biased, depending on the nationality of the writer. Needless to say, they are excellent sources of information on the day-to-day political changes. Unfortunately, almost all of these works are found only in Japanese or Russian language periodicals.[7] Works on this topic in English have been limited the past half-decade, apart from occasional journal articles.[8]

Although these studies all have their particular strengths, in the past few years various new factors have emerged in the bilateral relationship that for the most part have been overlooked in the literature. These factors, though they may have existed in the past eras of Japanese-Russian *rapprochement*, must be re-examined in the context of international relations of the early years of the twenty-first century. The traditional analyses tend to return to oft-repeated and oft-examined issues (the Cold War, territory, etc.) while missing the changes that have influenced bilateral relations over the past few years. These changes associated with the post-Cold War era have significantly affected bilateral relations, but most of the works covering Japanese-Russian relations during the 1990s are still influenced by Cold War thinking, and Cold War terminology.[9]

Unfortunately, the territorial dispute has become the beginning point and the ending point of much of the analysis undertaken to date. Utilizing historical examinations that begin with the earliest explorations of the disputed islands, many authors attempt to establish who had the earliest territorial claims, thus from the very beginning focusing on a dispute which has really only existed since the end of the Second World War.[10] In fact, the malaise in relations predates the territorial dispute. The dispute does not define the relationship, as is often the mistaken assumption. The territorial dispute is currently the outstanding issue dividing the two nations, and it serves as the primary negotiating point for policymakers. Nevertheless, the territorial dispute is more a symptom of the general malaise in relations than the source of the abnormal relationship.

The roles of domestic actors in both countries have been transformed because of the dramatic political changes of late in both nations. Throughout much of the Cold War the Japanese Ministry of Foreign Affairs (MOFA) was a unitary actor that dominated policy toward Moscow. Today MOFA is no longer as unitary as it once was, and it has lost much of its influence. MITI (today renamed METI, for the Ministry of Economics, Trade, and Industry), has gained more influence, and there are new actors in the Japanese political arena – such as the Japan Defense Agency, the Ministry of Finance, the Maritime Safety Agency – who have gained more influence over Russia policy.

Within the 'Russia School' of the Japanese Ministry of Foreign Affairs there has been tremendous discord over Russia policy over the past decade. As mentioned, the Russia School had once been a unified core in the formulation of Japan's Soviet and Russia policy over the Cold War decades and into the 1990s. Differences within the ministry have surfaced and have complicated Japan's Russia policy over the last decade. This is an issue that has received little attention in even the most recent studies, due to the controversial and sensitive nature of the subject.[11] Elected politicians in Japan (especially among the Liberal Democratic Party (LDP) elite) have also increased their role in the formulation of Russia policy. They have taken advantage of discord within the Ministry of Foreign Affairs to leverage their power and influence. Little attention has been given to the role of bureaucracies in the formulation of Japan's Russia policy since Donald Hellman's work on the Japanese-Russian *rapprochement* of the 1950s.[12] Hasegawa gives extensive coverage to the role of the Japanese Ministry of Foreign Affairs, but much less attention to other ministries. Gilbert Rozman also touches on the subtleties of Japanese domestic policymaking in one of his works, but it covers only the Gorbachev years.[13]

In Russia, too, the various governmental ministries and agencies tend to view relations with Japan differently from one another as they vie for control over the decision-making process. It is vital to point out that the collapse of the Soviet system resulted in the collapse of centers of policymaking and the fragmentation of the policymaking process. The Russian Foreign Ministry seems less influential, while other government ministries (such as the Ministry of Fuel and Energy, the Ministry of Natural Resources, the State Fisheries Committee, and others) have increased their clout. The Ministry of Defense has retained most of its power in the inter-governmental ranking in Russia, especially concerning relations with Japan. Perhaps the most important change in Russian foreign policymaking has been the increase in the role of local governments in the Russian Far East, especially their influence on Japan policy. To date there has been little analysis of the intricacies of internal Russian policymaking in the late 1990s and early 2000s and how this has affected the formulation of Japan policy.

Most of the work to date published on this topic does reference economic relations, but it is never a major focus. Indeed, except for a brief period of time during the First World War and in 1970, neither country has ever been a principal trading partner for the other.[14] During the latter half of the 1990s and into the 2000s energy cooperation began to take on an important role in the relationship. This is especially significant since cooperation is no longer a bilateral affair, but now includes the United States, China, Great Britain, India, and South Korea. Only in the last few years has production begun on the Sakhalin Island shelf, and only recently has there been activity in multi-national extraction and supply (or pipeline) projects elsewhere in the region.[15] This is an area that deserves more examination, particularly against the background of rising oil demand across East Asia, the growth of the East Siberian energy projects, and the effect this has on both Russian and Japanese policy. The last decade has been a time of great activity in the energy field and recent studies have neglected this. Energy and resource development has been important to the Japanese-Russian relationship since the early days of the twentieth century. But now the issue has become more internationalized, with multinational consortia taking the lead in energy development in the Russian Far East. Additionally, strategic factors are important to consider in analyzing energy trends. Since 2002 Japan and China have been engaged in an intense competition over the course of oil and gas pipelines coming from Siberia to the Far East. Japanese commercial and diplomatic activity in Central Asia is focused squarely on energy, and is also linked to its Russia policy. Tied in with the energy factor, environmental issues began to assume more importance in the late 1990s, as is evidenced with the recent Sakhalin imbroglio. Nuclear non-proliferation initiatives have also come to figure prominently in the bilateral relationship. Together these issues have become an area of major concern, particularly for Japanese policymakers, and already they have had an effect on the formulation of policy in Japan. All of these issues merit more attention than has been given recently.

The end of the Cold War and the demise of the Soviet Union have dramatically affected the situation in Northeast Asia. The specter of the 'Northern Threat' (i.e. the Soviet Union) had been one of the driving forces of Japanese defense and security policy during the Cold War (in fact, this threat had been a major factor in Japanese strategic planning as early as the eighteenth century). Suddenly, with the end of the Cold War, the primary rationale for Japan's national defense strategy disappeared. The rise of China and questions about the US commitment in the region, along with various other events of the mid-1990s, prompted new Japanese policy initiatives. Russian leaders, also concerned about China's intentions in the Russian Far East and in Central Asia, as well as US intentions in Eastern Europe and Eurasia, also realized that a dramatic shift in their nation's strategic calculus was in order. Consequently, during the late 1990s, the territorial debate became part of a broader picture.

Korean Peninsula issues provide interesting insights into the recent diplomatic efforts in Japan and Russia. The Democratic People's Republic of Korea's pursuit of nuclear weapons and development of ballistic missiles drove Japanese policymakers to reconsider existing security policies in the early- to mid-1990s. Russia also was concerned about conflict on the Korean Peninsula along its border, and had to reevaluate its position in the region. Additionally, early US-led international efforts to bring about peace and reconciliation on the Korean Peninsula aroused concerns in both Moscow and Tokyo. This was not because a lessening of tensions would be adverse to national interests, but because the initial dialogues did not fully include either Japan or Russia. Despite the inclusion of the two nations in the Six-Party talks on Korean Peninsula nuclear issues, diplomats in both nations often feel that they are on the outside looking in. The two nations realize that they now play a minimal role in events on the Korean Peninsula, which had been the traditional crossroads of Japanese-Russian confrontation and diplomacy early in the twentieth century. Japan and Russia have called for a wider dialogue in Korea that would include their participation. Japan wants to demonstrate that it has the capability to influence the regional political agenda; Russia wants to show that it is still a player in East Asia. Both sides hope to regain lost diplomatic clout and to enhance their political positions in Northeast Asia. For Japan events on the Korean Peninsula took on additional meaning. Were Korea to reunify (whether on southern terms, or even in the unlikely event that North Korea would dictate terms), Japan would be faced with a potentially hostile neighbor.[16] Analyses of Japanese-Russian relations and Korean Peninsula issues tend to be treated separately but, in fact, there is a direct linkage.

International factors also influence non-traditional factors such as national identity and perception. Since the middle of the nineteenth century the ups and downs of Japanese-Russian bilateral relations have often reflected concerns of regional hierarchy and political-economic development for both countries. Both nations strove to modernize their economies and develop a catch-up strategy vis-à-vis perceived threats from the West at the same time, in the mid-1850s. They have been on a parallel development course since that time, and have directly competed as latecomers to industrialization and in the game of power politics in East Asia.[17] Traditionally, when one country had the upper hand over the other, it reflected badly on the other nation both at home and internationally. This is of particular importance for these two nations because they each have a tendency for prioritizing status, or hierarchy, in international relations. This has become crucial at the dawn of the twenty-first century as both nations attempt to define a new national identity (or mission). Both Japan and Russia, amid global and domestic change, seek a higher place for themselves in the emerging international hierarchy.

Furthermore, the studies on Japanese-Russian relations have to date been too narrowly focused on the territorial dispute, failing to place it in the context of broader international issues. An example is the role played by the Soviet Union in the formulation of policy by conservative forces in Japan both before and after the Second World War. Conservative forces in Japan utilized the Soviet specter to cement themselves in political power for almost four decades. Interestingly, during the Cold War, many conservative Japanese politicians favored recognition of the People's Republic of China. Although China is a communist country, many conservative policymakers in Japan saw great advantage in dealing with their closer socialist neighbor.[18] It can be deduced, therefore, that the Russian form of socialism was seen as a bigger evil to many Japanese policymakers. This suggests that the image or perception of the Russian threat has played an important role in

Japanese policymaking, especially among conservative leaders and even in socialist political circles.[19] This attitude has existed for a better part of the last century. Robert Jervis has noted that, "it is often impossible to explain crucial decisions and policies without reference to the decision-makers' beliefs about the world and the images of others."[20]

Long subsumed by Cold War political priorities, the importance of the role of historical memory cannot be too strongly emphasized. Japan has still not come to terms with its wartime actions. While many nations in Asia demand that the Japanese government make amends for its checkered past, Japan clings to the idea that it was also a victim in the war. This 'victim mentality' (*higaisha ishiki*) is particularly stressed in reference to the actions of the Soviet Union at the end of the war in August 1945.[21] Japan can conveniently point to this experience when decrying its own suffering in the Second World War.

Similarly, Russia has not truly come to terms with its history. In particular, Russia has a hard time accepting its reduced role in East Asia. Russia clings to Soviet notions of being an East Asian power. It suffers from a double identity when it comes to its Asian neighbors. In the past Russia (and the Soviet Union) viewed itself as a bridge between an advanced Europe and a backward Asia. Now that much of East Asia has developed to economic levels higher than that of Russia, the nation grapples with traditional feelings of superiority, and a newfound inferiority complex.[22] In no other relationships is this more obvious than in Russia's relationships with Japan and China. These issues deserve more attention, and indeed are crucial to understanding the course of relations over the past decade.

A major new factor in the bilateral relationship is the growth of democratic politics and the free market economy in both countries, and the effect this has had on policymaking and public opinion. For the first time in its modern history Russia has a democratically elected government. In the 1990s Japan also underwent a major transition when the Liberal Democratic Party (LDP) lost absolute power and the Japanese 'economic miracle' came to an end. Earlier this century, Japan and Russia were two imperial powers on the rise in Asia. Today, these two democracies are experiencing growing pains, and they are both answerable to the general public. This undoubtedly affects policymaking. Japanese interest groups and regional governments also play a role in influencing policy decisions. In Russia, the roles of interest groups and particularly of regional governments have grown substantially. How have these new roles changed the relationship? What new actors are influencing policy in these two democracies?

Along with the process of democratization, leaders in both countries must adjust to the economic and political forces associated with 'globalization.'[23] In both Japan and Russia politicians and the public are having a difficult time assessing their nation's place in this newly emerging system. These new forces affect economic, political, and social institutions, and consequently they affect foreign policymaking. Although their respective experiences are quite different, both Japan and Russia are finding it difficult to adjust to the new changes associated with the shifting dynamics of the global political economy.[24]

By examining the historical pattern that has pervaded Japanese-Russian relations it will be possible to assess the factors that have remained constant in the relationship, and the factors that have changed. Past studies have traced the history of relations chronologically, but they do not make a comparative historical study. As a result they have neglected the historical/psychological forces that have definitively shaped this relationship over an entire century. An historical study can help to shed light on what transpired during the last days of the tumultuous twentieth century. It is equally important to identify new or transformed factors, and to explain how these factors might set the period 1996–2007 apart from the general pattern of relations that has existed throughout the twentieth century.

1 The patterns begin

1907–45: Japanese-Russian relations over four decades, and among four wars

Japan and Russia are two countries that are on paper vastly different. One is a multiethnic nation that is also a vast continental power, with seemingly no maritime tradition.[1] The other country is a homogeneous island-nation that by geographical necessity lives by maritime trade and by what the sea produces. Japan is known for its innovation and adaptability; Russia is often viewed as being slightly behind the times (especially in comparison to its European neighbors) and perhaps too closely tied to its history and traditions. While Japan had the luxury several times during its ancient and modern history of shutting itself off from the rest of the world, Russia has often been at the mercy of hostile neighbors, and has never had the luxury of remaining isolated from the world. Russia has in the last few centuries been preoccupied with events in Europe. Japan has, and always will be, first and foremost an Asian country. Little seems to tie the two nations together other than geography: Russia is Japan's closest neighbor.

History demonstrates, however, that the two nations, though often at odds over one issue or another, can come together when political necessity dictates a *rapprochement*. Two prominent historians of Japanese-Russian relations, George Lensen and Peter Berton, have argued that bilateral relations have not been as bad as is commonly perceived. These two historians point to the pre-Second World War period in particular to demonstrate that relations are relatively friendly when the needs of politics and strategy intersect. They have both argued that the periods of *rapprochement* offer "proof of the possibility and feasibility of close cooperation" between the two countries.[2] Former Russian Deputy Foreign Minister Georgii Kunadze feels that although Japanese-Russian relations have been less than ideal since the Second World War, they have generally been better than many other bilateral relationships in Northeast Asia, such as the Sino-Russian, Sino-Japanese, or Japanese-Korean relationships. He points out that since the end of the Second World War there has never been a threat of war between the two, and that a certain "stagnant cordiality" has existed among the leadership in both countries since 1956.[3] Kunadze generally agrees with the Berton-Lensen argument that cooperation between Moscow and Tokyo can and does exist when the situation calls for cooperation.

Both before and after the Second World War, the two nations did agree to cooperate economically and politically at various times, even as they were agreeing to disagree on issues such as the territorial dispute. Although it might be argued that the international situation after 1991 and the collapse of the Soviet Union is different from that of previous periods, the brief political *rapprochement* that marked the years 1996–2007 echoed earlier periods of bilateral cooperation in certain ways. An examination of these earlier periods of *rapprochement*, therefore, would seem to be a logical step toward understanding not only the evolution of bilateral relations, but also the constraints on full cooperation and normalization. What do the earlier periods of *rapprochement* have to tell us about the latest period of cooperation? Furthermore, and more importantly, what do earlier failures to completely normalize relations tell us about the most recent failure to sign a peace treaty after the year 2000?

1907–16: From wartime enemies to wartime allies

In February 1904, after a significant deterioration of relations that dated from the early 1890s, war between Japan and Russia broke out. The war was essentially fought over control of port and railroad facilities in Manchuria and over the political domination of the Korean Peninsula. The war was a costly affair – economically, militarily, and politically – for both countries.[4] By the summer of 1905 both Japan and Russia were already exhausted from the short war. In Japan questions arose over whether the government could meet its loan repayments to third-party nations, due to the rising costs of the war. In Russia the public was growing weary of the bad news streaming in from the front, and the government feared a revolutionary movement would grow from the bloody demonstrations in St. Petersburg. Both had also suffered enormously on the battlefield. Besides losing more than one hundred thousand lives each, Russia lost almost its entire Baltic and Far Eastern surface fleet, and Japan was forced to call up more and more reserves as battlefield casualties mounted. After 18 months of combat both nations were ready for peace.

In a remarkable turn of events within two years of the end of the war, Japan and Russia had signed the first of four different treaties, culminating in an alliance aimed ostensibly toward Germany in 1916 at the height of the First World War.[5] In 1907 the Japanese and Imperial Russian government, concerned about a renewal of tensions in Northeast Asia, signed a treaty known as the Russo-Japanese Convention of 1907. This treaty came into effect on July 30, 1907 and it had both public and secret clauses. The public clause recognized the territorial integrity of China, while the secret clause reaffirmed Japan's 'special' position in Korea and Russia's 'special' position in Outer Mongolia (today the sovereign nation of Mongolia). The secret clause also reaffirmed the divisions of respective spheres of influence in Manchuria (in rough terms, the northern half lying in Russia's sphere and the southern half in Japan's).[6] An agreement was also reached on commercial fishing rights around Sakhalin and the Kuril Islands.

Just three years later on July 4, 1910 the two nations yet again signed a pair of treaties, public and private. The public treaty reaffirmed the 1907 convention, while a secret clause outlined the right of each nation to undertake any measures necessary to maintain the status quo in Manchuria (i.e. to guard and maintain each nation's "special interests").[7] Article 5 of the secret treaty went so far as to state:

> In the event that these special interests should come to be threatened, the two High Contacting Parties will agree upon the measures to be taken with a view to common action or to the support to be accorded for the safeguarding and the defense of those interests.[8]

The treaty took on the form of a quasi-alliance, already five years after the Russo-Japanese War and four years before the First World War.

Yet again on July 8, 1912, a third secret treaty was signed between the two nations. This third treaty recognized each nation's sphere of influence in Inner Mongolia (part of China proper today). Unlike the previous two conventions this treaty had only a secret clause and was unknown to the rest of the world. Japan also recognized yet again Russia's special interests in Outer Mongolia.[9]

The last of the four conventions was signed on July 3, 1916 during the First World War. The 1916 convention dealt with all the areas outlined in the three earlier treaties. The public convention announced that the two nations, if necessary, would undertake joint measures to maintain the status quo in China, Korea, Manchuria, and Mongolia. Several articles stipulated for the supply of Japanese munitions to Russia's beleaguered armies on the Eastern Front in Europe. In return Russia also extended new river navigation and railroad rights to Japan in northern Manchuria. In a secret clause the two nations promised that no third power should come to dominate China, and that in the event of such a threat the two nations promised to cooperate militarily, even if only one of the two nations should be drawn into a war initially with a third power.[10]

The international context

It is important to understand the rapidly changing international situation as the context for the series of public and secret Japanese-Russian conventions from 1907 to 1916. It should be remembered that in 1907, the year of the first Japanese-Russian convention, Great Britain had patched up its differences with Imperial Russia, and had joined Russia and France in the Triple Entente. It should also be remembered that Britain had been Japan's ally since 1902, and so there was an intersection of interests between the three nations. Britain in no way urged Japan or Russia to seek a *rapprochement*, but for leaders in both St. Petersburg and Tokyo, it was clear that their interests coincided strategically. Leaders in Japan and Russia were also concerned that misunderstandings stemming from the war in 1904–05 could get out of hand and cause a second war between the two. Japanese and Russian leaders were consequently interested in shoring up the bilateral relationship, in order to concentrate on their particular areas of interests without raising the suspicions of the other. Japanese generals in particular were concerned about Russia launching a possible "war of revenge."[11] In fact, even after the 1907 convention there was still talk in both Japan and Russia about a second war, and in Japan the national defense plan still outlined Russia as Japan's biggest threat.[12] It was no coincidence that Japan's final annexation of Korea was carried out on August 22, 1910, six weeks after the July 4 Japanese-Russian convention.[13]

In 1911 the Chinese Revolution and the overthrow of the Ch'ing dynasty put a question mark over the issue of all foreign concessions in China. Japan and Russia wanted to make sure that there would be no confusion about the status quo in Manchuria. In addition, Japan was increasingly nervous about Britain's commitment to the 1902 Anglo-Japanese alliance, whose utility was being questioned in Britain by 1911.[14]

Perhaps most important to Japanese and Russian leaders was the newfound American interest not only in China proper, but especially in Manchuria. In 1899, US Secretary of State John Hay and his staff had formulated a US plan for an 'open-door' policy in China. Although Hay formulated this policy with respect to trade, it was also seen as a tacit call for the political equality of the Western powers (and Japan) in China. The international response to the Boxer rebellion in 1900 seemed to advance Hay's idea even further. In 1902 Hay addressed Russia regarding the open-door policy, and questioned Russian designs in Manchuria.[15] Meanwhile relations between Japan and the United States (which had been tense since before the US annexation of Hawaii in 1898) began to deteriorate further while the Portsmouth peace negotiations were still under way. During the 1905 Russo-Japanese peace negotiations (which US President Theodore Roosevelt had offered to mediate), Japan insisted on an indemnity of a sum between $600 million and $1.2 billion from Russia. The United States strongly opposed this and soon Japan was forced to relent. The United States also opposed Japan's desire to "temporarily" occupy all of Manchuria after the war, and insisted on a neutral Manchuria with an 'open door.' The 1907 US Immigration Amendment Law prohibited further Japanese immigration from Hawaii into the United States. These events combined with the beginning of the construction of the Panama Canal (completed in 1914), and the tour of the American Great White Fleet in the Pacific left the Japanese wary of the growing assertiveness of the United States in the Pacific and Northeast Asia.[16]

In November 1909 the new US Secretary of State, Philander C. Knox, encouraged by American railroad executives such as E.H. Harriman and a group of investment bankers led by J.P. Morgan, called on Japan and Russia to permit their railroad concessions in Manchuria to be placed under control of an international consortium in accordance with the open-door policy.[17] Knox's railroad neutrality plan was a potential blow to Japan and Russia's ambitions in Manchuria. In addition, the Russian-built China Eastern Railroad (which crossed Manchuria from the Lake Baikal region into the Russian Maritime Province and on to Vladivostok) was seen in Russia as an actual national railroad whose function was to succor and protect the vital and outflanked Maritime region that hosted the Russian Far Eastern fleet. Japan was also loath to part with Japanese-constructed railroads in a region over which they had spilled blood to acquire. What made Japan even more nervous was British support for Knox's plan. American, Japanese, and Russian historians are virtually unanimous in their agreement that the 1910 Japanese-Russian convention was signed in response to the stepped up US political presence in northeastern China. Two Japanese historians claim that Knox's proposal demonstrated America's ambition to expand her political and economic influence in the region.[18] A Russian historian concurs and wrote that the convention of 1910 was aimed solely at opposing the United States.[19] An American historian pointed out that with his plan Knox had nailed the 'open door' shut with the United States looking in from the outside.[20] Perhaps the final irony of this whole affair was that this 'anti-US' Russo-Japanese convention was signed on July 4, 1910.

Considerably more debate has been undertaken with reference to the 1916 treaty. The majority of historians feel that, like the 1910 convention, the 1916 wartime alliance was aimed primarily at the United States. In fact the new Soviet government, soon after taking power, published the contents of the secret agreement of 1916. Lenin himself declared that the treaty was aimed at "the United States, Great Britain, and the people of China."[21] In spite of this Soviet declaration it is still unclear against which "third" country the treaty referred to. The logical choice is of course Germany, since the two nations were fighting on the side of Britain, France, and Italy in 1916. Japan had been sending large amounts of weaponry to shore up Russia in its struggle with Germany on the Eastern Front. Already by the middle of 1915 (a year before the last treaty was signed), 500 cannons, nearly half a million rifles, and several million rounds of ammunition had been delivered from Japan to Russia via the trans-Siberian railroad. In a dramatic gesture that demonstrated the turnabout in relations, Japan returned to Russia the massive siege guns that had been captured from Port Arthur after the fall of that city in 1905.[22] The question arises, however, if Germany was the target of the secret clause of the 1916 alliance treaty, why did it have to be kept a secret? Peter Berton argues forcefully that the treaty was not simply aimed at the United States, but was a broader effort by the Japanese to avoid being opposed by a coalition of powers in Northeast Asia, whether it be Anglo-American or German-American. Japanese leaders dreaded the idea of an 'all-white' alliance aimed at carving up China and opposing Japan in East Asia.[23] Another historian feels that Japan and Russia themselves could not agree against whom the alliance was aimed.[24] As far as this work is concerned it is really of no consequence what the purpose of the alliance was. The decision by Japan and Russia to sign both public and secret conventions between 1907 and 1916 was undoubtedly influenced greatly by numerous international events during this highly fluid period in world history.

The domestic political context

Domestic politics, as is usually the case, played a strong role in policy formulation in both countries during the period of the four treaties. Both countries were in periods of internal transformation. In Japan, the passing of the Emperor Meiji in 1912 and the accession of his weak-minded son were, and still are, seen as a major milestone in modern Japanese history. In Russia after the 1905 revolt in St. Petersburg, the Tsar was experimenting with a representative legislative body for the first time in the history of the Romanov dynasty.

In Japan pro-British and pro-Russian factions battled for influence. These two factions were also divided on views concerning the United States, the pro-British faction being seen as more conducive to working with the United States, and the pro-Russian faction being more bellicose toward the United States. Count Hirobumi Ito, one of Japan's original modern *genro* (or oligarchs) was seen as one leader of the pro-Russia faction. After his assassination by a Korean nationalist in Harbin in 1909 (where he was on the way to a meeting with Russian officials), the leadership torch of the pro-Russia faction passed to Prince Aritomo Yamagata, also a *genro* and the founder of the Imperial Japanese Army. Ironically, Ito and Yamagata did not see eye to eye on most issues, but on this one they did. In fact, many Japanese leaders saw the advantages in shoring up relations with Russia. Among them were Baron Shimpei Gotō, Marquis Kaoru Inoue, General (later Prime Minister) Giichi Tanaka, and Prime Minister (1901–06) Tarō Katsura. The fact that these conventions were signed demonstrates the power of the pro-Russia faction and the growing disillusionment (even before the First World War) with the Anglo-American powers.[25] The power struggle in domestic politics between the two factions was reflected in the power struggle between Yamagata and Baron (and Foreign Minister) Takaaki Katō. A Russian historian has pointed out that Japan's search for national autonomy and a way out of the impending post-war economic slump began as early as 1916, with special attention given to Russian oil deposits in northern Sakhalin.[26] Although this seems rather early for Japan to have been contemplating a slump that would not come until 1919–20, it is plausible that Japanese economic planners foresaw the drop-off in demand for Japanese goods in Europe and North America after the termination of hostilities in Europe.

In Russia the decision to move closer toward Japan after 1905 was seen by at least one historian as part of a broader struggle within the Russian government between pro-British and pro-German elements. The pro-British faction won out. The decision was made not only to join the Triple Entente with Britain and France, but also to improve relations with Japan, a British ally.[27] Russian leaders also undoubtedly wanted breathing space in the Far East to be able to complete construction of the trans-Siberian railroad (especially the Amur mainline around Manchuria) and to develop the Russian Far East both economically and militarily. At the same time, French loans started to pour into Russia, overtaking American loans. Consequently, an anti-US treaty with Japan would not sound a death knell for Russia's developing economy. Finance Minister Sergei Witte, who initially supported the 1904–05 war with Japan, later was to become a proponent of shoring up relations with Japan in order to concentrate on domestic economic reform, particularly in the Far East. Although many in Russia feared that another war might ensue in the Far East with Japan, war with Germany and Austria in Europe was deemed a greater overall strategic threat to Russia.[28]

The ideational context

Did something beyond mere political convenience push the governments in both Japan and Russia to undertake efforts to cement a political *rapprochement* so soon after the war in 1904–05? As mentioned, Japan and Russia were two nations undergoing dramatic political transformations at home. In addition, at the turn of the century both Japan and Russia were still developing economically and playing a catch up game with the Western powers. In essence, the two nations were in search of a national identity and national mission at the dawn of the new century. And because of the brutal nature of the recent war the image of each nation played a large role in shaping the national psyche of the peoples in Japan and Russia.

Japan's internal political struggle that began after the Emperor Meiji's death in 1912 was, in the words of one historian, "a battle for national identity."[29] Aritomo Yamagata wanted a return to the Meiji glory, which involved oligarchic rule backed by the support of the Imperial Army. His opponents wanted a more liberal, Western-leaning form of government. Yamagata's vision was more similar to that of Nicholas II than to that of David Lloyd George or Woodrow Wilson. Yamagata's view prevailed for the next several decades in Japan, but the collapse of the Tsarist government in 1917 had to be seen as an ominous sign of the shortcomings of such a system of governance. When, in fact, the Bolsheviks did take power in Russia it came as a grave shock to the Japanese. As one Japanese leader proclaimed, "No one imagined that the [Russian] Imperial Household, which possessed an enormous army and international renown, would meet in an instant such a fate."[30] The demise of the Ch'ing Dynasty – as well as the Houses of Hapsburg and Hohenzollern – was seen as a worrying trend to those in the Japanese government who were dedicated to the continuation and the integrity of the national polity (or *kokutai*), represented by the emperor. Some Japanese were even more worried about Woodrow Wilson and democracy than they were about Bolshevism.[31] Russia's collapse gave Japanese military leaders the opportunity to do what they had wanted to do for a while – go into the Russian Far East. The Siberian Intervention began in 1918 and lasted until 1925 (although Japanese forces were withdrawn from the mainland in 1922, remaining only thereafter in northern Sakhalin).

Russia was perhaps less driven by ideas in its attempts to repair relations with its erstwhile enemy in the Far East. Nevertheless, the movement toward a *rapprochement* with Japan between 1907 and 1916 played to the imagination of the Eurasianist movement in Russia, which had argued that Russia looked too attentively toward Europe and in doing so had ignored its Asian heritage. As Aleksandr Block wrote in his famous 1918 poem 'The Scythians': "Yes, we are the Scythians. Yes, we are Asiatics/With slanting and greedy eyes!"[32] The Eurasianist movement (or perhaps more properly, idea) had begun earlier in the nineteenth century, but undoubtedly Japanese exploits and events in China at the turn of the century captured the imagination of the proponents of Eurasianism in Russia.

It should also be pointed out that in the first two decades of the twentieth century cultural contacts between Japan and Russia saw a period of flowering which has perhaps never been surpassed in the history of their bilateral relations. Although they were political rivals in Northeast Asia, there was a profound mutual admiration and fascination among the peoples of the two nations. This mutual admiration was of course concentrated in the educated and elite urban classes, but it was a strong feeling. In Japan the love for Russian literature, especially Chekhov, Dostoyevsky, and Tolstoy, was profound.[33] The same admiration extended to Russian classical music. A popular Japanese song in the days after the Russo-Japanese War said: "Yesterday's enemies are today's friends." [*Kinō no teki ha kyō no tomo.*][34] In

Russia, too, a growing admiration for Japanese culture became apparent after the Russo-Japanese War. France is partly responsible for introducing the movement known as *le japonisme*, which influenced painting, fashions, and literature of widespread circles of artists throughout Europe, before moving into Russia.[35] Russian fascination, not only with Japanese art and culture, but also with Japan's political accomplishments, grew through the decade, even after the Bolshevik revolution. Karl Radek, a well-known Soviet journalist and communist party insider, later insisted that after the 1905 war, the Russian public felt antipathy for the Tsarist government, not Japan.[36] Although the role of culture should not be overplayed, it was nonetheless a very important factor in shaping national perceptions. "Culture is not independent of politics, but it does have a life of its own, and in nations like Russia and Japan, where cultural accomplishment is esteemed, it can cast a strong and persistent shadow."[37]

But is must not be forgotten that the war had a tremendous psychological impact on Russia, whose condescending and racist attitude toward Asian nations made the defeat to Japan that much more difficult to stomach. Although attitudes among some elite may have been more balanced, the majority of people in Russia undoubtedly felt shame at the defeat at the hands of an Asian nation.

Since the beginning of efforts at modernization and national revitalization in the 1850s both Japan and Russia have always felt the necessity of 'catching up' with the West.[38] As Western nations scrambled to find their 'place in the sun,' Japan and Russia also battled to find their own place in the Darwinian struggle of international power politics. The two nations viewed one another as rivals in political and economic development, not just in Northeast Asia, but also in regards to relations with both Great Britain and the United States. Not only were Japan and Russia in competition for sources of investment, but also for sources of international political support. The two nations' search for allies led them in parallel directions, toward Great Britain and away from the United States (ironically after the United States had helped broker a peace in 1905). When Great Britain and the United States drew together before the First World War this helped to push Japan and Russia together.

Another perspective argues that Japan and Russia were *not* two powers on the rise at the turn of the century. In fact, "Japan and Russia were both backward, brittle societies, compared with the rapidly industrializing Americans and West Europeans … both [Japan and Russia] were being driven outward by weaknesses and divisions at home."[39] Whatever was the ultimate cause of the Japanese-Russian *rapprochement* of 1907–16, the fact is the two nations did come together at a time of national reckoning and national transformation. How far-reaching the *rapprochement* might have been is up for speculation. Soon the Bolshevik revolution put an end to it.

The breakdown

When the new Soviet government decided to withdraw from the war against Germany in early 1918, the era of Japanese-Russian *rapprochement* came to an end. The provisional government under Aleksandr Kerensky had kept alive hopes that cooperation could be maintained, but after its fall from power cooperation proved impossible. Not only was the Bolshevik government anathema to the conservative Japanese leadership, but the crumbling of authority gave Japanese military leaders something they had longed hoped for: a free hand in Manchuria, Mongolia, and the Russian Far East. In July of 1918, more than seven thousand troops from the Japanese Imperial Army were deployed in the Maritime Region up to Khabarovsk. By November 1918 there were almost fifty thousand Japanese troops deployed across the Russian Far East from Sakhalin to the Baikal region. Soon this number would expand to seventy thousand.[40]

The breakdown of the Japanese-Russian *rapprochement* was a foregone conclusion once the Bolsheviks seized power. It is not assured that the era of goodwill would have continued much longer, even if Imperial Russia had emerged unscathed from the First World War. There were clear signals of suspicion from the leadership in both countries during the entire period from 1907 to 1916. Nevertheless, both countries did make a conscious effort at all levels to improve the relationship. Political contacts proliferated, economic interaction grew to the extent that by 1916 Japan was Russia's largest trading partner,[41] territorial issues were settled to mutual satisfaction, and cultural contacts flourished.

The two nations' domestic political structures and elite attitudes and perceptions were unprepared to consider the *rapprochement* anything more than an expedient measure. Additionally, elite rulers in Japan, grounded in the imperial, oligarchic system, were loath to consider a partnership with a nation of 'Bolsheviks.' The mere idea would have been repugnant, although this would change soon enough.

1923–25: Japan recognizes the Soviet Union

Japanese forces withdrew from the Russian Far East in October 1922, except for a contingent that remained on northern Sakhalin Island until 1925. Southern Sakhalin Island had been annexed by Japan after the Russo-Japanese War in 1905. This created some of the mistrust that lingered in Russia even during the period of *rapprochement* from 1907 to 1916, because the Russians viewed Sakhalin (unlike concessions in northeastern China) as Russian soil. Japanese forces crossed over into northern Sakhalin in the summer of 1918, as other troops landed in the Russian Far Eastern Maritime Province. Within Japan, the Imperial Army leadership, formerly enthusiastic supporters of *rapprochement* and friendship with Tsarist Russia, held a deep suspicion toward the new Soviet Union. Ideologically, Bolshevism was seen as a threat to the imperial system and, hence, the Japanese *kokutai*.

The First World War economic boom in Japan was coming to an end by 1918–19 and horrible deflationary pressures caused great economic dislocation. Army leaders were worried that events such as the rice riots that hit every major Japanese city in 1918 (caused by a spike in the price of rice) portended serious civil unrest and promised trouble for the conservative leadership.[42] Bolshevism, they felt, could throw gasoline onto the fire. Consequently, there was little support for recognition of the new Soviet regime. During the Russian Civil War, the Japanese government supported anti-Bolshevik, white Russian leaders such as Dmitri Horvath, Grigorii

Semyonov (both located in Harbin, Manchuria), and ultimately Admiral Aleksandr Kolchak.[43]

In July 1920, as the Russian Civil War wound down (and after Aleksandr Kolchak had been executed) the Japanese government partially 'recognized' the short-lived Far Eastern Republic, but not the government in Moscow. In Tokyo, however, the Far Eastern Republic (a rump republic founded by the Soviet Union whose territory extended from east of Lake Baikal to the Maritime Province) was seen as no more than an appendage of the Bolshevik government in Moscow. A temporary buffer zone was convenient for Japan while Japanese troops occupied large parts of the Russian Far East.[44] Fighting between Japanese troops and Red partisans, as well as Red Army forces, occurred almost from the very beginning of the Siberian intervention and lasted until Japan's withdrawal from the mainland areas. In many cases the fighting was of a particularly savage and bloody nature, and involved Japanese and Russian civilians residing in the region. Both sides committed atrocities, against civilians and soldiers alike. There were reports that Japanese soldiers routinely executed captured Russian partisans and soldiers.[45] The event that captured the most attention was the so-called Nikolaevsk Massacre, which occurred in March 1920 in the town of Nikolaevsk near the Amur River estuary on the Sea of Okhotsk. In this battle a group of Red forces with partisans took the town and allegedly slaughtered 700 Japanese prisoners (both soldier and civilian), including the resident Japanese counsel and his family. The Soviets claimed that the Japanese had shown the white flag and then opened fire on advancing Red troops; the Japanese claimed that they had surrendered and were needlessly butchered.[46] Whatever both sides might have claimed, the four-year 'Siberian intervention' was a sordid affair that cast a long shadow on bilateral relations.

Yet, as early as mid-1921, it appeared that Japan was prepared to make a deal with the new Soviet government. A bilateral conference was convened in Dairen, China in August 1921 to discuss mutual recognition. The conference broke down over the issue of Japanese troops in Russia and over the Nikolaevsk Massacre. A second conference was convened in Changchun, China in September 1922 after Japanese troops were withdrawn from the mainland. Like the Dairen conference the Nikolaevsk affair was a stumbling block, as was the disposition of Japanese troops in northern Sakhalin. The Far Eastern Republic was incorporated into the Soviet Union on December 30, 1922. From 1923 Japan began dealing exclusively with the Soviet government in Moscow.[47]

Baron Shimpei Gotō (Lord Mayor of Tokyo at the time, as well as former Foreign Minister of Japan) took it upon himself to begin negotiations with the Soviet government. Gotō's counterpart was the able Soviet diplomat Adolf Ioffe. Ioffe had experience in Asia, and was actually in China at the time in January 1923 as plenipotentiary representative extending Soviet promises of material and moral support to the *Kuomintang* (or KMT) government of Sun Yat-sen.[48] Gotō and Ioffe agreed to meet on an informal basis in Yokohama, Japan in early 1923. As noted earlier, Gotō had been a proponent of the 1907–1916 Japanese-Russian *rapprochement*, and so his involvement with the Bolshevik representative suggested that Japan was treating the negotiations sincerely. The Japanese government decided to begin formal negotiations with Ioffe in Japan. Representing the Japanese Foreign Ministry was Toshitsune Kawakami, Japan's Minister to Poland. The negotiations, which lasted through the summer of 1923, once more ground to a halt over the issues of the Nikolaevsk Massacre (Japan demanded a formal Soviet apology) and northern Sakhalin (which Japanese troops still occupied). Japan wanted the Soviet government to sell it the remaining half of Sakhalin. Moscow refused to entertain this idea.[49]

In the spring of 1924 negotiations were resumed in Beijing. This time the representatives were the Japanese and Soviet ministers in China, Kenkichi Yoshizawa from Japan and Lev Karakhan from Russia. The negotiations ran through the end of the year, and finally on January 20, 1925 a treaty of recognition was signed. In return for Japan returning northern Sakhalin, the Soviet government recognized the validity of the 1905 Portsmouth Treaty and granted Japanese companies oil and coal concessions on Sakhalin. Japanese troops were withdrawn from northern Sakhalin on May 15, 1925. Subsequent commercial negotiations yielded settlements on fishery issues in 1928.[50]

The international context

The period that witnessed the brief *rapprochement* culminating in Japanese recognition of the Soviet Union in 1925 was another time of great ferment in world politics, especially in Northeast Asia. China was still the center of attention in East Asia for all nations involved in this region, especially Japan, Russia, the United States, and Great Britain. The Washington Conference of 1921–22 was convened in large part to settle the China issue. Both Britain and Japan agreed to give up concessions in lower China (outside of Manchuria). The issue of tariffs and extradition treaties was to be addressed, and the policies described in the US open-door notes of the early 1900s were once again announced as the principles on which to base relations among the powers in China, and between the powers and China. This meant equal opportunity for foreign trade and the preservation of China's territorial integrity. Japan was less than happy with the results of the conference agenda on naval armaments in the Pacific region, as it was not given parity in tonnage allowed with either the United States or Great Britain. The Soviet Union and the KMT, of course, were even less thrilled with the results of the Washington Conference as each had been denied the right to participate (as was the case at Versailles in 1919).

The Chinese Revolution that began in 1911 was still an unfinished chapter. Sun Yat-sen was desperately trying to unify the country against the growing power of regional warlords. Meanwhile a communist movement was born in the southern regions of China and, though still weak, it was attractive enough to invite the interest of Sun and several of his top lieutenants, including Chiang Kai-shek, who was to leave China briefly in 1923 to spend time training in the Soviet Union. The Soviet Union rejected all past Tsarist policies and promised to give up all claims on China, including Boxer indemnities and the Chinese Eastern Railroad in Manchuria. Minister Lev Karakhan's charm offensive in China was also backed by increasing Soviet material support not only for the KMT, but also for communist groups in China. The KMT also openly supported the Chinese Communist Party and the Comintern at one point.[51] Early on Soviet leaders recognized China as a fertile breeding ground for communism.

Japan, meanwhile, was concerned about both Western and Soviet designs on China. As the Japanese economy worsened in the early 1920s, China was seen as a possible elixir for the country's economic ills. Just as Japanese leaders feared Western domination in China, they also feared Chinese-Soviet collusion. The Washington Conference had further left Japan with a feeling of isolation, because it called for the revocation of the 1902 Anglo-Japanese Alliance. Meanwhile, US-Japanese relations continued their downward spiral, and immigration issues became an area of serious contention. Exclusionary policies had existed among local governments in the western United States, and attacks on Japanese immigrants there were a common occurrence. By 1924 a Japanese Exclusion Act was passed by the US Congress.[52] Japan was unhappy that immigration and race issues were not covered either at Versailles or at the Washington Conference. Japanese foreign ministry documents from the period that cover the Japanese-Soviet normalization talks are full of negative references to the United States. The Japanese government was not only concerned about tensions in the Western Pacific, immigration policies, and China issues; there was also a reawakening of concerns regarding US intentions in Manchuria and in the Russian Far East. Japanese leaders were also worried that the United States might recognize the Soviet Union and reap windfalls in the Far East, including on Sakhalin.[53] *The Japan Weekly Chronicle* quoted Gotō as saying: "The restoration of Japanese-Russian relations will bring about a considerable change in the policy of Britain and America toward Russia, Japan, and China, and the relations between [the four] will become more complicated than ever."[54]

The Soviet Union was also somewhat preoccupied with the United States, although for different reasons. The leadership in Moscow wanted an active US presence in Northeast Asia to keep Japan occupied until the Soviet Union was able to rebuild its capabilities in the region. Keeping China in turmoil was another Soviet tactic designed to keep Japan involved elsewhere. Playing Japan and the United States off of one another was not hard to do given the state of US-Japanese relations at the time. When Japan did finally recognize the Soviet government, Karakhan was quoted in *Izvestia* saying, "The agreement with Japan, by strengthening our position in the Pacific, serves as a warning to America which, by not making a treaty with us, only makes its own position worse."[55] A cartoon which accompanied the article in *Izvestia* showed a lonely Uncle Sam standing on the street as a Japanese soldier marches into a door with a sign overhead reading, "Recognition of the USSR." The caption reads, "Uncle Sam remains alone."[56] Lenin had also advocated letting the United States and Japan eventually come to blows, something he believed was 'inevitable.'

International factors continued to greatly influence the bilateral relationship between Japan and the Soviet Union. Press reports from the day in Europe, Japan, the Soviet Union, *and* the United States were full of references to the anti-US tilt of Japanese-Soviet normalization.[57] Unlike the unsteady transformations under way in Japan and the Soviet Union, the only change in progress in the United States with regard to Northeast Asia was its increasingly powerful profile there. In Japan this was seen as a cause for great concern; in the Soviet Union it was seen as an opportunity.

The domestic political context

The transformation of the international system during this second period of *rapprochement* was also reflected in the respective domestic arenas of Japan and Russia. Japan was still recovering from the effects of the passing of the Meiji Emperor and domestic political struggles continued throughout the 1920s. The new Soviet government in Russia was also still in a formative stage. Not only did Soviet leaders have to contend with shifting international issues, but also during the early 1920s the domestic political hierarchy was still being sorted out in Moscow.

In the early 1920s, led by Kijuro Shidehara (Ambassador to the United States and then later Foreign Minister), Japan flirted briefly with an accommodating policy toward the Western powers *and* the Soviet Union. The Japanese government also began working more closely with international organizations such as the League of Nations. Shidehara took the new policy of equitable relations with Japan's neighbors to such an extreme that even the United States government criticized Japan's new flirtation with the Soviet Union.[58] The Japanese Imperial Navy was also a backer of normal relations with the Soviet Union. Indeed, the Prime Minister in 1923, when Japanese-Soviet negotiations began, Tomsaburō Katō, was an admiral and served concurrently as Minister of the Navy. The navy was not only concerned about a stable flank for Japan to the north, but it was also interested in the oil of northern Sakhalin. The navy leadership was particularly interested because the Soviet government had granted oil concessions in northern Sakhalin to an American oil company *during* the Japanese occupation of the early 1920s.[59]

But resistance to recognition of the Soviet Union was strong in Shidehara's own Foreign Ministry. Shidehara's predecessor as Foreign Minister, Yasuya Uchida, was particularly concerned about the Comintern and the potential for subversive activities in Japan.[60] The issue of fishing in the waters off Sakhalin and the Russian coast became a controversial matter in Japan as various groups connected to the fisheries industry lobbied hard for an agreement with the Soviet government.[61] Meanwhile Shimpei Gotō, whose influence remained strong, continued to lobby for good relations with the Soviet Union.[62] But suspicion toward the new Soviet government was deep-seated among the Japanese army leadership. Not only was communism seen as a threat in China, but it was also seen as a threat to the *kokutai* of Japan itself. By the late 1920s the groups favoring accommodation with the West and with the Soviet Union began to lose influence to a group whose sole concern was preserving Japan's 'special interests' in China. The Imperial Army leadership dominated this group.

The Soviet government's teething process was dramatically upset with the incapacitation and death of Vladimir Lenin in early 1924. By 1922 Lenin's first of a series of strokes had already taken him somewhat out of the picture. As for policy in East Asia, Soviet intentions (at least for the short term) were to try to create a peaceful border to give the Russian Far Eastern provinces time to recover from the ravages of war and foreign intervention (which mirrored Soviet policy throughout the country). Some Soviet leaders and diplomats (such as Ioffe, Karakhan, Blukher, and Borodin) felt that East Asia afforded the young Soviet nation opportunities to expand the revolutionary wave. China was seen as offering the greatest of these opportunities, and historian Adam Ulam has suggested that China, along with Germany, was the focus of the new government's diplomacy.[63] At this stage of the Soviet government's formation there was little luxury in choosing foreign policy priorities; nevertheless, Ioffe, Karakhan, Blukher, and Borodin, as well as the head of the People's Committee of Foreign Affairs (*Narkomindel*), Georgii Chicherin, favored a new policy toward Asia. While this group still held favor in Moscow (many of them started out as Mensheviks), relations with Japan benefited. Already by the late 1920s, the influence of this group had begun to wane, as Joseph Stalin's influence and power rose. Even had new leaders in Moscow wanted better relations with Japan, by the late 1920s events elsewhere might have forced them to take a hard look at their preferences.

The ideational context

In spite of the fact that the Siberian intervention was carried out by forces from France, Great Britain, and the United States, as well as from Japan, memories of the Japanese intervention in Russia were particularly fresh. Up until 1925 Japanese forces occupied all of Sakhalin Island. In contrast, Soviet relations with Germany had been normalized with the Rapallo Treaty in 1922. Although most diehard Bolsheviks viewed Great Britain as a principal evil in the world, Britain was the first Western power that had intervened in the civil war to eventually recognize the Soviet Union in January 1924 (commercial relations had been established as early as 1920). The Soviet leadership also desired relations with the United States.[64] Relations with Japan were a more difficult matter. As mentioned, the 1904–05 war was still a fresh memory, as was the long Siberian intervention and the continued occupation of Sakhalin.

In Japan the feeling seemed to be no different. Communism was viewed as a cancer that threatened both China and Japan. The day the recognition protocol was signed in January 1925, Prime Minister Kato, addressing a group of his fellow party members (the *Kenseikai*), said: "The Soviet government at Moscow, which is only another name for the Third Internationale, never misses an opportunity to propagate communism in every country of the world, whether or not there are diplomatic relations."[65] As mentioned before, ideologically the Soviet Union opposed everything Japan stood for. Japanese leaders were still nervous about what had happened to the imperial families in Russia, Germany, Austria, and China.

No matter how desirous the Soviet Union was of peaceful relations with Japan in order to buy time to settle the situation in the Russian Far East, the distrust of Japan was deep-seated among the leadership and the people, especially in the provinces occupied by Japanese troops. Chicherin, in a letter to the Soviet ambassador in Tokyo, cautioned:

> Not one state, after recognizing our government, was so friendly in its expressions toward us as the Japanese one. Your reports ... point to a strikingly, even exceptionally strikingly underlined friendliness. What is the meaning of this? That is what one must decipher. What do they expect? Do they want territory for immigration, do they want concessions, or do they want a safe rear for the coming war with the United States? There is a clash [of interests] between us and the Japanese ... This question is most serious. If the Japanese have further designs on Eastern Siberia, there can be further friction with them and they will be beset by disappointment.[66]

This feeling of mutual suspicion was further complicated by psychological issues, such as 'saving face.' This came up time and again during the protracted normalization negotiations between 1922 and 1925. Ioffe mentions this in a 1923 letter to Gotō, saying that Russia, too, has a pride and is loath to see it wounded.[67] In a gesture that would repeat itself in the early 1990s, the Japanese government refused Soviet food and material aid that was transported to Yokohama in the aftermath of the Great Kantō earthquake of 1923. The relief ship, *Lenin* (the first foreign relief vessel to arrive in Japan), was forced to turn around fully laden with supplies after it had arrived. Japanese army units responsible for distribution in the stricken areas were suspicious of Soviet motives and would not allow Soviet workers to do the off-loading and distribution. This gesture deeply offended the Soviet government.[68]

Mutual recognition in 1925 did not bring about a drastic change in the way both governments and both peoples viewed one another. Mutual perceptions were perhaps at an all time low. The wounds received in 1904–05 were reopened in 1918–22, and they would leave deep scars that showed for decades to come.

The breakdown

It was not long before the latent mistrust surfaced once again after diplomatic relations were established. Spy scandals and expulsions began almost immediately after the treaty of recognition was signed. Soviet support of the Japanese Communist Party was increased, prompting increased surveillance of Soviet citizens in Japan. Soviet diplomats were constantly harassed by Japanese nationalists, and often Japanese citizens working at the Soviet embassy in Tokyo were attacked by their fellow Japanese. Japanese citizens in the Soviet Union were also kept under close surveillance. The respective diplomatic delegations became virtual house prisoners. "The Japanese trusted the Soviets as little as the Soviets did them," commented one historian.[69] The Japanese government rejected Soviet proposals for a neutrality pact in August 1926. The following May a Soviet request for a non-aggression pact was also turned down. In addition, negotiations over fishing rights bogged down.[70]

In 1929 when Chinese Nationalist (KMT) troops seized the Russian-controlled Chinese Eastern Railroad, Red Army forces counter-attacked and routed the Chinese. The Japanese looked on as neutral observers, but they had sympathy for the Soviets. They understood that attempts by the KMT to nationalize railroads in northeastern China were not in Japan's best interests. The speed and effectiveness of the Soviet counter-attack, however, gave some Japanese concern that Russian power in the Far East was greater than had been assumed.[71] Two years later when the Kwantung Army subjected Manchuria to *de facto* Japanese control, the Soviet Union also looked on as a neutral observer. It was clear to leaders in Moscow, however, that Japan now no longer looked to its neighbors for cooperation in establishing a peaceful region. The events of 1929–31 heightened the mutual distrust. In 1930 the Japanese government commemorated the twenty-fifth anniversary of the victory in the Russo-Japanese War, with much more enthusiasm than in the past. An elaborate kabuki play was staged at the national Kabuki Theater in Tokyo, reenacting the fall of Port Arthur. A plethora of newly published books, magazines, and paintings celebrated the victory.[72] In 1932 the Japanese government rejected the final Soviet request for a non-aggression pact after the Japanese takeover of Manchuria. In 1935 the Soviet Union sold its rights to the Chinese Eastern Railroad to Japan, primarily because Stalin saw the vulnerable stretch of railroad as more of a liability than an asset.[73]

Leaders of the Kwantung Army and elements in the Imperial government in Tokyo clamored for a war against the Soviets. Japan became even more nervous when the United States finally recognized the Soviet Union in 1933. Fears of American-Soviet collusion against Japan in the Far East were reawakened.[74] Japan was also concerned about Soviet aid to the KMT, which continued even after the Japanese invasion of China-proper in 1937. Soviet aid to the KMT lasted until the United States became Chiang Kai-shek's principal backer in 1941. Until then Soviet aid was not insubstantial (loans amounted to more than $200 million between 1937 and 1939). Red Army officers trained Chinese soldiers, and Soviet 'volunteer' pilots took to the skies and fought against the Japanese for several years (over 100 gave their lives).[75] Japanese militarist circles were already openly talking about the upcoming war with the Soviet Union when in 1936 the Japanese government signed an anti-Comintern Pact with Nazi Germany.[76] There was concern in the Soviet Union that Japanese interests in Outer Mongolia extended to the far western Chinese province of Xinjiang. Japan, it was reasoned, had designs on Siberia's industrial centers that could be outflanked from China's Far West.[77] In fact, Japan's interest in Xinjiang was probably linked more to Central Asia and the resources of the Middle East than to Siberian factories.[78] Japanese-Soviet border skirmishes along the Manchurian and Mongolian borders began escalating by the late 1930s. In 1938 a sizeable battle was fought by Lake Khasan near the Korean-Soviet border (known as the Changkufeng Incident to the Japanese). In the summer of 1939 a huge battle involving one Japanese division of the Kwantung Army and armored units of the Soviet Far Eastern Army took place along the Mongolian-Manchurian border at Khalkhin-Gol (Nomonhan to the Japanese). Red Army units (and units of the Mongolian People's Army) under the leadership of Georgii Zhukov (later to become famous as the conqueror of Berlin) delivered a stinging blow to the proud Kwantung Army, destroying an entire division (Japanese casualties totaled close to 17,000). The devastating defeat helped to finally convince Japanese leaders to turn southward to attack Western colonial possessions in Southeast Asia in 1941, rather than teaming up with Germany to attack the Soviet Union.[79]

It was probably clear to leaders of both sides long before the harsh rhetoric and pitched battles of the 1930s that any broad efforts at cooperation between Moscow and Tokyo were doomed to failure due to the unhappy combination of international, domestic, and ideational factors.

As in the 1907–16 period, the brief Japanese-Soviet *rapprochement* of 1923–25 was a product of international structural forces. Recognizing potential dangers in the Far East, the leaders of both nations sought to shore up their strategic position. Moving toward one another was a temporary tactic designed to shield vulnerabilities. The geopolitical nature of the *rapprochement* was attested to by both sides. As for the breakdown in relations, it seems to have been due to a combination of domestic ideational factors and international structural factors. Growing political conservatism in Japan, along with a growing Soviet agenda in the Far East, ruled out a long life for the *rapprochement*. The political ideologies and the negative mutual perception of the two societies killed off this short period of amicability between the leaders of both sides.

1941–45: A neutrality doomed to failure

George Lensen, quoting a Japanese historian, referred to the state of relations between Moscow and Tokyo during the Second World War as the "Strange Neutrality."[80] Perhaps a better moniker would have been the 'Temporary Neutrality.' The two nations did a remarkable job maintaining the neutrality for as long as they did. Japan was abetting the Soviet Union's principal enemy, Germany, while the Soviet Union was allied with Japan's principal enemy, the United States. Yet somehow they maintained a civilized cordiality. The neutrality was a mere stop-gap measure meant to provide relief while greater evils were combated. Almost from the beginning it was clear that neither side trusted the other and that the neutrality would last no longer than each nation's enemies remained standing on the battlefield. Unlike other periods of *rapprochement*, the neutrality of 1941–45 was not accompanied by efforts of both governments to truly improve relations. More pressing matters were at hand, namely the stake for national survival.

As mentioned, by the early 1930s it seemed not a question of *whether* Japan and the Soviet Union would go to war, but *when* the two would go to war. After the two nations did engage in combat in 1938–39, both sides were forced to reassess the practicality of continuing to aggressively oppose one another and engage in small-scale wars that might escalate into a general war in Northeast Asia. In the late summer of 1939 the Japanese received a double blow, which caused them to reassess the situation along the Manchurian-Mongolian-Soviet border. First, the Kwantung Army had been humiliated at Khalkhin-Gol/Nomonhan. Then came the news that the Nazi-Soviet non-aggression pact had been signed on August 23, 1939. This 'duplicitous' act was a rude awakening for the Japanese government. After the Russian Civil War Japanese military leaders had assumed that it would be a long time before the Russian threat from the north would make its reappearance.

The Red Army purges of the late 1930s, which shook the Far Eastern forces every bit as much as those in the western Soviet Union, seemed to vindicate this line of thinking.[81] On the battlefield in 1938–39 Japanese army leaders realized they were sorely mistaken in these assessments of the Red Army. When Japan's ally Nazi Germany signed a non-aggression pact with the Soviets, Japan's leaders felt that a rug had been pulled from under them. Japanese leaders considered the 1936 Anti-Comintern pact as primarily an anti-Soviet agreement. Suddenly Japan was forced to re-evaluate the changed situation. It was as much of a shock to Japanese leaders as the 1941 German invasion of the Soviet Union would be. The twin thunderbolts of Nomonhan and the Nazi-Soviet Non-aggression Pact forced a shake-up in Japanese thinking.

Japan, which had twice rejected Soviet overtures for a non-aggression pact earlier in the decade, now approached Moscow with its own set of overtures. The cabinet of Prime Minister Kiichirō Hiranuma was forced to resign in late August 1939 because of the humiliation of being left in the dark by Hitler about the Molotov-Ribbentrop Pact. General Nobuyuki Abe formed a new cabinet and began working almost immediately on improving relations with the Soviet Union. Japan's ambassador in Moscow, Shigenori Tōgō, was instructed to begin working with Soviet counterparts to secure some kind of entente.[82] Now the tables were turned and unlike in the early 1930s the Soviet Union had the strategic upper hand. Stalin was somewhat cool to Japanese approaches. But as the wars in Europe and China heated up, the two sides were pushed to reach an understanding. A border agreement signed on June 9, 1940 eased tensions along the Soviet frontier in the Far East. As for a general agreement, Japan held out for the sale of northern Sakhalin, whereas the Soviet Union desired a withdrawal of Japanese companies from oil concessions in northern Sakhalin (and even intimated that *they* desired to purchase southern Sakhalin). Finally, in early 1941 Japanese Foreign Minister Yōsuke Matsuoka took it upon himself to personally sign a neutrality agreement with the Soviet Union and he traveled to Moscow in March to do so. On April 13, Matsuoka got what he (and the Japanese government) desired: a neutrality pact. Matsuoka and the head of the *Narkomindel*, Vyacheslav Molotov, signed the pact in the presence of Joseph Stalin. Japan and the Soviet Union agreed to remain neutral in the event that either one became engaged in a war with any third party. The Japanese side was unable to acquire northern Sakhalin, and Japanese oil companies were to withdraw from their concession areas at a later time to be specified.[83] In a final gesture of friendship, Stalin accompanied Matsuoka to the train station to see him off on his return to Tokyo. Among wishes of goodwill and toasts aplenty, Stalin remarked that the two countries could now solve the world's problems.[84] Both sides consistently observed the neutrality for more than four years in spite of whatever ill will may have existed in the minds of both leaders.

The international context

Little explanation is necessary to understand that the international situation during the years 1941–45 was extremely complex and fluid. Japan and the Soviet Union were, however, remarkably successful in treating their relations almost as if they existed in a vacuum. The Soviet Union fought its successful Great Patriotic War against Nazi Germany in Europe, and eventually triumphed. Japan was less successful in the war for its Greater East Asia Co-Prosperity Sphere in the Pacific.

The most complicating factor in maintaining the 'temporary' neutrality was the United States. Nazi Germany was a factor in some respects, but cooperation between the Nazis and the Japanese never existed to the extent of the cooperation between the Americans and the Soviets. Japanese leaders were fearful that Stalin might be persuaded to allow the United States to establish air bases in the Russian Far East from which American aircraft might bomb Japan and interdict Japanese shipping. Indeed, the Roosevelt administration lobbied the Kremlin extensively for such rights.[85] The United States also repeatedly requested Soviet participation in the war against Japan.[86] American fliers that had flown bombing raids against the Japanese homeland and naval targets sometimes landed in Soviet territory due to mechanical and weather-related problems. This created diplomatic incidents for the Soviet Union, when Japan asked the Soviets to either not allow landing rights or that they hand over any American pilots that landed on Soviet territory. The Soviets instead interned approximately 300 of these American pilots, but then stage-managed their 'escapes,' normally through Central Asia to British forces in Persia (Iran).[87] Japanese leaders also frowned upon American shipping of Lend-Lease goods to Soviet Far Eastern ports. But there was little they could do as the goods either arrived on Soviet transports or on US Liberty ships flagged with the Soviet hammer and sickle. Occasionally the Japanese sank some of these ships, either by design or accident. US submarines operating around the Japanese home islands and the Kuril Islands at times also inadvertently sank several of these ships.[88] American-Soviet naval cooperation in the Far East went beyond Lend-Lease. Soviet naval aviators began receiving training in the United States in 1944. In the spring of 1945 American naval forces in Alaska secretly began training and equipping Soviet naval forces for amphibious operations against Japan.[89]

By 1944 it should have been clear to Japan's leaders that it was only a matter of time before the Soviet Union joined its American and British allies in the fight against Japan. Nevertheless, many leaders in Tokyo tried their best to convince themselves and others that the Soviets would abide by the neutrality pact, and that they would have a year's notice if the Soviets decided to withdraw, as the treaty stipulated. Japanese diplomats in Moscow and Tokyo constantly sought Soviet reassurances of neutrality as the war wound to its conclusion. By the end of 1944 circles within the Japanese government vainly hoped for Soviet mediation to an end to the war against the United States. It was, however, becoming painfully clear to the Japanese which way the wind was blowing. In a November 1944 speech, Stalin referred to the Japanese as one of the "aggressor countries." On April 5, 1945 Molotov informed Japanese Ambassador Naotake Satō in Moscow that the Soviet Union would not renew the neutrality pact. The Japanese frantically tried to restore relations or to head off an impending catastrophe, but all efforts failed when the Soviet Union attacked Japanese forces in Manchuria and Korea on August 8, 1945.[90]

The domestic political context

Within the Japanese leadership there was a split between two groups regarding Soviet policy. Deep rifts over Soviet/Russian policy had become a tradition within the Japanese government dating to the end of the nineteenth century. In the Soviet Union, by the end of the 1930s there was little opposition to the wishes of Joseph Stalin, though the turnover in leadership at the head of the *Narkomindel* during this time portended some change.

Prime Minister Nobuyuki Abe, who was asked to form a cabinet after the Nazi-Soviet non-aggression pact in August 1939, was inclined to improve relations with the Soviet Union, but his Foreign Minister, Kichisaburō Nomura, favored improved relations with the United States and Great Britain.[91] Abe's cabinet lasted in power only until the end of 1939. The Foreign Minister of the new Yōnai cabinet, Hachirō Arita, was an avowed anti-communist and favored an Anglo-American tilt in Japanese policy. But three influential groups within the government favored a *rapprochement* with the Soviet Union. The first group was led by Japan's ambassador in Berlin, Toshio Shiratori, who wanted to follow in step with the Germans. A second group originated in the 'Soviet School' of the Foreign Ministry and was led by Ambassador Shigenori Tōgō in Moscow. A third group came from within the Imperial Army General Staff, and it wanted to end Soviet aid to the KMT in China.[92]

Within the Japanese Army, the Imperial Way (*Kōdō*) Faction and the Control (*Tōsei*) Faction, who struggled for the good graces of the emperor, were also at odds over Soviet policy. The *Kōdō* Faction, led by Sadao Araki (one time Army Minister), was deeply opposed to Bolshevism as an ideology, because of its complete incompatibility with an imperial system. The *Tōsei* Faction (which would one day be led by Hideki Tōjō) wanted to accommodate the Soviets to have a free hand in China. The pro-Soviet group had its victory, and serious negotiations with the Soviets began in the summer of 1940. In July 1940 the Konoe cabinet was formed, and Yōsuke Matsuoka was named Foreign Minister. Matsuoka had served with the Foreign Ministry in Russia before the revolution and had grand ideas about Japan's place in the new world order. In his mind this meant cooperating with the Soviets. After lengthy negotiations, the neutrality pact was signed by Stalin and Matsuoka in the spring of the following year.[93]

In May 1939 Stalin replaced Maxim Litvinov with Vyacheslav Molotov as People's Commissar at the head of the *Narkomindel*. Litvinov was not very favorably disposed either toward Germany or Japan, and so his dismissal paved the way for the Nazi-Soviet Non-aggression Pact in August. Litvinov tended to lump Germany and Japan together, which Stalin did not do.[94] Litvinov, like some Japanese leaders, preferred working with the West (particularly Great Britain and France), and was against cooperating with Nazi Germany and its fascist allies. Stalin did not seem averse to cooperating with Germany or Japan, and evidently he felt that Molotov was the one to carry this policy out. When negotiations with Japan appeared to have become bogged down in March 1941, Stalin's personal intervention secured the neutrality treaty.[95] Of course, history shows that this was but a temporary phenomenon.

The ideational context

There was little room for ideas and cultural exchange during the tumultuous years of the Second World War. The exigencies of war demanded that both countries formulate policy and strategy based purely on national interests and national survival.

In Japan the group of leaders opposed to the Soviet Union and the communist ideological threat remained powerful throughout the war. After 1940, however, the group favoring *rapprochement* with the Soviet Union actually won out. In spite of the hatred for the Soviet Union, the national strategy was based on meeting the immediate needs of the war. In spite of the ideological gulf there was some admiration in Japan toward the fighting qualities of the Red Army soldiers (instilled no doubt after Nomonhan). Japanese generals were even told by American and British observers in the late 1930s and early 1940s that the Japanese Imperial Army was the purest form of communism.[96] The influence of Russian literature still lingered, and in one of his most famous novels (*Sasame Yuki* – or, 'The Makioka Sisters,' in its English translation) Junichirō Tanazaki touched briefly on the lives of one Russian émigré family living in the Osaka area during the war. One son of Russian émigrés in Tokyo even became a famous baseball player in the Japan pro leagues both before and during the war.[97] As mentioned, by the end of 1944 certain circles in Japan were in favor of bringing in Soviet mediation to end the war. Japanese diplomats made unsuccessful efforts throughout the first half of 1945 to bring this about.[98] Up until the bitter end the Japanese leadership (including the emperor) convinced itself that the Soviet Union could be an impartial arbiter.[99]

Policy in the Soviet Union was quite simple. Ideology ceased to have the power in the Soviet Union that it did prior to the war. Stalin reached back into Russian history to find heroes and traditions to rally the people of 'Mother Russia'. The Soviet leadership dealt with fascists and democrats alike in order to keep from being overwhelmed from all directions. Stalin made a deal with the Japanese to quiet the east while the Soviet Union dealt with a growing crisis in the west. Although the crisis in April 1941 did not seem quite as dire, this would change for the Soviets in June 1941. Citizens and politicians in the Soviet Union had little time to think about Japan during the war. Stalin's November 1944 speech (cited above) identifying Japan as an aggressor nation, however, was a clear signal that it was time to start thinking about Japan as an enemy.

The breakdown

The ultimate breakdown in Japanese-Soviet relations occurred when Red Army forces invaded Manchuria on August 8, 1945 and smashed Kwantung Army forces stationed there. The *rapprochement* of 1941–45 was a brittle one. Its demise was evident as early as two months after the signing of the neutrality pact when Germany invaded the Soviet Union on June 22, 1941.

It was in late June 1941 when the Japanese government began debating whether it should take advantage of the opportunity afforded by Germany's invasion to settle the score once and for all with an old foe. Ironically, one of the most forceful advocates for an invasion of the Russian Far East was the very man who had signed the neutrality pact for Japan, Foreign Minister Matsuoka.[100] A July 2, 1941 imperial conference tentatively sanctioned a build-up of forces in Manchuria for a possible push northwards, should the Soviet Union collapse from the Nazi onslaught. But a group led by Prime Minister Fumimaro Konoe insisted on watching events closely in the Soviet Union and in Southeast Asia before any hasty action was taken. Matsuoka's insistence was such that Konoe decided to replace him as Foreign Minister with Admiral Teijirō Toyoda, himself a cautious figure.[101] The decision not to move north was made final when Japanese forces occupied southern Indochina and the United States froze all Japanese assets and imposed an embargo on oil and gasoline exports to Japan in late July 1941. The decision was made then to attack American, British, and Dutch possessions in the Asia-Pacific region.

The roles were significantly reversed as the fortunes of war turned to the allies in 1943. At an allied foreign ministers conference in Moscow in October 1943, Soviet leaders verbally assured the US Ambassador Averill Harriman that the Soviet Union would enter the war against Japan after Germany's defeat. Stalin reiterated this promise to Roosevelt at the Teheran conference in November 1943.[102] In 1944 Japanese fishing rights around northern Sakhalin were restricted and oil and coal concessions were shut down. In a meeting in the fall of 1944 with British Prime Minister Winston Churchill, Stalin again gave clear assurances that the Soviet Union would enter the war against Japan after the surrender of Germany. As planning for the final phases of the war with Japan developed in the United States, both sides began discussing conditions for Soviet entry. By the Yalta conference in February 1945, it was agreed that the Soviet Union would enter the war against Japan three months after Germany's surrender. In return southern Sakhalin and the Kuril Islands would become Soviet possessions. The Soviets would also have lease of certain concessions in China, including the Chinese Eastern and South Manchurian Railroads, and a naval base on the Liaotung Peninsula (close to the former Russian base at Port Arthur).[103]

When Soviet Army Forces attacked the Japanese it probably should have come as no great surprise in Tokyo, but Japanese leaders had hoped to the very end that war could be avoided. Kwantung Army forces had been stripped over the years and this once proud army was a shell of its former self. Although there were pockets of spirited resistance, Japanese forces simply collapsed after a week.[104] And thus, too, collapsed the neutrality whose days were numbered from the very beginning.

The *rapprochement*, or 'temporary neutrality,' was a factor of the international structure during the turbulent years of the Second World War. So was the breakdown. Geopolitical interests and war exigencies dictated policy in both nations. International factors dominated the relationship during this time. Additionally, ideational factors continued to weigh heavily, in a negative sense. The militarism of Japan and communism of the Soviet Union were simply fated to be incompatible.

2 Cold War patterns

The Cold War and its influence on Japanese-Soviet relations

The series of wars fought between 1904 and 1945 left a legacy of deep suspicion and mistrust in Japan and Russia that the opposing sides of the Cold War were able to use for maximum benefit in their respective propaganda campaigns. Although the fighting subsided after 1945, the rhetoric at times was every bit as harsh and the hostility remained just below the surface in both nations. There were periods of brief *rapprochement* during the Cold War that reflected the pre-war pattern of fluctuating relations, but by the early years of the twenty-first century Japan and Russia were no closer to normalization than they had been in 1945.

1955–60: Diplomatic recognition

For ten years after the end of the Second World War, Japanese-Soviet relations remained in a state of suspension. The Soviet Union maintained a diplomatic mission in Tokyo during the occupation years (1945–52), but the mission spent more of its time dealing with the US occupation authorities than with Japan's leadership.[1] The Soviets had representatives on the two Allied consultative bodies that ostensibly governed occupation policy in Japan, but they had no say in the way policy was carried out. When the Allied occupation ended in 1952, the Japanese government asked the Soviet mission to leave, and most members returned to the Soviet Union, but a skeleton staff was maintained at the embassy building in Tokyo.

As early as 1952 it was clear that at least some leaders on both sides were interested in establishing diplomatic relations, despite the fact that the Soviet Union had refused in 1951 to sign the San Francisco Peace Treaty that ended the Pacific War.[2] On New Year's Day 1952 in a radio address Joseph Stalin sent his greetings to the Japanese people, hinting at the Soviet Union's desire to establish friendly relations. But six months later the Soviet mission in New York vetoed Japan's application for admission to the United Nations.[3] In Japan political factions within the conservative leadership and certain business circles also were interested in establishing diplomatic relations with the Soviet Union immediately after the occupation.[4] Yet, domestic political events in Japan and in the Soviet Union kept the two nations from being able to deal with this issue for the next few years. In Japan, post-occupation political rule was still being sorted out. In the Soviet Union, Stalin's death in early 1953 set off a political struggle and it was a year before the picture became somewhat less muddled, and several years before Nikita Khrushchev was truly able to establish himself at the top of the Soviet hierarchy. The political leadership in both nations was simply unable and unwilling to deal with the relatively minor issue of diplomatic recognition while questions of political hierarchy were being wrestled with at home. Admittedly, in Japan, because of the territorial issue, UN membership, and the repatriation of Japanese POWs imprisoned in the Soviet Union, diplomatic recognition with the USSR was of greater importance. As long as the Soviets were unwilling to negotiate, however, there was little the Japanese leadership could do.

The Kremlin made the first move toward Japan, establishing a pattern that would also repeat itself throughout the Cold War. Soviet Premier Georgii Malenkov called for the normalization of relations with Japan in an August 1953 speech. A year later in September 1954 Soviet Foreign Minister Vyacheslav Molotov announced that the Soviet Union was ready to establish diplomatic relations with Japan, if Japan were ready to reciprocate. The first formal initiative did not take place until January 1955 when the Soviet government sent a note to the Japanese government requesting negotiations leading to a peace treaty, diplomatic recognition, and normalization.[5] The new Japanese government, led by Prime Minister Ichirō Hatoyama of the Democratic Party, agreed to negotiations despite the hesitancy and initial opposition of his Foreign Minister, Mamoru Shigemitsu, and bureaucrats in Japan's Ministry of Foreign Affairs.

Hatoyama appointed Shin'ichi Matsumoto as Plenipotentiary Ambassador to the negotiations. Matsumoto was a career diplomat and had the confidence of the Prime Minister. For Japan the negotiations centered not just on the twin issues of a peace treaty and normalization, but also on the more pressing matters of the repatriation of Japanese POWs (of whom there were an estimated 170,000 still in Soviet camps in the early 1950s), fishing rights in the disputed northern areas, and UN membership. The first round of negotiations was held in the summer of 1955 in London. Matsumoto's Soviet counterpart was Iakov Malik, the Soviet ambassador to Great Britain. The negotiations were cordial but became bogged down over the territorial dispute. Japan initially asked for the return of the entire Kuril archipelago, as well as the southern half of Sakhalin Island.[6] The Soviets countered by offering to return the two southernmost of the Kuril Islands, Shikotan Island and the Habomai group of islands, which in fact met the original initial minimum Japanese requirement for a peace treaty. The Soviets also intimated that they would not insist on a US military withdrawal from Japan upon the signing of a peace treaty. Matsumoto indicated that these terms were acceptable but that he would need to check with Tokyo. Back in Tokyo Shigemitsu and the Foreign Ministry insisted that Matsumoto demand the return of the 'Northern Territories' (i.e. the four southernmost of the Kuril Islands) and that the Soviets agree to address the status of the rest of the Kuril Islands and southern Sakhalin at a later conference that included Allied (British and US) participation. When Matsumoto presented these new demands to Malik the negotiations were broken off. The issue of POWs was left unresolved as well.[7]

The negotiations were resumed in London in January 1956 but no agreement was reached and the stalemate continued. Fishery talks were taken up in the spring after the Soviet Union published a set of restrictions on salmon and trout fishing in the northern Pacific waters contiguous to the Kuril Islands. The limit on fishing as well as the growing clamor to bring home the POWs brought a sense of urgency to the Japanese government. Normalization seemed more important than ever. Negotiations were continued in July 1956 in Moscow, this time with Foreign Minister Shigemitsu at the head of the Japanese delegation. After initially holding out for the return of Etorofu and Kunashiri, as well as Shikotan and the Habomai group, Shigemitsu appeared willing to agree to the return of only the latter two after seeing the hard line taken by the Soviet leadership (he conferred personally with Soviet Foreign Minister Dmitri Shepilov, chairman of the Council of Ministers Nikolai Bulganin, and Khrushchev).[8] Ironically, it was Shigemitsu's hard line back in 1955 that pushed the Soviets into a tougher negotiating stance. When Shigemitsu consulted with Prime Minister Hatoyama in Tokyo, he was instructed to hold out for all four of the southern island groups, as Japanese public opinion had hardened in the last few months, mainly due to the publicity efforts of Shigemitsu's own Foreign Ministry. As historian Tsuyoshi Hasegawa writes, "Shigemitsu became a victim of his own public relations [campaign]."[9] Still it was unclear whether or not the Japanese would accept the two-island formula.

The United States entered the picture at this point. Like a large number of Japanese politicians in the conservative ranks, influential policymakers in the United States did not want Japan to accept just Shikotan and the Habomai group. They expected Japan to hold out for all four of the southern islands. Shigemitsu traveled to London in August 1956 during a break in the negotiations to attend a foreign ministerial conference to discuss the Suez Crisis. There he met with US Secretary of State John Foster Dulles. Dulles informed Shigemitsu that he was not at all happy with the idea of Japan merely accepting Shikotan and the Habomai group in exchange for a peace treaty. He stated that the United States was prepared to invoke article 26 of the San Francisco Treaty, which promised equal treatment to any signee of the treaty should Japan subsequently sign a new peace treaty with another nation. He brought up the issue of article 26 with the understanding that should Japan acquiesce in the Soviet annexation of the Kuril Islands, the United States might not be obliged to return Okinawa and the Ryukyu archipelago to Japan. Dulles urged Shigemitsu to demand the return of all four southern island groups.[10] Shigemitsu, taken aback by what he viewed as a veiled threat of his American counterpart, had to then return to Moscow and inform the Soviets that Japan could not accept the two-island formula. Most Japanese and Russian historians have blamed the United States for sabotaging the peace talks through the 'Dulles threat incident.'[11] It is not entirely clear what Dulles meant with his warning to Shigemitsu. Even Dulles' own State Department seemed taken aback by his words.[12] Three American historians have argued that Dulles' intentions were benign and that he was hoping to convey more leverage to the Japanese side during negotiations. An article authored by these historians based on recently published archival material argues that Dulles urged Shigemitsu to use the specter of a permanent American occupation of Okinawa as an inducement with which to convince the Soviets to return the four 'Northern Territories'.[13]

Whatever Dulles' intentions may have been, the discussions between Japan and the Soviet Union were once again stalemated. At the eleventh hour, Japanese Prime Minister Hatoyama decided to lead a delegation to Moscow to reach some sort of agreement. Prior to his departure Hatoyama had cabled Bulganin with a list of five matters that he proposed to discuss. These were: (1) the technical ending of the state of war; (2) the establishment of diplomatic relations; (3) the repatriation of Japanese POWs; (4) the implementation of a fishery agreement; and (5) Soviet support of Japan's UN membership.[14] The territorial issue would be discussed only after these issues had been settled. The Soviets accepted and in October 1956 the Hatoyama delegation arrived in Moscow. The two sides were able to hammer out a Joint Declaration based on the five points outlined above. The Soviet Union agreed that upon the signing of a peace treaty it would return Shikotan Island and the Habomai group to Japan. The legislatures in both countries duly ratified the Joint Declaration by the end of the year. Negotiations culminating in a peace treaty were to be held in the near future, and then the territorial issue could be solved once and for all.

Upon the exchange of permanent diplomatic missions, trade volume increased, Japanese UN membership was confirmed, and, most importantly, the repatriation of Japanese POWs began, the last returning in 1958. In January 1960, the United States and Japan renewed and revised the security treaty that had been signed in 1951, allowing US bases to remain in Japan. Massive protests in Japan signaled widespread public discontent over the new treaty. Khrushchev took this opportunity to attempt to sow public discontent in Japan, and to push Japan and the United States apart by unilaterally declaring that the Soviet Union would only sign a peace treaty and return two of the disputed islands when the last foreign (i.e. US) soldier left Japanese soil. Khrushchev was in fact reneging on an agreement that his government had already ratified. The Japanese response was terse – Japan would determine its own security policy independent from outside counseling. Khrushchev continued to insist on a US withdrawal.[15] From that point on over the next decade diplomatic relations were barely cordial. Thus by the end of 1960 the brief era of Japanese-Soviet *détente* had come to an end.

The international context

Not coincidentally the brief *rapprochement* between Japan and the Soviet Union came about during an era of international flux. The events transpiring between 1953 and 1956 were momentous in scale, and a quick glance can provide an understanding of how the events of this period influenced decision-making in both Japan and the Soviet Union.

The major event of early 1953, the death of Joseph Stalin, helped to bring to a close the bloody Korean conflict that had so destabilized Northeast Asia. Both Japan and the Soviet Union had actively participated, in auxiliary roles, in that conflict. Japan was America's major supply base; the Soviet Union was the major supplier of hardware to both the North Korean and Chinese forces fighting on the peninsula. In some cases both Japanese and Soviet individuals participated in combat or direct support operations.[16] Another major event affecting international relations in Asia was the ending of the war in Indochina and the French withdrawal from Southeast Asia in 1954. Obviously, the revolution in China (1949) and the independence of Indonesia (1949) affected the international climate in East Asia during the early 1950s as well.

In Europe major changes were also under way. In 1955 the Soviet Union negotiated neutrality treaties with Austria and Finland, in both cases withdrawing troops from those countries. In 1955 the Soviet Union and West Germany established diplomatic relations, indefinitely postponing a territorial settlement (in the same year West Germany joined NATO). The 'Spirit of Geneva' resulting from the 1955 American-Soviet summit meeting in that city signaled a thawing in East-West relations across the globe. Khrushchev had announced that international relations should be based on the principles of 'peaceful coexistence.' His speech at the twentieth Party Congress of the Communist Party of the USSR in which he denounced Stalin and the cult of personality was a further indication of the newfound quest for global normalcy after the dark early days of the Cold War.

It is against this background of global change that the events surrounding the Japanese-Soviet *rapprochement* should be understood. To be sure, there was still conflict and strife across the globe (Guatemala, Hungary, Indochina, Suez, the Taiwan Straits, etc.), but both sides in the Cold War wished to step back from what they feared could become a nuclear nightmare in Northeast Asia and Central Europe. It was in this context that Nikita Khrushchev claims in his memoirs that he wished to approach the Japanese leadership and normalize relations. Of course, he also intimates that he wished to sway public opinion in Japan to somehow drive a wedge between Tokyo and Washington.[17]

The cabinet of Prime Minister Hatoyama in Tokyo was also in search of a new direction for Japan's foreign policy. Partly for domestic political reasons (to be discussed below), and partly out of conviction, Hatoyama felt that Japan needed to wean itself from its over-reliance on the United States. As early as 1952, even before he became Prime Minister, Hatoyama advocated for the normalization of relations with the Soviet Union and the People's Republic of China.[18] Early in 1954 while Yoshida Shigeru was still Prime Minister Japan had recognized the Chinese government on Taiwan. In fact, Yoshida had been in favor of recognizing the communist Chinese government in Beijing but his hand was forced by the United States.[19] The American government, though wary of a Japanese-Soviet *rapprochement*, could not pressure the Japanese government to *not* establish relations with the Soviet Union, because Washington maintained relations with Moscow. Hatoyama was eager to move Japan in a more independent direction, and establishing relations with Moscow was a start (that ultimately was to become an end in itself for the Hatoyama cabinet).

The United States government, meanwhile, kept a vigilant watch on the Japanese-Soviet peace negotiations. US leaders were wary of Soviet intentions, and were nervous that the establishment of Japanese-Soviet relations might be followed by Tokyo's establishment of relations with the communist government in Beijing. Leaders in the United States were also concerned that should the Soviet

Union return the disputed 'Northern Territories' to Japan, demands for the reversion of Okinawa would intensify, and that the Soviet propaganda machine would use this and the continued presence of US troops in Japan to maximum effect. There was in fact a growing opposition in Japan to the US military presence after the occupation, and a series of bloody labor riots on May Day in 1952 added to the American disquiet.[20] Therefore when the Japanese-Soviet normalization talks were under way in London and Moscow, the United States kept a close eye on events. In this context it is much easier to understand the 'Dulles' threat incident.'

By the early 1950s it seemed to most casual observers that Japan was firmly entrenched in the American camp, with China firmly in the Soviet bloc. Fissures in the latter relationship, however, were already apparent to some even before the start of the Korean War. Mao's dislike for the Soviet leadership (and for Stalin in particular) dated back to the 1920s and 1930s, when the Soviet Union and the Comintern's involvement in the domestic politics of China was often at the expense of the young Chinese communist party.[21] Mao's enmity was further enhanced during his highly publicized visit to Moscow in early 1950, when he spent most of his time waiting in a guest dacha for 'audiences' with Stalin.[22] It was probably already clear to the leadership in Moscow in the early 1950s that China would not be satisfied playing the role of a mere client state. The fissures in the relationship became even more apparent in 1956 with China's lukewarm response to Khrushchev's speech at the twentieth Party Congress, where he attacked Stalin and the 'cult of personality.' Mao undoubtedly read into this a veiled criticism of his leadership in China. The Soviet leadership felt that its policy in East Asia was in need of a new focus. Khrushchev and Bulganin had already visited India in 1955 and had been successful in getting Indian leaders to look to the Soviet Union for assistance and guidance. In Northeast Asia, however, relations with China were proving difficult. Khrushchev and the Soviet leadership, meanwhile, were optimistic that Japan could somehow be separated from the American-led camp.[23]

The impact of the international situation in the early and mid-1950s was every bit as large as it had been in past eras of Japanese-Russian *rapprochement*, and much like the earlier eras the role of the United States continued to be a major one.

The domestic political context

As mentioned earlier, the changes in the international system in the mid-1950s mirrored changes under way in the respective domestic political arenas of both Japan and the Soviet Union. Japan was still grappling with the post-occupation political issues in Tokyo. Similarly, the Soviet Union was dealing with its own political infighting in the Kremlin.

By the mid-1950s the post-occupation political landscape in Japan had become relatively polarized. The two political forces competing for control were not left versus right, but rather two competing forces within the conservative leadership. Socialist political forces had been strong in the early days of the occupation, and a socialist prime minister had even formed a coalition cabinet in 1947–48. But the onset of the Cold War and the 'reverse course' in US occupation policy doomed the chances of success for the Japan Socialist Party in its earliest days. Instead, by the early 1950s two conservative parties, the Liberal Party (led by Shigeru Yoshida) and the Democratic Party (led by Hatoyama) vied for control of the government. The two leaders of these parties had originally been political allies. Yoshida was Hatoyama's political protégé when US authorities purged Hatoyama early during the occupation, barring him from holding public office. When Hatoyama re-emerged on the political scene in 1952, Yoshida refused to give up the reins of the conservative movement that Hatoyama had bequeathed to him in 1946.[24] Even when the two parties merged to form the Liberal Democratic Party (LDP) in 1955 the struggle between the two did not die off. When Hatoyama pushed for normalization with the Soviet Union in that same year, Yoshida's faction fiercely opposed him.[25]

Consequently, one must consider whether Hatoyama's decision to re-establish relations with Soviet Union was pushed through to a conclusion before Japan was ready to arrive at an agreement mutually satisfactory to both sides. The five points agreed to in the Joint Declaration were amicably adhered to, but in the end Japan settled for the return of no territory. As Hatoyama's health failed and as his faction's power base became shaky, the move to normalize relations took on an extra urgency for his cabinet. It was in fact the last major act of his career. He died shortly after his return from the peace talks in Moscow. In spite of the political infighting in Japan it must be remembered that the normalization talks with the Soviet Union did have an urgency of their own, due to the issue of UN membership, the matter of fishery agreements, and, most importantly, the grave problem of repatriating Japanese POWs from Siberia and elsewhere in the Soviet Union.

In Japan the prime minister and top political leaders dominated the formulation of Soviet policy in the early post-war years. The Ministry of Foreign Affairs (MOFA) was still relatively weak, in spite of the strong leadership of Shigemitsu. Many bureaucrats in the ministry were still very loyal to Yoshida (himself a career diplomat), and this forced Hatoyama and his followers to work as much as possible around the Ministry of Foreign Affairs. In some ways, MOFA officials were seen as more of a hindrance than an asset to the Prime Minister.[26] The Ministry of Agriculture and Forestry played a big role, due to the strong leadership of its minister Ichirō Kōno, who had a close relationship with Hatoyama. Hatoyama used Kōno at times to bypass the Foreign Ministry. For example, during Japanese-Soviet fishery talks in early 1956 Kōno was given a green light to discuss the territorial issue by the Prime Minister. The role of Japanese big business (*zaikai*) also helped speed Hatoyama along in making his decision. Certain businesses were anxious to import Soviet products – mainly in the lumber and fishing industries – and they let this be known to Hatoyama and the rest of the government.[27] At the same time, other business groups more closely linked to trade with the West wished Hatoyama to go slow so as not to antagonize the United States.[28] Hatoyama also had to take into account the growing clamor among the Japanese public to bring home the POWs.

The Soviet leadership was still dealing with issues of succession in 1955–56. Though the Bulganin-Khrushchev clique had seemingly gained the upper hand over the Malenkov-Molotov faction, Khrushchev probably felt himself on shaky ground given the history of internal Soviet politics over the preceding three decades. Nevertheless, the general atmosphere of *détente* afforded the Soviet Union opportunities in areas around Asia where it had previously been shut out. Japan was one of them.

As for the internal driving forces behind the Soviet Union's new Japan policy, there are several possible theories. Soviet foreign policymaking was strictly centered on the upper echelons of the government hierarchy. The International Department of the Central Committee of the Communist Party and the Ministry of Foreign Affairs (MID) were thought to have some influence over Japan policy. Indeed, they were probably important in fleshing out details given larger guidelines by the Politburo, but the ultimate power rested at the top level. It has been assumed that the final decision to approach Japan was made by Khrushchev himself. In his memoirs Khrushchev insists that he had to convince Molotov that the Soviet Union had to normalize relations with Japan: "Since we had absolutely no contacts with Japan, our economy and our policy suffered."[29] Khrushchev probably did not really believe the Soviet economy suffered dramatically due to the lack of trade with Japan. He wrote his memoirs in the late 1960s when Japan was establishing itself as an economic dynamo. But in the mid-1950s, Japan's war-shattered economy was anything but dynamic. He did point to the potential for trade in raw materials, but Khrushchev was undoubtedly more eager to exploit political opportunities. This was particularly the case given labor problems in Japan and anti-US sentiment in some segments of Japanese society.

Others are not so sure that Khrushchev had a firm grip on Soviet foreign policy. Russian historian Aleksei Zagorsky attributes the change in Soviet policy toward Japan not to Khrushchev but to Kremlin insiders Lavrenti Beria and Georgii Malenkov. Zagorsky ridicules Khrushchev's grasp of matters pertaining to foreign affairs, and argues that the initiatives to improve relations with Japan came about sometime in 1953 after the death of Stalin but before the accession of Khrushchev. Although Japan was not a central focus of the new Soviet policy, it could be seen as an offshoot of the newer, relaxed attitude toward certain countries. According to Zagorsky, Khrushchev merely followed through on initiatives that were already in the pipeline. Unlike proposals that, for example, envisioned the reunification of Germany and which would have aroused great opposition in the Soviet leadership, establishing relations with Japan was an easy, safe route for Khrushchev to follow.[30]

The historian James Richter feels that Khrushchev's foreign policy was more of a compromise between that of Malenkov and Molotov, and that Khrushchev made gestures toward other nations to offset the differences between the two. Furthermore, to counteract the US move to ring the Soviet Union with bases, Khrushchev attempted to make diplomatic inroads into these nations on the periphery of Soviet territory.[31]

The ideational context

After the war ideational factors came to play a large role in bilateral relations. Arguably the Japanese and the Soviet peoples suffered more than any other in the war, apart perhaps from the Chinese.[32] In this sense the war and the memories associated with it came to play an enormous role in the minds of both societies. The war still influences both countries today, perhaps more than anywhere else. Perceptions of self and of other nations are still colored by associations with the war. Japan's relations with its Asian neighbors and its close ally the United States reflect experiences of the war years. Russia still clings to the past glories associated with the war, and at the same time it is still scarred by the traumatic events of those years – the injustices carried out by its enemies, and the injustices the Soviet leadership perpetrated on its own citizens during the war (such as the Chechens, the Volga Germans, or the Koreans in the Far East). The bilateral relationship between Moscow and Tokyo is still heavily influenced by the war, evidenced by the fact that the two sides have still not signed a peace treaty ending the official state of war.

In Japan, in spite of the fact that for almost four years the people fought a battle to the death with the United States, the 'betrayal' by the Soviet Union with its last-minute attack was treated with more anger and indignity. Many Japanese leaders viewed the US occupation as eminently preferable to the introduction of Soviet troops on Japanese soil.[33] All through the occupation and afterwards there was a sincere concern among Japan's conservative leadership and among top US policymakers that the communist/socialist movement in Japan was gaining in strength and might actually win an electoral victory and control of the Diet. As mentioned a socialist prime minister (Testu Katayama) had formed a coalition government in 1947 which lasted barely a year. There was also concern that the split in the conservative camp would allow the leftist forces to gain victory. In fact, Hatoyama had used the aid of two socialist parties to help bring down Yoshida's Liberal Party in the 1954 election.[34] The merger of the two conservative parties to form the LDP, ironically, was undertaken a year later in response to the socialist 'threat.'[35]

Conservative politicians feared that re-establishing ties with the Soviet Union would open up Japan to potential communist subversion. The communist party in Japan had been legalized in 1946 and the wretched economic conditions proved a fertile recruiting ground for members. In addition, the repatriation of Japanese POWs who had been in either Soviet or Communist Chinese captivity added more fuel to the mix. Frequently, such soldiers returned with newly implanted ideas about 'socialism' and 'revolution.'[36] The Japanese Communist Party (JCP) was an outlet for many of these individuals, although a split within the JCP in 1950 was disastrous for the future electoral chances of this party. The United States was concerned enough about communism and socialism that in the 1950s the CIA reportedly began funneling cash to top politicians in the LDP to help keep them in power.[37]

After the 1955 merger the LDP became firmly entrenched with a solid majority in the Japanese Diet. The power of the opposition Japan Socialist Party slowly faded. The conservatives were able to garner even greater support from Japanese big business (*zaikai*) and the United States. As incomes began to rise in the early 1960s, the chances of a socialist electoral victory became ever slimmer.

In addition to concerns about the rise of communism and socialism in Japan, and lingering public anger over the sudden Soviet attack and seizure of territory in

1945, the matter of Japanese POWs fueled anti-Soviet sentiment. Some 640,000 POWs had been kept in the Soviet Union after the war, and though the majority was returned in 1947–49, there were still approximately 170,000 unaccounted for. Furthermore, of 2.7 million Japanese civilians in territory captured by Soviet forces at war's end, 150,000 were missing. Soviet information on the number of prisoners was at variance with Japanese statistics, and this led to further tension. The last POWs were returned in 1958, but close to 60,000 had died in captivity.[38] In spite of the cruel treatment meted out to Japanese prisoners in the Soviet Union, and the harsh attitude of Japanese citizens to this bit of history, the Japanese government might be seen as somewhat complicit. According to Herbert Bix in his biography of Emperor Hirohito, the Japanese government considered using the forced labor of Japanese POWs in the Soviet Union as a form of reparations payment for war damages, and hence turned a blind eye for a while.[39] Furthermore, certain officials in Japan insisted that Japan could sacrifice its POWs in order to uphold 'national prestige.' One of these was *Keidanren* (Federation of Economic Organizations, a powerful business lobby and quasi-governmental organ) President Taizō Ishizaka who stated that, "neither Japanese captives nor … fishing rights should be traded for our national prestige nor our territory [sic]."[40] It should also be remembered that Japan was not a signatory of the Geneva Convention on the treatment of war prisoners, and its treatment of captured soldiers and civilians (especially in China) was probably no better than that meted out by the Soviets. This is not to condone what the Soviets did to the Japanese POWs, but the Japanese government could be seen as partially responsible for what happened. During the war the Japanese viewed prisoners of war as those without honor and indeed without country; now ironically these POWs had become a *cause célèbre*.

The issues associated with these wartime experiences left a feeling of extreme dislike for the Soviet Union among the Japanese. Although an anti-American sentiment existed in various segments of Japanese society, it was superseded by a reinvigorated hatred for the Soviet government. The LDP's ability to capitalize on this feeling (with US concurrence and assistance) helped it to retain uninterrupted political control of the Japanese government for 38 years.

After the war years and Stalin's obsession with *Realpolitik*, the Soviet Union once again turned to ideology as a tool in foreign policy. Soviet leaders felt that Japan offered fertile ground for the spread of communism. In his memoirs Khrushchev stated that the Soviet Union should take advantage of any anti-American feelings still lingering because of the destruction wrought on Japan by the US atomic bombings. He went on to say:

> As soon as our embassy reopens in Tokyo, it will act like a magnet, attracting all those who are dissatisfied with Japan's current policies. These elements will begin to establish ties to our embassy, and we will begin to exert some influence on Japanese politics.[41]

Khrushchev viewed Japan through the lens of the growing third-world non-alignment movement. His goal was not to set up Japan as a Soviet client, but merely to pull it out of the US orbit through peaceful means. As it turns out, Khrushchev misread the extent of anti-American feeling in Japan. It existed, but it was tempered by widespread support for the United States, its occupation policies, and its anti-communist policy in East Asia.

By 1960 Khrushchev felt that Japan was an even better target for Soviet attentions. The 1960 anti-treaty riots in Japan were seen in Moscow as bringing Japan closer to a "revolutionary situation."[42] The plan to denounce the 1956 Japanese-Soviet Joint Declaration until all US troops were withdrawn from Japan ended up backfiring, as Moscow's actions did nothing to speed along the process of 'revolutionary socialism' in Japan. Japan actually grew closer to the United States and considered the Soviet actions as a breach of a ratified pact and of good faith. By the early 1960s the Soviet Union's attention was turned more and more toward events in Europe, the Middle East, and the Western hemisphere.

The perception of Japan in the Soviet Union was different than it had been before the war. The people of the Soviet Union had viewed Imperial Japan as an aggressor of the worst sort. After the war, Japan barely registered in the minds of the Soviet people. Later, it had taken on a quaint, exotic image.[43] This view is visible in Khrushchev's memoirs, and he probably shared more in common in his views and opinions with the average Soviet citizen than any Soviet or Russian leader has before or since. Nevertheless, neither Khrushchev nor the average Soviet citizen was amenable to returning territory that they felt was earned by the nation's travails during the war.

The breakdown

Both international and domestic political factors helped bring Japan and the Soviet Union to a *rapprochement* in 1956. By 1961 Khrushchev had admitted that relations, though on a sounder footing than after the war, were still not completely normalized.[44] After Khrushchev's denunciation of the 1960 US-Japan security treaty, political relations between Moscow and Tokyo could no longer be said to be amicable. Although trade volume did increase dramatically, this was from a zero baseline. Japan did become an important trade partner for the Soviet Union, but overall bilateral trade figures were still relatively low as a percentage of total trade.[45] Meanwhile, Soviet leaders continued to insist on a Japanese renunciation of the security alliance with the United States.[46]

The international situation following the *détente* of the mid-1950s deteriorated into a second phase of the Cold War. Domestically, in Japan the growth in conservative strength did nothing to improve the bilateral atmospherics. More importantly, the negative images associated with Soviet actions in the war continued to poison the public perception in Japan. The threat of communism spreading throughout Northeast and Southeast Asia not only frightened Japan's conservative leadership but it also frightened the Japanese public.[47] Although China was seen as a concern, the bigger threat seemed to emanate from the north. And once Soviet missiles could reach Japan in the late 1950s this shaped the negative public perception even more. In the end, territorial issues combined with the issues surrounding the war to doom any chance for a meaningful *rapprochement* with the Soviet Union. As historian Tsuyoshi Hasegawa wrote:

> To the Japanese the demand for the Northern Territories represented more than a wish to recapture lost lands. It was rather a moral imperative to right all the wrongs that had been perpetrated by the Soviet Union in the summer of 1945. The petulance and tenacity with which the Japanese government clung to the return of the Northern Territories can be explained by the self-righteous belief that few people in Japan have questioned – a belief that all justice rested on their side .... What separated them [from the Soviet Union] was therefore not merely a difference in views on the territorial dispute, but also the more profound differences in historical memory about the war and the psychological needs to cling to the myths they had created.[48]

The Soviet leadership, meanwhile, viewed Japan more and more as an American appendage, especially when it became clear that the social unrest of the 1950s was a thing of the past. No attempt was made by the Soviet leadership to truly cement relations and build a firm foundation. Improving relations with Japan was seen as nothing more than an opportunity to poke the Americans in the eye. There seemed to be little more substance to Soviet motives than this. They did not truly see Japan as a potential trade partner, or as a nation that could and would formulate an independent foreign policy course. Therefore once resistance to normalization was felt in certain areas, it was easier to turn away than to stay the course.

In spite of the international and domestic factors that pushed Japan and Russia together briefly in the mid-1950s, the two nations could not get over certain biases and ideas. Public perceptions of one another were often very far from the mark, and in the end it was in this graveyard of misperception that normalization attempts foundered. In this period of *rapprochement*, structural factors were again key in bringing the leaders of the two nations together. But both domestic issues (especially in Japan) and ideational factors were also important in bringing about the *rapprochement*. But these same ideational factors were part of the breakdown. The Japanese leaders and the Japanese people were still scarred from the war and still viewed the Soviet Union through this lens of the war experience. There could be no true normalization, in their mind's eye, with the Soviets.

1972–73: The search for partners in an uncertain time

Amid the great international tumult of the late 1960s and early 1970s both Japan and the Soviet Union began to experience a certain feeling of isolation. Although Japan had strongly backed the United States and its effort in Vietnam throughout the 1960s, the US announcement of troop withdrawals, Nixon's surprise China visit in 1972, and increasing US-Japanese trade friction created a concern in ruling circles in Tokyo that the United States was drifting away from Japan.[49] The Soviet Union was also feeling somewhat estranged from former friends, not only in East Asia but also in Eastern Europe. The most disconcerting issue for Soviet leaders was the Sino-Soviet rift that had exploded into armed clashes and confrontational posturing along the length of their several thousand-mile border. And although the US announcement of a withdrawal from Vietnam was undoubtedly viewed with joy in the Kremlin, it was very quickly tempered by the sudden and warm Sino-American embrace. In one fell swoop the Soviet Union's two gravest enemies were fraternizing like two long-lost friends. Practically overnight East Asia became simultaneously the scene of a great Soviet moral victory over the United States (in Vietnam), and the arena posing the greatest threat to Soviet homeland security. Because of the mutual feelings of isolation and the new bond between the United States and China, leaders in Tokyo and Moscow decided that normalization between the two countries was a worthy goal.

In spite of the fact that the previous attempt at normalization in 1955–56 had fallen short, diplomatic relations had been re-established, and the 1960s were by no means a lost decade in Japanese-Soviet relations. In fact, the two nations had taken advantage of a 1957 treaty of commerce and trade to greatly expand trade relations. In 1956 the two-way trade volume equaled approximately $3.6 million. The next year (1957) it jumped to roughly $21.6 million, and by 1972 the bilateral trade level had already reached $1 billion.[50] Soviet exports to Japan of raw materials such as lumber, marine products, raw cotton, coal, non-ferrous metals, gold and other items soon outpaced Japanese exports to the Soviet Union of steel, machinery, machine tools, textiles, chemical products, and other materials; until 1975 Japan ran a consistent trade deficit with the Soviet Union. Japan became one of the Soviet Union's largest capitalist trading partners (the largest in 1970), but in comparison with overall trade figures the bilateral trade level remained a fraction of each nation's total trade volume. The Soviet Union, for example, never ranked among the top ten trading partners of Japan. By the end of the decade, however, Japanese businesses and government leaders became more and more interested in the prospect of Siberian energy development.[51]

During the 1960s there were no bilateral summits between the heads of state. The respective Foreign Ministers did exchange visits in 1966–67, while other high-ranking government officials and business leaders visited either Moscow or Tokyo, and cultural relations were somewhat reawakened. Nevertheless, political negotiations were non-existent. Japanese leaders still insisted on discussing the territorial dispute, while Soviet leaders refused to acknowledge that a territorial dispute even existed.

By the early 1970s, however, it became apparent that leaders and policymakers in both nations were interested in a *rapprochement*. Not only was the international situation in East Asia in a period of transition, but internal political and economic factors were aligning to the point that policymakers in both countries were able to contemplate more freely a substantive *rapprochement*. Attempts at *rapprochement* prior to 1972 had been greeted with suspicion in both capitals. Japanese Prime Minister Eisaku Satō had hopes in the late 1960s to completely normalize relations with Moscow, but his overriding concern was concentrated on the disputed islands, and he had no vision for a working relationship beyond the issue of territory. Similarly, Soviet General Secretary Leonid Brezhnev's 1969 proposal for a collective security system in Asia (which included overtures to Japan) was seen by most Japanese policymakers as a ploy to pull Japan away from the United States.[52] But the dramatic international transformation described above, as well as disconcerting economic trends in both countries, forced a change in thinking.

The first fruit of this new line of thinking was Soviet Foreign Minister Andrei Gromyko's visit to Tokyo in January 1972, the first such visit since 1966. Prior to the visit the Soviet press had begun painting a friendlier picture of Japan, no doubt to prepare the public for the upcoming change in Soviet East Asian policy.[53] Gromyko had also met with the chief of the Japanese delegation to the United Nations in the fall of 1971. During this meeting in New York Gromyko reportedly sensed the Japanese nervousness toward the Sino-American *rapprochement*, and it was here that he received the official invitation to Tokyo.[54] Gromyko, who had earned the nickname 'Mr. Nyet,' could not have approached his 1972 visit any differently than his previous visit in 1966. In 1966 he assumed a dour countenance and maintained it throughout his trip, failing to even acknowledge that a territorial dispute existed. By contrast, in 1972 he arrived bringing a different attitude that helped create a positive atmosphere. In fact, so changed was Gromyko that the Japanese ambassador to Moscow, Kin'ya Niizeki, referred to the visit as "Gromyko's Smiling Diplomacy."[55] During his visit Gromyko, though careful to avoid acknowledging the territorial dispute, did not dismiss it out of hand as he and other Soviet leaders had done in the past. In fact, he hinted that the Soviets would be willing to return to the 1956 promise to return two of the disputed islands upon the signing of a peace treaty. Gromyko also requested Japanese economic assistance in developing energy resources in Siberia. Specifically, he asked for a Japanese loan to build a pipeline linking Irkutsk (located near the Mongolian border and already connected with oil and gas pipelines to fields in Western Siberia) to Nakhodka on the Pacific coast just east of Vladivostok. At the end of his visit a promise was secured to open negotiations toward a peace treaty and an invitation to visit Moscow was extended to the Japanese Prime Minister (Eisaku Satō, at the time).[56]

In October of 1972, Japanese Foreign Minister Masayoshi Ohira reciprocated Gromyko's visit and arrived in Moscow to represent the new cabinet of Prime Minister Kakuei Tanaka, who had succeeded Satō in July of that year. Tanaka's first major foreign policy act was his recognition of the People's Republic of China in September 1972. Now he wished to move his country closer to the Soviet Union. The atmosphere during Ohira's visit was decidedly less warm than during Gromyko's visit earlier in the year. This was in large part due to Japan's normalization of relations with Beijing, which the Soviet Union was less than happy about. Ohira spent more time answering questions about Japan's China policy than about Japan's Soviet policy.[57]

Tanaka's visit to Moscow finally took place in October 1973. Just prior to his visit, the Yom Kippur War broke out in the Middle East. Like Ohira's visit a year before, the atmospherics were much less warm compared to Gromyko's 1972 Tokyo visit. During his earlier visit Gromyko had invited the Japanese Prime Minister to visit Moscow, but at the time the Japanese prime minister was Satō. Tanaka had sent a letter to Brezhnev in January 1973, proposing a Moscow visit and negotiations toward a peace treaty. Because Tanaka had in a sense invited himself, and because he had directly mentioned the territorial dispute in his overture, the Kremlin put him on ice for a while. But the summit was eventually held in October. Tanaka arrived and by the second day substantive negotiations were under way. Tanaka made it clear from the beginning that a resolution of the territorial dispute was Japan's primary goal, though the expansion of economic contacts should not be held hostage to this issue. The principle of the 'inseparability of economics and politics' (*seikei fukabun*) that was to define Japan's Soviet and Russia policy in later years had still not become the abiding principle in Tokyo's policy with Moscow. Brezhnev made a case for Japanese investment in the development of Siberia, and an interest existed on the Japanese side (although other factors beyond the mere question of economic feasibility were complicating these prospects). The Soviet leadership also reiterated its desire for Japan's participation in an Asian collective security system. Japan was amenable to the former offer, but lukewarm to the latter. The two sides parted, agreeing to continue discussions toward a peace treaty, although there was a miscommunication about the wording of the final joint communiqué summing up the summit meeting. Included in the communiqué was a promise by both parties to strive their utmost to resolve 'outstanding issues' (or 'unresolved problems') dating from the Second World War. The Japanese side believed this was a tacit affirmation that the Soviets agreed to resolve the territorial dispute before signing a peace treaty, but the Soviets adamantly denied that this communiqué even hinted at the territorial dispute. Later, participants of the meeting would argue whether Brezhnev had uttered 'Da' or had merely cleared his throat in response to Tanaka's query whether the 'outstanding issues' referred to the territorial dispute.[58]

The optimism that had arisen during the two years of awkward dancing around a *rapprochement* and the blind groping for some sort of normalization vanished after the October 1973 summit. Another visit to Tokyo by Gromyko in 1976 proved unsuccessful in mitigating the mutual acrimony. Economic contacts, however, continued to flourish and the bilateral trade volume grew at an accelerated rate through the end of the 1970s.

The international context

As has historically been the case, seismic shifts in the international system helped to trigger Soviet and Japanese overtures to one another in 1972. Although 1972–73 is generally seen as the beginning of the second era of *détente*, it must be remembered that the several years just prior to this were years in which all three great powers were engaged in conflict in East Asia. The United States was warring in Vietnam, while the Soviet Union and China came to blows with one another along their border in Northeast Asia. And although the powers stepped back from conflict in the early 1970s, the situation was still fairly tense in East Asia. High concentrations of battle-ready American, Chinese, and Soviet armed forces existed in close proximity to one another from Sakhalin to Saigon. Soviet and American proxies continued to do battle from Southeast Asia, through the Middle East and Africa, into Latin America. Not only did Chinese and Soviet troops engage in conflict, but leaders in the Kremlin also considered strikes against China's atomic assets and even reportedly solicited American support in this effort. In the Middle East in the fall of 1973 US and Soviet forces were on a hair-trigger alert that threatened to expand the Yom Kippur War into a global conflict.[59]

One prominent Japanese historian has claimed that the events that transpired in 1972–73 were of more significance for the nations of East Asia than the fall of the Berlin Wall in 1989.[60] The events of this period were certainly of enormous consequence for Japan. Japan's feeling of shock over Nixon's secret overtures toward China cannot be overstated. It was even more of an unexpected shock to the Japanese leadership than had been the announcement of the Molotov-Ribbentrop Pact in August 1939. Unlike Nazi Germany in 1939, the United States was in 1972 (as it is today) Japan's guarantor of security. What had made Nixon's action even more galling to the Japanese was that it came in the midst of serious bilateral trade friction, and signaled, many felt, a concerted US effort to move away from, and even isolate, Japan in the East Asian region.[61]

The effects of the changing world situation on Japan were not confined to East Asian events. The first oil shock in 1973 in the wake of the Yom Kippur War left an indelible imprint on the Japanese. At the time, Japan relied on imported oil for close to 75 percent of its domestic energy needs, and 90 percent of this imported oil originated in the Middle East.[62] Oil prices skyrocketed and the Japanese gross national product (GNP) recorded negative economic growth in 1974 for the first time since the Second World War. It was also the first time since the Second World War that Japan found itself at a *disadvantage* because of its close ties to the United States. The Arab oil exporting nations tagged Japan as an uncritical ally of the United States and as a tacit supporter of Israel. Japan staggered, politically and economically, under the weight of these events.

The events of the early 1970s imparted to the Soviet leadership a paradoxical mood of newfound confidence, tinged with a bit of unease. The Soviet Union had achieved strategic parity with the United States, had seen its long time nemesis get a bloody nose in Vietnam, and the halcyon days of the Brezhnev era reached their peak before the stagnation of the command economy became apparent. At the same time, among the leadership there was knowledge of the beginning of economic troubles, while the Sino-American *rapprochement* was the worst of Soviet nightmares that seemed to be realizing itself. Furthermore, Soviet-American global confrontation in the Third World, either direct or through proxy, continued apace. Soviet leaders could not afford to rest easy. Most troubling of all for Soviet leaders, of course, was China. The only situation that could be seen worse than the current one that was unfolding would be a trilateral US-China-Japan alliance aimed at the Soviet Union. Such a scenario was to be avoided at all costs. The Soviet leadership moved quickly, and when it was established that the Japanese leadership felt every bit as uneasy about the Sino-American *rapprochement* (during Andrei Gromyko's trip to the United Nations in the fall of 1971) an overture to Japan was extended. Another factor that gave the Soviet leadership hope that an arrangement could be worked out with Tokyo was the August 1970 treaty signed by the Federal Republic of Germany recognizing the status of Europe's post-Second World War borders.[63]

The Soviets hoped to outdo the United States and China by cementing some sort of agreement with Japan, whether a bilateral peace treaty, or Brezhnev's idea for an East Asian collective security organization. Gromyko visited Tokyo in January 1972 one month before Nixon visited China, hoping to achieve progress in diplomatic relations that could lead to a substantial breakthrough.

At subsequent meetings between Japanese and Soviet diplomats, the topic of China frequently arose, whatever issue the negotiations might be covering. As had been the case prior to the war and after, the United States was still a primary factor in Japanese and Soviet bilateral discussions. But now a familiar element had re-entered the picture, this time on a different level. China had always been a factor in Japanese-Russian/Soviet relations, dating from the 1894–95 Sino-Japanese War. Now, however, China had emerged as a political player, capable of affecting the strategic balance between Moscow, Tokyo, and Washington. China was no longer nothing more than booty to quarrel over; it was now a major actor. Beginning in the 1970s the KGB made Japan its primary base for gathering intelligence on China and recruiting Chinese agents.[64]

The domestic political context

As prominent as international factors may have been to the brief Japanese-Soviet *rapprochement* of 1972–73, domestic factors were closely linked to these large-picture issues and in combination they proved crucial to the changes in thinking at the top levels of leadership in both nations. Although there was no great transition under way politically in either country in the early 1970s, subtle changes could be detected at the very summit of leadership. Additionally, economic factors re-entered the picture prominently in the early 1970s. Economic relations had always been somewhat important in Japanese-Soviet relations, but they were often seen as secondary to political and strategic factors. By the 1970s, economics were directly linked to strategic issues.

In Japan in the early 1970s the domestic changes under way were primarily in response to world events. The oil shock had caused a serious disruption in Japan's economic machine. As it turned out, the Japan economy rebounded and responded quite well to the changes associated with higher oil prices. Nevertheless, even prior to the 1973 oil shock, Japanese leaders had been concerned about the long-term security of oil supplies and had wanted to find alternative sources in order to lessen dependence on Middle Eastern sources.[65] Japanese plans to exploit sources in Siberia had existed for years, and Japanese Prime Minister Kakuei Tanaka was a particular proponent of Siberian energy development. Prior to assuming his post at the head of the cabinet in July 1972, Tanaka had shown great interest in a project to develop petroleum resources in the Tyumen region of western Siberia as head of the Ministry of International Trade and Industry (MITI). The project also involved the construction of a pipeline linking the Tyumen fields via Irkutsk to the Pacific port of Nakhodka. Later, upon assuming the premiership, Tanaka became even more attracted to the Tyumen project, to the point of completely separating this issue from the territorial dispute. In a letter to Brezhnev in March 1973, Tanaka stressed the importance to both sides of the Tyumen project. Original estimates for the project were around $1 billion, but as cost estimates expanded Japanese interest contracted, especially when the Soviets announced that they would transport the oil via a new northern branch of the trans-Siberian railroad whose costs were to be covered by Japanese financing. Although Japanese businesses continued to show interest in the project, government opposition seemed present in all the concerned ministries.[66]

Japanese Prime Minister Kakuei Tanaka was a member of the LDP, as all of his predecessors had been since 1955, but this is where the similarities ended. Unlike most of the LDP leadership, Tanaka had not been a politician before the war, and had no connection to the LDP old-boy network. Tanaka was a self-made millionaire with no formal education beyond high school. He rose up through the business world to become a construction magnate, profiting from contracts both during the war and during the American occupation. Tanaka drew his strength not from party contacts but from his popular appeal to the Japanese people. Tanaka's rise and his ability to raise large amounts of money through business contacts partially transformed the way the LDP did politics.[67] It should be emphasized that Tanaka came into office viewing the relationship with the Soviet Union as a businessman might: what can we get out of this? His first thoughts were of the oil industry. As Minister of International Trade and Industry Tanaka knew that Japan badly needed to diversify its energy sources, and the Soviet Union, he felt, was a potential source to exploit. Tanaka also had another motivation. Almost every one of the LDP prime ministers beginning with Shigeru Yoshida had scored some sort of major foreign policy success. Yoshida had negotiated the signing of the San Francisco Treaty and the ending of the American occupation. Ichirō Hatoyama had re-established diplomatic relations with the Soviet Union. Nobusuke Kishi signed and cemented the US-Japan Security Treaty. Tanaka's predecessor Eisaku Satō was responsible for bringing about the reversion of Okinawa to Japanese political control in 1972. Tanaka wanted his own legacy to secure his place in history. The two most glaring holes in Japanese foreign policy were China and the Soviet Union. Tanaka pursued them both. He visited Beijing in September 1972, and it was agreed that relations would soon be normalized. He hoped to further this diplomatic breakthrough with a far-reaching energy agreement with the Soviet Union. Like Hatoyama in 1956, Tanaka wished to devise a more independent foreign policy with regard to the United States, the country that many felt had gone behind Japan's back during the so-called 'Nixon shocks.' Tanaka was one of many Japanese politicians angry with US actions in regards to recognizing China without first having consulted Tokyo.[68]

Tanaka received unqualified support for his overtures to Moscow from Japan's business world (*zaikai*). The *zaikai* was partially backed up by MITI, but this support slowly waned as Soviet financial demands for energy development became more exorbitant. Like MITI, the Ministry of Finance was initially favorably disposed toward the Tyumen project, but became more disillusioned as time went by and the details grew clearer. Even the normally cautious Foreign Ministry favored the use of Japanese bank loans to finance the Tyumen project, but this support quickly waned as well.[69] The Japanese Defense Agency (JDA), interestingly enough, was concerned about the impact the Tyumen project might have on Japan's relations with China. The JDA was also opposed to any project that might further Soviet strategic capabilities in the Far East.[70] China had all the while been warily eyeing Japanese-Soviet negotiations on the Tyumen project, and this was ultimately to have an impact as will be seen. By the beginning of 1973 the enthusiasm over the Tyumen project had disappeared. When Tanaka visited Moscow in the fall of 1973 the priorities of Japan had changed and the territorial issue became once again Japan's overriding concern.

One of the Soviet Union's primary strategies with regards to Japan was economics. Soviet leaders hoped to lure Japanese investment into Siberia and the Russian Far East. As mentioned, by the early 1970s it was becoming clear to some in the Soviet leadership that the economy was slowly grinding to a halt. Soviet leaders were particularly concerned about the growing technological gap between Moscow and the West. The race to achieve strategic parity had exhausted the Soviet economy, and many in the Soviet leadership recognized this.[71] The annual average growth rate of the Soviet GNP, which averaged 6.4 percent in the 1950s and 5.4 percent in the 1960s, had slowed to 3.7 percent by the early 1970s.[72] The Soviets hoped to rectify the worsening economic situation with a massive development plan for Siberia and the Russian Far East, involving not only energy but also industrial and infrastructure development. In their eyes, Japan was the perfect partner to carry out such an ambitious project.[73]

Under Brezhnev a range of more diverse players had emerged within the Soviet policymaking ranks. In the case of Japan policy, certain officials in the International Department of the Central Committee of the Communist Party and in the Ministry of Foreign Affairs had carved out positions of influence. Ivan Kovalenko was chief of the Japan section in the International Department of the Central Committee from 1963 to 1983 (and deputy chief of the department thereafter). He became one of the primary advisors to the Politburo on East Asian policy. After the Second World War he had been responsible for overseeing and re-educating Japanese POWs in Siberia. His experiences left him with a strong impression of Japanese servility in the face of strength and pressure. He maintained a hard-line stance toward Japan his entire career until he was forced to retire from the International Department by Mikhail Gorbachev in 1988. Deputy Foreign Minister Mikhail Kapitsa supervised East Asian policy in the Foreign Ministry and shared Kovalenko's views. He, too, maintained his position and influence for many years into the 1980s when Gorbachev removed him as well.[74] With these two men exerting influence on Japan policy there would be no softening of the Soviet position on the territorial issue.

As was the case though the 1960s, several fishery agreements were signed in the 1970s. Often the Soviet leadership used the fishery card as a tool for leverage against the Japanese when something was wanted, or when negotiations came to an impasse on other issues. This was a tactic used during the 1955–56 negotiations and was used again during the early 1970s. The Soviet Union was able to do this because most of the discussions over fishery issues focused on Japanese rights to fish in Soviet territorial waters. From the end of the war until Tanaka's 1973 Moscow visit, the Soviets allegedly seized close to 1,400 Japanese fishing boats and detained almost 12,000 Japanese citizens. The issue became even more pressing for the Japanese after 1976 when the Soviet Union declared an exclusive 200-nautical mile fishery zone. Incidents at sea between Japanese and Soviet boats happened with increasing regularity. The Japanese public, backed by the fishing industry, demanded resolute action from the government. The Japanese government was forced to listen because the political influence of the fishery groups in Japan was something that could not be ignored. Some Japanese today argue that the fishery issue is every bit as important as the territorial issue.[75]

The ideational context

The brief period of the Japanese-Soviet *rapprochement* in 1972–73 rose quite suddenly, and just as quickly faded. Consequently, there was little increase in contacts beyond the official level. The societies of the two countries were as distant from one another during this time as perhaps any other time over the past century. Furthermore, the citizens of each country were as ignorant of one another as they were distant.

By the early 1970s the citizens of Japan were feeling good about themselves. The decade of the 1960s was an era of unprecedented growth in Japanese history. Between 1965 and 1970 the economic growth rate was a staggering 12.2 percent per year.[76] The successes of the 1960s, though due to a myriad of factors, were brought about in large part by the hard work of the Japanese people themselves. By 1970 they deserved to feel good about themselves and their country, and this also engendered an unavoidable feeling of smugness. This was perhaps translated into a growing feeling of superiority toward other countries, including the Soviet Union. Nowhere was this attitude better personified than in the person of Prime Minister Tanaka, himself an economic success story. According to several commentators, Tanaka's brusque, confident style put off Brezhnev and the Soviet leadership during his 1973 visit to Moscow.[77] Tanaka was sure that the Soviet appeals for Japanese investment were out of necessity, while Brezhnev seemed to go out of his way to tell Tanaka that Japanese investment was not necessary, merely desired.[78] This misperception of the other side's motivations and desires has continued to dog the relationship up to the present.

In the Soviet Union by the early 1970s, the people still lived in the ignorant bliss of the Brezhnev years, which were rapidly approaching their twilight. The average Soviet citizen had a feeling of superiority when it came to Japan. This was not due to recent events as was the case in Japan, but more to the smug nature peculiar to citizens of a great empire. Of course, no one could be fully aware in the 1970s that the Soviet Union's days were numbered. In 1972 the Soviet Union was at the zenith of its power and influence. If Tanaka personified the attitude of the Japanese people in the 1970s, then Soviet Foreign Minister Andrei Gromyko perhaps best personified the attitude of the Soviet people during the Brezhnev years. Gromyko's diplomacy was based squarely on the bilateral Cold War confrontation with the United States. Any diplomacy was undertaken with the United States foremost in mind, and perhaps secondarily China. Japan and its leaders were viewed as politically reliant on the United States. As for Japan and its people, one historian points out that Gromyko's view was that of a quaint, vaguely exotic little country in the Far East. Japan was not yet seen as an economic superpower at this point in time. The views of the majority of the Soviet peoples were probably not much different, although Japan as a threat to the Soviet Union was an issue that was brought up time and again by propagandists.[79]

There were some cultural exchanges between the two nations in the early 1970s. The Bolshoi Ballet, as well as the Moscow and Leningrad Philharmonics, toured Japan and were given brilliant reviews. Japanese groups such as a kabuki troupe were also well received in the Soviet Union. Modern Japanese literature (such as works by Kōbō Abe and Kenzaburō Oe) found some popularity in Soviet intelligentsia circles when texts could be procured. Japanese consumer electronics were prized for their technology and were bought promptly whenever and wherever available in the Soviet Union (most frequently through the black market).[80] Japanese film director Akira Kurosawa made a wonderful film in the Russian Far East, *Dersu Uzala*, which was the first film made in the Soviet Union by a non-communist bloc director. It also captured the Oscar in Hollywood in 1975 as the best foreign film. It was slowly becoming known in the Soviet Union that Japan was more than a military base for the United States, and that it had an economy and culture to be admired.

But mistrust still ruled mutual perceptions. As the 1970s progressed, some in the Soviet leadership became more and more aware that Japan was a country that was to be reckoned with in its own right. Beginning in the early 1970s there was a proliferation of official Soviet writings portraying Japan as a nation bent on military revanchism.[81] Similarly, as the US commitment to the Asia-Pacific region came into doubt and as the Soviet military build-up in the Far East continued apace, Japanese leaders and the Japanese public came to view the Soviet Union more and more as a nation with sinister intentions.[82]

The breakdown

The initial impetus driving Japan and the Soviet Union together in 1971–72 had almost died out by the time Kakuei Tanaka visited Moscow in the fall of 1973. Nevertheless, there was still hope that economic cooperation and the emerging strategic situation would bring the two sides closer together. For the Soviet Union, this meant either Japanese participation in a collective security organization, or a peace treaty based on a territorial status quo. For the Japanese this meant a peace treaty and the reversion of the disputed territories. By the mid-1970s, it was clear that this brief attempt at *rapprochement* had faded even quicker than the earlier periods of bilateral warming.

For starters, both sides were on a completely different page. Although the Soviets grudgingly recognized that there were 'unresolved issues' dating from the Second World War, they were not about to return the disputed territories, as the Japanese continued to insist. The Soviets were prepared to agree to a peace treaty based on the 1956 formula (returning only Shikotan and the Habomai group), but this clearly did not go far enough for the Japanese. The publics in both nations were also unwilling to compromise and yield to the other side.

Economically, the Tyumen project, which had assumed such importance in bilateral discussions, was going nowhere. The initial bill for exploration, drilling, and a pipeline came in at $1 billion. The estimated costs suddenly expanded by the time Tanaka visited Moscow in the fall of 1973. Nevertheless, Japanese investors maintained their interest, while discussions continued through the year. By this time the Soviets were hoping to construct a second trans-Siberian rail line running parallel approximately 200 miles to the north of the existing line to transport the oil, and the costs were now estimated to run to $3.3 billion. The Soviets wanted the Japanese to finance this. As mentioned, the various ministries in Japan involved with the project had already begun voicing concerns about the economic feasibility. There were also concerns in Japan that this second railroad was meant to transport more than just oil. The Soviet Union was in the midst of a significant military build-up in the Far East, and the second rail line (known as the Baikal-Amur Mainline, or BAM) would provide the Soviets with some strategic depth along the Chinese border, because the original trans-Siberian line could easily be cut in the event of a war with China.

Japan suddenly had to take into account the political and military implications of such a project. Not only might his arouse the opposition of groups in Japan (such as the Japan Defense Agency), but also the opposition of the United States and China. Japan had hoped to garner not only American support for the project but to enlist the participation of American firms. From the beginning American companies were lukewarm on the project. Chinese officials had expressed concern about Japanese participation in the project, but made it clear that this opposition would cease to exist with the participation of American companies. But American firms were simply not interested.[83] Although economic factors were probably foremost in the minds of Japanese officials, some analysts have concluded that it was China's opposition that eventually drove Japan from participation in the Tyumen project.[84] In fact, China stepped up its oil exports to Japan in an attempt to lessen Japan's desire to develop Siberian resources. China's oil exports to Japan in 1972 amounted to 200,000 tons. By 1975 they reached eight million tons.[85] Although Japan was highly interested in resource diversification in the wake of the first oil shock, the Tyumen project never got off the ground.

Several bilateral issues beyond the territorial dispute were directly related to the failure to completely normalize relations in the 1970s. Some of these, as mentioned, had to do with the changing perceptions of one another that resulted from the growing confidence of the Japanese due to economic achievements, and the smugness of the Soviets because of the new status as a strategic equal to the United States. Although the summit meeting of 1973 was a failure, the Soviets did approach the Japanese again in 1975 with an offer of a treaty of friendship, but the Japanese side demurred. In September 1976 a Soviet pilot landed his MiG-25 aircraft in Hokkaido and requested political asylum, which was granted by the United States. US military experts promptly dismantled the MiG, giving it a thorough inspection over the protests of the Soviet government before returning it in crates to the Soviet Union.[86] Several months later the Soviet government unilaterally declared the 200-mile exclusive fishery zone. All bilateral existing fishing agreements were abrogated. It took several months of tense negotiations before a settlement was concluded to the satisfaction of both Moscow and Tokyo. The Soviet Far Eastern military build-up that had been under way since the early 1970s was extended in 1978 to include the disputed islands. After 1960 the Soviets had withdrawn most of their troops from the islands of Etorofu/Iturup, Kunashir/i, and Shikotan. In the spring of 1978 Soviet ground forces and military aircraft were redeployed on these islands. Additionally, Soviet SS-20 missiles, previously stationed only on Soviet European territory, were moved to the Far East. Stepped-up maneuvers in the Far Eastern districts, including landing exercises on the disputed islands, put the Japanese Defense Agency in a state of alarm.[87] In 1980 Japanese Prime Minister Masayoshi Ohira declared the Soviet forces in the Far East a 'threat' to Japan.[88]

The build-up of Soviet forces in the Far East (starting in 1966) was directly linked to the China factor, and Japan's disquiet was directly linked to the American factor. China's opposition to the Tyumen project has been noted. China was every bit as wary of Japanese-Soviet cooperation as the Soviet Union was suspicious of Sino-Japanese cooperation. Chinese attempts to come between Moscow and Tokyo, whether merely coincidental or planned, began as early as the mid-1960s when the Sino-Soviet rift was coming to a heated denouement. In 1964 Mao Zedong told a group of visiting Japanese Diet members from the Socialist Party that the People's Republic of China fully supported Japan's position on the territorial dispute: "As far as the Kuril Islands are concerned, there is no doubt in our view: they must be returned."[89] Prime Minister Zhou Enlai told another group of visiting Japanese parliamentarians in 1972, on the eve of Gromyko's Tokyo visit, that China would support Japan's demands for the return of the islands. The Chinese echoed this support at the United Nations in 1973, four days before Tanaka's Moscow visit. The Soviets reacted viscerally on each occasion, and Gromyko took every opportunity to denounce China's meddling. Gromyko was also interested in seeing that China and Japan did *not* sign a treaty of peace and friendship that had an anti-hegemony clause aimed, in the Soviet view, against Moscow. It has been speculated that the main purpose of Gromyko's January 1976 Tokyo visit was to prevent Japan from signing such a treaty with China.[90] If the Soviet Union felt that China was meddling in the internal affairs of Moscow and Tokyo, Japanese leaders were similarly insulted by the Soviets' veiled threats against the Japanese government not to sign a treaty with China.[91] The Sino-Japanese Treaty of Peace and Friendship was finally signed in 1978, and the anti-hegemony clause was included.[92]

The United States' direct involvement in Japanese-Soviet relations continued beyond the MiG incident. The Soviet invasion of Afghanistan took place in late 1979, and was immediately denounced by Washington. Japan joined the US sponsored sanction policy against the Soviets in early 1980. Japan also boycotted the 1980 Moscow Olympics. In 1980, Japan's Prime Minister Masayoshi Ohira put a freeze on several large economic joint projects under way in the Soviet Union at the time.[93] In 1981 the new Japanese Prime Minister, Zenko Suzuki, agreed to the US proposal for Japan to undertake defense of Japan's sea-lanes up to 1,000 miles. The Japanese government on February 7, 1981 established 'Northern Territories Day.'[94] The following year Japan accepted the deployment of US Air Force F-16s at an air base in northern Japan (Misawa). Prime Minister Yasuhiro Nakasone came into office in 1982 and enthusiastically took up the US call to become a more effective and responsible ally. He responded by saying that Japan would become America's "unsinkable aircraft carrier."[95]

Other events around the world contributed to the new animosity in the Japanese-Soviet relationship. These included the Soviet support for the Vietnamese invasion of Cambodia in 1979, the imposition of martial law in Poland in 1980–81, the shooting down of Korean Airlines flight 007 by a Soviet fighter off of Sakhalin in 1983, and the general hostility between the East and the West during the early years of the Reagan presidency. Though some would point to the events of the late 1970s as the demise of the Japanese-Soviet *rapprochement*,[96] it was clear by 1973 that the subtle warming of relations in 1972–73 was but a brief respite in the worsening of political relations that continued from the early 1960s through the 1980s.[97]

In spite of the transformation of the international situation in the early 1970s, and the attractive idea of the complementary nature of the Japanese and Soviet economies, the *rapprochement* of 1972–73 was dead long before the international tensions of the late 1970s. The hostile attitudes of both peoples that resulted from the Second World War continued to linger. This factor, combined with the new-found feelings of superiority in both nations, created a poisonous atmosphere, whose toxicity was heightened by the tense international situation. There was simply too much for the leadership of both nations to overcome in convincing their respective peoples of the benefits of a true and long-lasting Japanese-Soviet *rapprochement*. If the leaders were themselves unsure of the utility of a *rapprochement*, how could they convincingly persuade their people? No foundation for a strong bilateral relationship existed, and the leaders failed to build one. Once again, attempts at true normalization foundered due to an uncommitted leadership and an unconvinced people.

As in the 1956–60 period, structural factors both pulled the nations together and then ultimately kept them apart. The transformation of the international system in East Asia (i.e. the end of the Sino-American confrontation) was an important factor in the move by Moscow and Tokyo to explore areas of cooperation. But the worsening situation of US-Soviet relations later in the decade drove Japan and the Soviet Union apart. Ideational factors, however, were also key in keeping the two nations from ultimately normalizing relations. The Soviet image in Japan was still deeply negative. Meanwhile, Soviet leaders were unable to view Japan as anything more than a US appendage in the Far East.

1988–93: *Glasnost', perestroika*, and *demokratizatsiya*

The Gorbachev years of the late 1980s marked the end of the Cold War and a dramatic transformation of the international system. Yeltsin's accession to power after the collapse of the Soviet Union in 1991–92 breathed new life into the Japanese-

Russian relationship. Although relations never achieved normalization, the five-year period between 1988 and 1993 seemed to bring new hope to the relationship and marked a significant increase in bilateral interaction. The new era, however, scarcely brought about a decrease in distrust and suspicion between Japan and Russia. Mutual misperceptions were, if anything, heightened during this period. The newfound hope that a complete normalization might be reached only caused a greater disappointment when ultimately the two sides failed to bridge the differences that had divided them for an entire century.

Relations between Japan and the Soviet Union had reached a nadir in the late 1970s and early 1980s. Following the visit to Tokyo of Soviet Foreign Minister Andrei Gromyko in 1976, and the visit of Japanese Foreign Minister Sunao Sonoda to Moscow in 1978, there were no official high-level visits to either capital until 1986, when Gromyko's successor Eduard Shevardnadze visited Tokyo. Japanese Prime Minister Yasuhiro Nakasone had attended the funeral of Soviet Premier Konstantin Chernenko in 1985, and there he briefly met with Chernenko's successor, Mikhail Gorbachev. The ten-year period lasting until Gorbachev's accession to power was a lost decade in the political bilateral relationship. Like the 1960s, however, which also marked a period of political impasse, the 1980s were an active time in trade relations, debunking the conventional wisdom that increasing economic interaction necessarily accompanies warm political relations. The two-way trade level that stood at $3.3 billion in 1977 increased to $5.5 billion in 1982. After a decline to $3.9 billion in 1984, the two-way trade level reached $5.1 billion in 1986, hovering around this mark until it reached $6 billion in 1989. It did not reach the $6 billion level again until the year 2003 (coming close only once in 1995 at $5.9 billion).[98] During the five-year period 1988–93, trade relations actually stagnated, and then declined. By 1992 the trade level had fallen to $3.4 billion, though it increased again to $4.2 billion in 1993. Much of this, of course, was due to the political chaos surrounding the dissolution of the Soviet Union at the end of 1991. Nevertheless, the trade level failed to reach the $6 billion level even ten years after the Soviet collapse. Furthermore, two-way trade was but a mere fraction of each country's overall trade level. The trade in goods echoed earlier periods: Japan imported from the Soviet Union mostly raw materials (lumber, marine resources, cotton, coal, oil, minerals); Soviet imports from Japan consisted primarily of finished manufactured goods (though less centered on industrial equipment than previously).

In spite of Nakasone's positive assessment of Gorbachev after their private meeting at Chernenko's funeral in 1985, attitudes in Japan toward *glasnost'* and the 'new-thinking' of the Soviet leadership were slower to change than in the West.[99] Conditioned to turning a jaded eye to 'changes' in Soviet thinking and behavior, the Japanese leadership and the Japanese public first awaited concrete results. Gilbert Rozman has pointed out that as the world community debated the chances of success for Gorbachev and his *perestroika* in the late 1980s, the Japanese public merely debated whether or not he would return the disputed islands.[100] Nevertheless, the excitement Gorbachev produced around the world was contagious, and 'Gorby-mania' eventually caught on in Japan, although to a lesser degree than in Europe and North America.

Foreign Minister Eduard Shevardnadze's visit to Tokyo in January 1986 produced few results, but it did give the Japanese some room for optimism, because the Soviet foreign minister was actually willing to discuss the territorial issue. In the past Soviet leaders had even denied the existence of any such issue. It was also later learned that Shevardnadze had been prepared to discuss the 1956 joint declaration as a starting point for negotiations.[101] Japanese Foreign Minister Shintarō Abe followed up Shevardnadze's visit by making a trip to Moscow in May 1986. Although it was nothing more than a 'mop-up operation' to the first foreign ministerial meeting, it opened the opportunity to keep discussions ongoing.[102] In a speech he delivered in Vladivostok in July 1986, Gorbachev further hinted at his intention to put the Soviet Union on a new path of diplomacy in East Asia. The speech was primarily seen by most as a conciliatory gesture to the Chinese leadership, but it also gave hope to some in Japan that Gorbachev was a leader with whom they could work through the outstanding issues dividing the two countries. Most people in Japan, however, continued to doubt the credibility of the Soviet leader's words.[103] The Toshiba incident of April 1987, in which the Japanese company was found to have violated sensitive export regulations and to have sold technology that would make Soviet submarines quieter and more difficult to detect, caused a temporary, yet serious, setback in relations. In late 1986 the Soviets had told Japanese leaders that Mikhail Gorbachev would not be visiting Japan in 1987. The Japanese had hoped he would visit Japan after preliminary discussions. Historian Tsuyoshi Hasegawa believes that the cancellation of Gorbachev's tentative plan to visit Japan in 1987 marked the last best chance for the two countries to settle the territorial dispute. He reasons that only early on in his tenure as Soviet leader did Gorbachev have the power and political capital at home to make territorial concessions, whether just two or all four island groups.[104]

Most historians mark 1988 as the year that changes in attitudes in Japan resulted in a significant warming of relations. The Soviet leadership had been hinting that it desired a more rounded relationship with Japan since Gorbachev's accession in 1985. Gorbachev, in particular, felt this way. He was influenced in this regard by progressive thinkers at the Soviet Ministry of Foreign Affairs and by staff members of the Institute of World Economy and International Relations (IMEMO) of the Soviet Academy of Sciences.[105] In Japan it took a bit of time before the leadership, and the Ministry of Foreign Affairs in particular, were prepared to take the statements of the new Soviet leadership at face value. By 1988 more people in the Japanese leadership were prepared to listen to Gorbachev and move forward in an attempt to outline a new agenda on the path to improved relations.[106] Former Prime Minister Nakasone made a visit to Moscow in July 1988 in an attempt to kick-start the process of negotiation centering on the territorial dispute. During the visit Nakasone was granted an audience with Gorbachev and he took the opportunity to raise the territorial issue. In December of 1988, Shevardnadze made a second visit to Japan. This visit was followed up by a visit to Moscow by Japanese Foreign Minister Sōsuke Uno in May 1989. Once again, progress was slow, but at least the two sides were speaking openly about the territorial dispute, a change that was no doubt welcomed by the Japanese side. Uno and the Japanese side had unveiled its new policy of 'balanced expansion' (*kakudai kinkō*). This new policy was an attempt to get away from the singular focus on the territorial dispute, and to improve economic, political, and security relations with the Soviet Union.[107] In the fall of 1989 in a meeting with Japan's new Foreign Minister Tarō Nakayama, Shevardnadze announced that Gorbachev would visit Japan – but only in 1991.

The improved atmospherics received a setback at the end of 1989 when tensions began rising in a series of meetings held by a working group to discuss the signing of a peace treaty. This group was formed at the vice-ministerial level and was meant to discuss ways to settle the territorial dispute, the last barrier to a peace treaty. The group first met in December 1988 and had three more meetings in 1989 (in March, April, and December). One Japanese diplomat complained that the Soviets had begun speaking to their Japanese counterparts in a tone reminiscent of the bitter Gromyko years.[108] The Soviets undoubtedly felt that the Japanese sounded a bit like a broken record, as Japanese diplomats continued to harp on the territorial issue as the momentous international events of 1989 unfolded before the world's eyes.

This was also a difficult time domestically for both countries. Events in the Baltic Republics, in Georgia, in Armenia, and in Azerbaijan were unfolding with a disquieting rapidity for the Soviet leadership, as were events in Eastern Europe. Meanwhile Boris Yeltsin's power in Russia was growing, as evidenced by his overwhelming victory in an election for the Congress of People's Deputies.[109] In Japan, two different scandals brought down two different cabinets in a matter of months. First the Recruit financial scandal brought down the cabinet of Noboru Takeshita in April 1989. Then, his successor Sōsuke Uno was brought down in the wake of a sexual misconduct scandal in July 1989. The LDP lost control of the Upper House of the Japanese Diet in the July 1989 elections, partly as a result of these scandals and also because of an unpopular consumption tax.[110]

By 1990 it was unclear whether the positive momentum that had been gathering in the relationship could be maintained. A Tokyo visit Shevardnadze had scheduled for March 1990 was postponed. The cancellation was ostensibly to prepare for the Bush-Gorbachev summit, but it was also partially in response to Japan's perceived fixation with the territorial dispute. Politburo member Aleksandr Yakovlev backed up Shevardnadze's views on Japan after returning from a trip to Tokyo in December 1989. Although considered flexible toward Japan and generally pro-Western, Yakovlev was put off by the Japanese refusal to discuss other issues.[111] Although there were some individuals among the Japanese leadership that appeared flexible (including LDP kingmakers Shin Kanemaru and Ichirō Ozawa), bilateral talks and high-level missions repeatedly foundered on the territorial issue.

Shevardnadze did visit Japan again in September 1990, and met with Japanese Foreign Minister Tarō Nakayama and Japan's new Prime Minister Toshiki Kaifu. The revolving door of Japanese cabinets must have frustrated the Soviets. New faces meant that the initial groundwork of negotiations had to be redone time and again. Additionally, it allowed the Soviets little time to build trust with their Japanese counterparts. Each time a new foreign minister or prime minister appeared, they simply reiterated predictable arguments about the need to end the territorial dispute. Although Japan had supposedly abandoned its principle of the 'inseparability of economics and politics' (*seikei fukabun* – a euphemism for no economic aid until the territorial dispute has been settled) for a new 'balanced expansion' (*kakudai kinkō*) in its Soviet policy, nothing seemed to have changed. Shevardnadze was tiring of this merry-go-round. This attitude was shared around the world. Some Western commentators began criticizing the Japanese for their insistence on focusing on territory and turning 'deaf' ears to the calls for economic assistance to the sinking Soviet economy.[112] Foreign Minister Nakayama was caught on record saying that giving aid to the Soviets was like "throwing money into the gutter (*dobu ni kane wo tsuteru*)."[113] Such statements did not help Japan's cause, particularly around the time of the Gulf War, when Japan was being criticized heavily in the West for its lack of creative diplomacy and for its stinginess with the purse. Japanese leaders began feeling nervous that a failure to move forward in relations with the Soviet Union would result in Japan's isolation.

Preparations for the 1991 Gorbachev visit continued apace, and in spite of the polemics issued by both sides, the Japanese public was growing excited about the upcoming summit. Many within the Japanese government were also anxious for the summit to succeed, and began calling for a political settlement to be reached. These voices could be found both within the Foreign Ministry and within the ruling LDP.[114] As mentioned, two such individuals were LDP stalwarts Shin Kanemaru and Ichirō Ozawa. Kanemaru had been on record as advocating the acceptance first of a two-island solution, and then buying the other two. Ozawa was also reportedly interested in an 'Alaska solution' (i.e. a purchase of the disputed islands). In a visit to Moscow in March 1991 (one month before Gorbachev's Tokyo visit), Ozawa reportedly offered the Soviets $26–28 billion in economic aid for the return of the disputed islands.[115] This came on the heels of a report in the Soviet press in February 1991 that accused Gorbachev of conspiring to sell the disputed islands to the Japanese for $200 billion, in order to maintain his hold on power.[116] There were also those within the Soviet government who seem disposed to reaching some sort of settlement through compromise. Specialists at the Soviet Academy of Sciences had been advocating a return to the 1956 declaration as a starting point in negotiations. In a 1990 speech in Japan Boris Yeltsin advocated a five-point plan for settling the territorial dispute.[117] At that point in time, however, Gorbachev was in no position domestically for any compromise that could be construed by Soviet hard-liners or the Soviet public as backing down.

Gorbachev's visit to Tokyo took place as scheduled in April 1991. As had been predicted by most, the results of the summit were minimal. Although the Soviet government did officially recognize in a joint declaration the existence of the territorial dispute for the first time (this was seen as a coup for the Japanese side), neither nation could feel good about the results of the meetings. Territory dominated the agenda. Kaifu kept raising the issue, and Gorbachev kept ducking it. At one point an exasperated Gorbachev asked, "Isn't there anything new that Japan has to add to the arguments that I have heard so far?"[118] This was Gorbachev's last official foreign visit as head of the Soviet state (besides the London G-7 summit in the summer of 1991). In August the attempted coup d'état failed to bring about a hard-line dictatorship, but it did bring Gorbachev down and paved the way for the accession of Boris Yeltsin as head of the successor state to the Soviet Union, the Russian Federation. In the face of the quick moving events in the Soviet Union, Japan once again demonstrated a lack of resolve and was criticized from within its own ranks for not moving quickly enough to denounce the coup plotters. The Kaifu cabinet dithered several days before issuing a denunciation of the coup.[119]

Boris Yeltsin had been one of the chief thorns in Gorbachev's side. A bitter enemy in the ranks of the communist party, he had resigned his membership and was elected chairman of the Supreme Soviet of the Russian Socialist Federated Soviet Republic (the precursor to the Duma in Russia today) in May 1990 and on June 12 the declaration of Russian sovereignty was issued by the Supreme Soviet. While struggling in opposition to Gorbachev, Yeltsin had made various statements hinting that he was amenable to reaching a compromise with Japan on the territorial issue. But as the reality of his coming to power came closer Yeltsin backtracked significantly and then began criticizing those who advocated a reversion of territory. In a visit to Sakhalin and the disputed islands in August 1990, Yeltsin denied he intended to hand over any territory. In reaction to the Ozawa visit and the alleged offer of money for the disputed islands, Yeltsin angrily stated that no one in Russia could allow the islands to be sold as Alaska had been sold. Nor would he consider giving away any territory, a statement that he pointedly made in Kaliningrad, a former German possession that belonged to the Soviet Union.[120]

After the coup had failed and it was apparent that the days of the Soviet Union were numbered, hope again rose in Japan that a deal could be reached with the new government, notwithstanding Yeltsin's harsh statements.[121] The Japanese government responded to the coup and the threat of chaos in Russia by announcing an aid package totaling $2.5 billion. As late as the July 1991 London summit Japanese leaders had insisted that any aid dispensed was contingent on a settlement of the territorial dispute. Now aid was being issued with fewer strings attached.[122] Although the feeling in Japan toward the changes in Russia was still somewhat guarded in comparison to the rest of the world, a hint of goodwill could be detected.

Several officials in the new Russian Foreign Ministry were already voicing the need to reach a deal with Japan. Both sides moved quickly in the fall of 1991 in an attempt to sound the other side out. Ruslan Khasbulatov, the new chairman of the Supreme Soviet of the Russian Federation, visited Tokyo in early September along with the new deputy foreign minister, Georgii Kunadze. Kunadze had been one of the group at the Soviet Academy of Sciences that had urged Gorbachev to find a solution to the territorial dispute through the 1956 declaration (i.e. two islands and then further discussions). He was now the top Asia specialist at the Ministry of Foreign Affairs, working under Andrei Kozyrev. In Tokyo Khasbulatov handed to Japanese Prime Minister Kaifu a letter from Yeltsin enumerating Russia's desire to settle the territorial dispute based on the principle of "law and justice," leading to a relationship in which there were no "victors or vanquished." Khasbulatov reportedly also suggested a sum of $8–15 billion in financial aid. He also hinted that were a settlement reached, any territory would be returned sooner than the 15–20 years as originally proposed by Yeltsin.[123] Soon thereafter in an interview with *Pravda*, Yeltsin insisted that he sought a quick resolution to the territorial dispute.[124] Suddenly the Japanese were receiving warm signals from Moscow. In October 1991 Japanese Foreign Minister Nakayama visited Moscow. A few weeks prior to his visit Nakayama had made a positive speech at the United Nations expressing support for both the Soviet Union and the new Russian Federation. During Nakayama's visit to Moscow Yeltsin praised Japan's $2.5 billion assistance package.

By the end of 1991, however, it was once again clear that hope and reality are two different matters. The hope that had come about in the few months after the failed coup attempt was giving way to the reality of the domestic situation in Russia, in particular the Russian Far East. In the autumn of 1991 Deputy Foreign Minister Kunadze visited Sakhalin and the disputed islands and was treated to a dressing down by irate islanders opposing any territorial concessions to Japan. So heated were the discussions that Kunadze felt physically threatened.[125] Regional domestic elements would come to influence Russian policy toward Japan more and more as time wore on.

After the dissolution of the Soviet Union in December 1991, Russian Foreign Minister Kozyrev assured the Japanese that the Russian government would assume all the debts and obligations of the Soviet government, including the 1956 declaration.[126] In January 1992 at a UN summit in New York, Yeltsin met with Japanese Prime Minister Kiichi Miyazawa. Both sides expressed cautious optimism that an agreement could be reached and a peace treaty finally signed. On the occasion of the visit to Moscow by former Japanese Prime Minister Nakasone in March 1992, Yeltsin again thanked Japan for its economic assistance. Visa-free visits by islanders and former island residents between the disputed islands and Hokkaido were commenced in the spring of 1992, adding to the new positive atmosphere. In March 1992, on a visit to Tokyo, Kozyrev elaborated on his December statement (promising to honor all Soviet agreements) when he reportedly offered the Japanese a treaty based on the 1956 formula, with an immediate handover of Shikotan and the Habomai group, while agreeing to further negotiations over Kunashir/i and Etorofu/Iturup. The Japanese Foreign Ministry was willing to agree to such a deal, on the condition that Russia agree beforehand to a handover 'in several years' of the two additional islands.[127]

By the summer of 1992 political events in Russia again clouded the issue. Opposition to a territorial settlement with the Japanese existed in the Russian Supreme Soviet, in the military, and in various regional governments. Now Yeltsin was beginning to speak with an eye to these audiences. In an outburst in a June 1992 television interview Yeltsin lambasted the Japanese for not providing 'one cent' of assistance to Russia.[128] Only three months prior to this he had thanked the Japanese for their largesse. He may have been expressing the frustration of not being able to quickly access large amounts of the $2.5 billion assistance fund that was tied largely into trade insurance. He may also have been referencing Japan's *relatively* small level of assistance to Russia up to that point. During the crucial years of 1990–92 Japan ranked seventh in aid to the former Soviet states and Russia (behind even Saudi Arabia and South Korea).[129] But to have castigated the Japanese publicly was perhaps unfair.

The next month at the G-7 summit meeting in Munich, tensions mounted between Japan and her six partners. Japan was uneasy about supporting multilateral aid packages to the new Russian government through the auspices of the IMF and World Bank. Germany was most vocal in its criticism of Japanese hesitation. Much to Japan's consternation the United States, which had vacillated on the Russian aid issue in the first few years of the Bush administration, was now firmly behind the Europeans. Some leaders in Japan (especially among the LDP) even considered the new US position a betrayal of past support and tantamount to the 1971 'Nixon shock.'[130] The Japanese government insisted again (as had become common practice in the last few G-7 summits) that any final statement include support for Japan's territorial claims on Russia. In the end, Japan agreed to a Western aid package, but only if the other governments would include Japan's claims in the final communiqué. Much to the irritation of the European governments, the Japanese had their way. But these actions undermined the new Japanese-Russian relationship.[131] Parliamentary hearings in the Russian Supreme Soviet in July further hindered any positive developments toward a territorial settlement. Conservative groups called for the hearings in order to put a damper on Yeltsin's upcoming visit to Tokyo, scheduled for September 1992. Fear that Yeltsin might concede territory was put to rest in these hearings, which were nothing more than an attack on the 'Japan lobby' in Moscow.[132] A visit to Moscow by Japan's Foreign Minister Michio Watanabe in August was meant to be a last-minute preparatory meeting to sort out the details of Yeltsin's upcoming Japan trip. Instead, Watanabe's meeting with Yeltsin provided further negative publicity for both sides. The meeting was tense and the two sides exchanged heated rhetoric. The Russian side claimed that Watanabe was arrogant and rude and insistently pushed the territorial issue. The Japanese side insisted the blame lay with the Russians.[133]

Three days before Yeltsin was due to arrive in Tokyo, he called to postpone his visit. Besides the late nature of the warning, the insulting way Yeltsin had gone about it galled the Japanese. In the phone call announcing his intention not to come to Tokyo, Yeltsin put the Japanese Prime Minister Miyazawa on hold while he first called the President of South Korea (the second intended destination on his itinerary) to explain that he would not be coming. Furthermore, he rescheduled the visit to Seoul (for November 1992), and did not do so with the Japanese. The Japanese had arranged an audience with the Emperor and to cancel at the last minute went against all diplomatic protocol. Additionally, Yeltsin had said that part of the reason for the cancellation was so that the political situation in Japan could settle a bit more. Either Yeltsin was getting very bad information or he was fabricating excuses. Japan's domestic political situation was in no such state that a visit should be postponed. The Russians also posed questions about security details. The Japanese right wing was sure to protest Yeltsin's visit, but they protested the visit of every Soviet and Russian leader. He was looking for excuses, and not finding very good ones. Miyazawa took the cancellation diplomatically, but the Japanese side was utterly stunned.[134] The mercurial and unpredictable nature of the Russian president was a trademark that would come to define Yeltsin's career (and was not the last time he would postpone a visit to Japan).

Yeltsin promised to reschedule another trip sometime in 1993. It was already clear, however, that any such trip would fail to produce substantive results. At the April 1993 summit of G-7 foreign and finance ministers in Tokyo (the site of the upcoming G-7 heads of state summit in July 1993) the Japanese government announced a bilateral aid package for Russia as part of a larger package from all the nations. Japan would extend $1.5 billion in credit, much of it tied once again to export credits and trade insurance. Japan was slowly distancing itself from its past policy of the 'inseparability of economics and politics.' It was a conciliatory gesture given Yeltsin's behavior the previous fall. Nevertheless, Japanese diplomats began questioning the over-reliance on Yeltsin among the Western nations. No sooner had Yeltsin announced to everyone's surprise in April that he hoped to visit Japan in May, when he once again postponed the visit. It was just as well for the Japanese, for they were caught unprepared and needed time to develop what they hoped would be a fresh policy for Russia. The two sides had had little high-level contact since the fall 1992 cancellation, due to the unsettled situation in Russia in early 1993. At the time Yeltsin and the Congress of People's Deputies (of which the Supreme Soviet was part) were locked in a confrontation over a pending national referendum. Yeltsin did attend the July G-7 Tokyo summit, and his meeting with Miyazawa went off well. In September 1993 Yeltsin suddenly announced that he wished to visit Japan again in October. By now the Japanese side, understanding the delicate nature of the Russian domestic political situation, and used to the erratic nature of Yeltsin's decisions, was prepared to host a summit that fall.

In Japan there was not only a new cabinet, but also a new power-sharing coalition that excluded the LDP from power for the first time in 38 years. The new Prime Minister was Morihiro Hosokawa, head of the recently established Japan New Party. In September, Russian Foreign Minister Kozyrev met in New York with Hosokawa's new foreign minister, Tsutomu Hata. When Yeltsin announced his intention to visit it appeared that both sides, given a fresh start, might be willing to discuss new options to move the relationship along to a new level. Yeltsin's willingness to present a new agenda was partly reflected in his July visit to the G-7 Summit in Tokyo where he announced that partial troop withdrawals from the disputed territories had already begun.[135]

The fact that Yeltsin even visited Japan in the fall of 1993, fresh on the heels of his storming of the Russian parliament building (the stand-off ended only on October 4), was a gesture that this trip was important. Yeltsin further demonstrated his intention to give the relationship new energy upon his arrival in Japan on October 11, 1993. One of the first gestures he made was to issue an apology in the presence of the Emperor for the detention and death of the nearly 60,000 Japanese POWs in Soviet labor camps after the Second World War. Gorbachev had also issued an apology at a graveyard for Japanese POWs in Khabarovsk before his 1991 Tokyo visit, but Yeltsin's terminology went a step further than Gorbachev's (Yeltsin used the term *izvenenie*, while Gorbachev said *soboleznovanie*). The difference in 'apology' versus 'condolences' (thus implying guilt) was a big step forward for the protocol- and status-conscious Japanese.[136] The rest of the visit went well. The Japanese side made a conscious effort to not overstate the territorial issue.

Hosokawa and Yeltsin ended the summit with the signing of the so-called Tokyo Declaration in which the two sides promised to abide by all legal and historical precedents (i.e. treaties and agreements) in the pursuit of a peace treaty and a resolution of the territorial dispute, and that all agreements would be based on the principles of 'law and justice.' Furthermore, the names of the four disputed island groups were specified and Yeltsin orally affirmed that past agreements included the 1956 Joint Declaration. The Japanese also expressed support for the Russian democratic transition.[137]

Given Yeltsin's actions during the year leading up to the October 1993 Tokyo visit, the Japanese had every reason to believe that the summit had been concluded quite successfully. Yeltsin had managed to erase the negative image he had developed in the Japanese press and among the Japanese public. But no sooner was the Kremlin enjoying its newfound popularity among the Japanese than it was revealed that Russian naval ships had been dumping nuclear waste into the Sea of Japan for years (over 900 tons of waste). This revelation came three days after Yeltsin departed home from Tokyo. Although warnings had been issued in the Western press as early as July that a new round of nuclear dumping was imminent, the affair only garnered significant coverage in Japan on the heels of Yeltsin's visit. The Japanese government tried to downplay the issue, agreeing with Russian statements that the waste had a low radioactive level, but the anger of its citizens could not be held in check.[138] The Japanese government responded by opening discussions with the Russian government on the possibility of extending $100 million in aid to help stop the dumping of nuclear waste at sea. Nothing came of this first offer.[139]

With this latest outrage (in a series of diplomatic gaffes) perpetrated by the Russian government, the *rapprochement* of the late Gorbachev and early Yeltsin years sadly ended. The dramatic domestic transformations under way in both countries continued to occupy the thoughts and actions of the leaders of both countries. There was little room on either side for dramatic gestures and concrete steps.

The international context

The international situation changed during these five years in a more dramatic and fundamental fashion than at any time in the twentieth century apart from the 1914–20 and 1939–45 periods. Given that the change stemmed primarily from the transformation of the Soviet Union and Russia, it is no surprise that the relationship between Japan and Russia was affected. What is surprising is that given the degree of change domestically and in the international system, the Japanese-Russian relationship experienced no great breakthrough, such as occurred in US-Russian or Sino-Russian relations.

Although Gorbachev gained control in the Soviet Union in 1985, it was really only apparent by 1988–89 that change in the Soviet Union would translate to change in other areas of the world – particularly in the Soviet Republics and in the Soviet satellite states. Enumerating these events in detail is beyond the scope of this work. One need only keep in mind events such as the collapse of the Berlin Wall, the crumbling of the communist regimes in Eastern Europe and Mongolia, the ground breaking arms control agreements (conventional and nuclear) signed between the United States and the Soviet Union, the Tiananmen Square incident in China, the dynamic economic growth of East and Southeast Asia, UN membership for the two Koreas, US-Japanese trade frictions, the end of the war in Cambodia and the Vietnamese withdrawal from there, the US military withdrawal from the Philippines, the Soviet withdrawal from Afghanistan, the Persian Gulf War, the reunification of Germany, the independence of the Baltic Republics and the states of Central Asia, and finally the dissolution of the Soviet Union and the rise of Russia. Not only is this list not exhaustive, it also only takes into account the events on the Eurasian periphery and interior.

The positive effects of these changes on the Japanese-Soviet relationship were somewhat slow in coming about. Besides the complex issue of the territorial dispute, one reason Japan may have been less than excited about change in the Soviet Union was the extent of political and economic change going on in East Asia. While the Japanese government and commercial enterprises looked hungrily at Soviet resources in the wake of the 1970s energy shocks, now in the 1980s Japanese companies only had eyes for the booming markets in Southeast Asia and China. The 1980s and 1990s were a time of unprecedented growth for the Chinese economy, and also for Sino-Japanese trade and investment. Japanese economic aid and direct investment in China soared during that decade and numbered in the tens of billions of dollars. In 1980 Japan accounted for just 6 percent of all aid received by China; in 1990 this figure had risen to 29 percent. By 1990 Japan was disbursing close to $1 billion annually in Overseas Development Assistance (ODA) to China. In 1991 the Japanese government announced a loan package (separate from ODA) to the Chinese government for the amount of $4.6 billion from the Japan Ex-Im Bank. In 1992 Sino-Japanese two-way trade amounted to almost $30 billion; in 1999 it amounted to $66.2 billion (by 2006 it amounted to a staggering $207 billion). By comparison, Japanese-Soviet trade that had reached a peak of $6 billion in 1989 fell to $3.4 billion in 1992.[140] The figures in private direct investment are even more staggering. In 1992 Japanese firms invested just over $40 million in Russia; the same year they invested over $1 billion in China.[141]

Politically, China continued to play a major role in Japanese-Soviet/Russia relations. By this time China's official view of the territorial dispute had become somewhat ambiguous. Beginning in the mid-1960s and into the 1980s, China had supported Japan's claims to the disputed 'Northern Territories' (or 'Southern Kuril Islands'). China, too, had a territorial dispute with the Soviet Union along the river borders of Northeast China. Chinese leaders had been using the Japanese-Soviet territorial dispute for ammunition in Beijing's own war of rhetoric with Moscow. In 1991, however, the Soviets and the Chinese signed the first of several agreements legally establishing the border. This was the culmination of the dramatic warming in relations between the Soviet Union and China. In addition to the border settlement, all Soviet troops were withdrawn from Mongolia, and troop concentrations on both sides of the Sino-Soviet border were significantly decreased. China had also demanded that the Soviets withdraw from Afghanistan and that they withdraw support for the Vietnamese occupation of Cambodia. By 1991 both of these conditions had been met. Mikhail Gorbachev's visit to Beijing in 1989 and Chinese Premier Jiang Zemin's visit to Moscow in 1991 were seen as the final steps in putting behind the last vestiges of hostility that had divided the two nations since the early 1960s. China's relations with the new Russian government under Boris Yeltsin saw no drop-off in amicability in spite of ideological differences.[142]

In a 1991 visit to Beijing, Japanese Foreign Minister Tarō Nakayama asked Chinese Prime Minister Li Peng about China's opinion on the Japanese-Soviet territorial dispute and received an evasive answer. By 1991 China's views had evidently changed due to the dramatic improvement in Sino-Soviet relations.[143] China seemed now to be the primary focus of Moscow's Asia policy. Meanwhile, Japan seemed more preoccupied with China's rise than with the Soviet Union's dissolution. The Soviet Union/Russia was looking east – but to its 'other' eastern neighbor, China. Meanwhile, Japan was looking to its giant neighbor to the west – but to *its* 'other' giant neighbor, China. Thus the China 'factor' once again was a key influence on the policy formulation of Japan and Russia toward one another. It is not that leaders in Moscow and Tokyo chose to concentrate on relations with China to the exclusion of the other. Neither government could ignore the reality of China's rising economic, political, and military power. China simply afforded greater opportunities to both Japan and Russia than they could find in their own relatively stagnant bilateral relationship.

The United States continued to play an important role in Japanese-Soviet/Russian relations, but the nature of this role had changed. Long seen as playing an obstructionist role in Japanese-Soviet relations, leaders in Washington no longer feared a Japanese-Soviet *rapprochement*. In fact, a warming of relations between Moscow and Tokyo was seen as furthering the strategic interests of Washington. Initially, there was some concern in Washington that the demise of the 'Northern Threat' would have an adverse effect on the US-Japan security arrangement, which had long been based on a common hostility to the Soviet Union.[144] The United States had continued to voice support for Japan's claims to the disputed islands, even raising the issue at bilateral meetings with Soviet and Russian leaders. President George Bush raised the issue of the disputed islands in his meetings with Gorbachev at Malta in 1989 and Washington in 1990. At their last meeting in Moscow in July 1991, Bush privately and publicly urged Gorbachev and the Soviet leadership to reach a deal with Japan on the islands, stating that "this dispute could hamper your integration into the world economy, and we want to do whatever we can to help both sides resolve it."[145] Two high-ranking officials in the first Bush administration told this author that Washington had also extended an offer to Tokyo to mediate the dispute, but were politely told by Japanese officials that it was a bilateral affair for Japan and the Soviet Union. US leaders had become convinced of the necessity of Japanese-Russian economic and political cooperation. President Bill Clinton mentioned several times in early 1993 to Japanese leaders the importance of economic cooperation between Japan and Russia.[146] He would continue to do so throughout his presidency. Political cooperation among the three nations was also important given the rapidly changing situation in Northeast Asia. The Soviet Union had even stopped insisting that the US-Japan security alliance was detrimental to the interests of Moscow and her allies. In May 1989 the Soviet Union had officially de-linked the US-Japan security treaty from Japanese-Soviet negotiations.[147] Gorbachev on his visit to Japan in 1991 broached the possibility of trilateral US-Soviet-Japanese cooperation in confidence building measures that could one day form the basis of a new regional security structure.[148] This was a marked change from Cold War days, when even into the late 1980s the US and Japan trained jointly to fight a war against the Soviets in the Far East. So eager were some politicians in the United States for a political *rapprochement* between Moscow and Tokyo that by the early 1990s they often expressed frustration at the glacial pace in relations. While the United States and its European allies moved to extend massive amounts of credit to the Soviets and Russians during the various G-7 summit meetings, Japan was slow to embrace the remarkable changes brought about by Gorbachev and then Yeltsin.[149]

The late 1980s witnessed the emergence of another player in the Japanese-Soviet/Russian equation. Korea had been the central focus of Japanese-Russian rivalry one hundred years before. Now, the two Koreas had re-emerged as factors in the Japanese-Russian relationship. The influence of the two Koreas in Japanese-Soviet/Russian relations was peripheral, but both Seoul and Pyongyang assumed a role which neither Moscow nor Tokyo could ignore. When Mikhail Gorbachev formulated his new strategy in Asia, one of the most ground-breaking concepts was Soviet recognition of South Korea. The Soviet leader cultivated relations with Seoul for political and economic reasons. By moving closer to South Korea, Gorbachev hoped to place the Soviet Union in the unique position of being the only country with influence on both sides of the demilitarized zone. Gorbachev also saw Seoul as a source of economic assistance and capital for the beleaguered far eastern regions of the Soviet Union. Seoul could be an alternate, more reliable source of economic assistance in the event of continued Japanese adherence to the principle of withholding economic aid until the return of the disputed territories. Diplomatic relations between the Soviet Union and South Korea were established in September 1990.[150] In late 1990 South Korean president Roh Tae Woo visited Moscow and met Gorbachev for the second time (they had met in San Francisco in June 1990). Gorbachev's strategy paid off: Roh announced that South Korea would extend $3 billion in economic assistance. This was almost a year before the Japanese extended their loan of $2.5 billion.[151] Boris Yeltsin essentially followed Gorbachev's Korean policy, also hoping to cash in on Seoul's economic largesse. He became the first Soviet or Russian leader to visit Seoul in November 1992 (a year before he visited Japan).

Gorbachev's policy toward North Korea (the DPRK) was not quite so successful. Any remaining leverage Moscow had over Pyongyang dried up along with Soviet financial assistance. Meanwhile, Japan had begun making overtures to the North Korean leadership in the late 1980s. Ironically, it was South Korea's *Nordpolitik* initiative that served to temporarily bring Japan and North Korea closer together. South Korea's drive to establish trade relations, and then mutual diplomatic recognition, with China and the Soviet Union in the late 1980s and early 1990s served to drive a wedge between the DPRK and its two most important allies. An increasingly isolated North Korea was forced to look elsewhere for economic and political support. Kim Il-sung chose to utilize his contacts within the *Chosen Sōren* (the Korean community in Japan whose sympathies lie with the DPRK) in an attempt to improve relations with Japan. A delegation of Japanese politicians led by LDP leader Shin Kanemaru visited Pyongyang in September of 1990.[152] Normalization talks between the DPRK and Japan began in January 1991, and eight sessions were conducted until talks were suspended in November 1992. It is doubtful that Japanese attempts to improve ties with the DPRK were directly linked to relations with Moscow, but some feel that it was an attempt to play a different 'Korean' card in the face of the growing warmth in Sino-Soviet-South Korean relations.[153] In March 1993 North Korea announced that it was withdrawing from the Nuclear Non-proliferation Treaty (NPT) in response to requests by the International Atomic Energy Agency (IAEA) to conduct special inspections of two facilities suspected to be linked to an incipient nuclear weapons program. Condemnations of the DPRK were issued in both Japan and Russia. Two months later North Korea test fired a Rodong-1 missile, launching it into the Sea of Japan. Thereafter there would be no more attempts from either side to woo North Korea until the late 1990s.

As for Tokyo's relations with Seoul, the Russia factor did make a brief appearance when it was revealed in the fall of 1992 that South Korean firms were looking at business opportunities and possible joint ventures in the disputed 'Northern Territories.' Japanese Prime Minister Miyazawa took the occasion of a visit to Seoul in November 1992 to warn South Korean president Roh that Korean companies should not do business in the regions or waters Japan deemed the 'Northern Territories.' The advice was evidently heeded.[154]

The domestic political context

The international changes of this period almost all resulted directly from the domestic changes under way in the Soviet Union and Russia. Although domestic change in Japan was widespread, it was much less fundamental in nature than in Russia, and its link to international change was minimal. What is most important in this context is that domestic change in both countries caused a shift in the roles of traditional players and institutions that had influenced bilateral relations during the Cold War.

Japan was enjoying the dizzying heights of economic success in the late 1980s. This gave Japanese leaders much more confidence in their dealings with the Soviet Union. But, as the evidence shows, this did not translate into a successful strategy. In fact, the top leadership in Japan became divided on the territorial issue, and as to how much it should affect relations with the Soviet Union, and then Russia. Cracks in the unified edifice could be discerned as early as the late 1980s. The new policy of 'balanced expansion' (*kakudai kinkō*) had been unveiled in 1989, but the concept of the 'inseparability of economics and politics' (*seikei fukabun*) never seemed to die off. Whether one of these policies was meant to preclude the other was never clear to many – most of all it was unclear to the Soviet and Russian leadership. Gorbachev complained of this after his visit to Japan in 1991.[155]

In fact, within the Japanese Ministry of Foreign Affairs, which had dominated Soviet policy for almost two decades, there was a divide opening up. Certain personnel within MOFA were making a new appraisal of the Soviet Union by the late 1980s. Some within MOFA felt that little change was likely to result from Gorbachev's reforms (particularly with regard to the territorial dispute); others felt that change was inevitable. What caused this partial change in thinking? At least one historian feels that it was due to *gaiatsu*, or outside pressure, resulting from international change and the change in the stance of Japan's ally, the United States.[156] By 1992 the split within the Japanese Foreign Ministry's 'Russia School' had become even more pronounced. A group of more hard-line bureaucrats in the ministry centered on Hisashi Owada, Nagao Hyōdō, and Minoru Tamba were in favor of withholding any aid to Russia until the territorial dispute was settled (i.e. in Japan's favor).[157] Another group led by Ambassador Sumio Edamura, Kazuhiko Tōgō, and Kyōji Komachi was more amenable to expanding relations with Moscow in all areas, including economic assistance.[158] The former group represented the *iriguchi-ron*, or the 'entry-point' strategy insisting on the return of the islands as the key to improved relations. The latter group represented the *deguchi-ron*, or the 'exit-point' strategy that insists on better relations first, which will eventually lead to the return of the islands. These two different schools of thought within what had once been a unified 'Russia School' were to remain divided throughout the 1990s and into the 2000s. Through the early Yeltsin years, however, the division was only just emerging and thus still had little effect on policy.

Leaders in the still influential LDP interested in expanding contacts with Russia were eager to take advantage of the wavering MOFA line. LDP influence on Soviet/Russian policy increased to perhaps its highest level since the early 1970s. Through the 1970s and the 1980s the Ministry of Foreign Affairs had dominated Soviet policy. By the late 1980s LDP leaders such as Yasuhiro Nakasone, Shintarō Abe, Ichirō Ozawa, and Shin Kanemaru were engaging in clumsy attempts to formulate new policies. Professional diplomats were unhappy about this development and were no doubt satisfied to see the 1991 Ozawa mission to Moscow fail in its attempts to formulate (or 'buy') a new Soviet policy. The political corruption scandals that erupted in 1992, and that brought an end to LDP rule in 1993, emasculated all LDP influence for the next few years. By 1992–93 the Miyazawa cabinet was too weak domestically to accomplish anything new with Russia policy. In any case opposition parties from the left to the new reform parties on the right all supported the Ministry of Foreign Affair's principle of *seikei fukabun*.

Within other Japanese ministries the changes in the Soviet Union and the seismic shifts in the international situation were causing a reassessment, but it was a very cautious one. The economic ministries were still unsure about the Soviet Union. As mentioned, so much focus was being placed on China, the Soviet Union was but a mere afterthought. Some Japanese business leaders were very interested in new possibilities in the Soviet Far East. Foreign Minister Michio Watanabe insisted in 1992 that Japanese companies were not only interested but ready to take action.[159] The economic ministries were hesitant, however, due to the political situation involving the disputed islands, and to the unsettled domestic situation in the

Soviet Union, and later Russia. The Ministry of International Trade and Industry, the Ministry of Finance, and the *Keidanren* were all wary about the Soviet Union for political and economic reasons. Finance Minister (later Prime Minister) Ryutaro Hashimoto announced his support for the principle of *seikei fukabun* in relations with the Soviet Union.[160] By 1990 all 34 joint ventures involving Japanese firms in the Soviet Union (most of them begun in the early 1970s) had failed to make a profit.[161] In addition, Japanese trading houses (*shōsha*) were still waiting on unpaid debts for the purchase of goods during the Soviet era amounting to $1.36 billion.[162] Business leaders and economic bureaucrats were clearly less enthusiastic than they had been in the 1970s.

The military establishments in both countries were beginning to explore new avenues of cooperation. The Japan Defense Agency's White Paper of 1990 significantly downgraded the Soviet threat. The differences from the 1989 report were marked.[163] In spite of the JDA report, the Ministry of Foreign Affairs continued to insist that the Soviet threat had remained undiminished.[164] The Soviet military establishment was also surprisingly open to new contacts with Japanese counterparts. Undoubtedly, they were interested in Japan's defense modernization that had continued apace since the mid-1980s. Although as late as 1990 some Soviet generals pointed out Japan as a potential threat, in 1991 there was talk of military exchanges. That same year for the first time a Soviet general stationed in the Soviet Far East visited Japan.[165] These military contacts were still very low-key; Cold War prejudices still existed on both sides.

The domestic situation in the Soviet Union and Russia in the Gorbachev and Yeltsin years overshadowed concern about policy toward Japan. Beginning in the late 1980s Gorbachev met with an impressive number of Japanese diplomats, politicians, and citizens. This was a testament to his 'new thinking' about Japan policy. There was still hope that Japanese economic assistance could help transform the crashing Soviet economy. But by the early 1990s, as domestic events unfolded, Soviet and Russian leaders were less and less prone to give Japan much thought. Matters of political survival were much too pressing.

During the early days of Gorbachev's premiership, 'new thinking' on Japan was influenced by newer players, such as experts from the Soviet Academy of Sciences, and younger officials at the Ministry of Foreign Affairs. Gorbachev removed the hard-line apparatchiks Ivan Kovalenko and Mikhail Kapitsa from positions of influence by the end of 1988. Within the Politburo there were individuals such as Shevardnadze, Aleksandr Yakovlev, as well as Gorbachev himself, who seemed keen on formulating a more flexible Japan policy. But as *perestroika* began to falter and as the economy rapidly crumbled, Gorbachev became less prone to heeding the advice of these advisors. Although the Soviets seemed at one point interested in exploring the 1956 joint declaration as a starting point for discussions, by the early 1990s Gorbachev had become almost as obdurate about the territorial issue as his predecessors had been. He was constrained in his Japan policy by several factors: domestic political opposition in Moscow, regional opposition in the Far East, opposition from the military, and public opinion. Ironically, Gorbachev's *glasnost'* had been responsible for this new, last source of opposition.

The hard-liners that had been lurking behind the scenes gained more power with Gorbachev's 'turn to the right' in late 1990 and early 1991. By this time Eduard Shevardnadze resigned as Foreign Minister and Aleksandr Yakovlev had lost his influence. Men who would later lead the coup such as Gennady Yanaev, Boris Pugo, and Valentin Pavlov were appointed to higher posts.[166] Another source of opposition was of course Boris Yeltsin. Yeltsin wavered back and forth about Japan, but as Gorbachev's hold on power grew tenuous, Yeltsin sensed the national mood and denounced any calls for a turnover of territory to Japan. Gorbachev and other Soviet leaders were suspicious that Yeltsin would play the Japan card in order to get exclusive economic benefits for the Russian Federation rump state before the dissolution of the Soviet Union in 1990–91.

Although regional opposition to territorial concessions under Gorbachev was much more subdued than under Yeltsin, it did exist. One event that seemed to have a great impact on Gorbachev was the visit he made to Khabarovsk just prior to his visit to Japan in April 1991. On a walking tour of the city center people called out to him by his patronymic 'Mikhail Sergeyevich,' and pleaded with him to not turn over any territory to Japan. According to Gorbachev's new Foreign Minister Aleksandr Bessmertnykh, on the onward plane flight to Tokyo Gorbachev sat with a care-worn look on his face and muttered how he "could not do this." Bessmertnykh also added that Gorbachev may have wanted to make some sort of deal with Japan at this late date, but the opposition of fishery groups and the military was just too strong.[167] The Soviet press had taken on a lively debate on Japan policy and public opinion had come out strongly against any territorial concessions. Gorbachev frequently alluded to the new role of public opinion and his inability to ignore it.[168]

Boris Yeltsin had to live with these same opposition groups once he became president of the Russian Federation. The opposition had matured and grown even stronger by 1992–93. Yeltsin's opposition in the Russian parliament (the Congress of People's Deputies until autumn 1993; then the Federal Assembly of which the Duma was a part) extended to every area of his rule. In addition, the regional governments of the Far East, incensed by Gorbachev's 'handover' of territory to China in 1991, were determined to not let this happen again. The clout of the military, which had supported Yeltsin during the coup, had also grown. And now that Russia was a democracy, Yeltsin had to keep public opinion in mind. Yeltsin had assiduously cultivated public support and utilized it in the 1991 national referendum that contributed to his rise to power. He could not afford to alienate the masses now.

Andrei Kozyrev headed Yeltsin's foreign policy team. Deputy Foreign Minister Georgii Kunadze was by background a China, Japan, and Korea specialist. He was known for his controversial views on the relationship with Japan. Kunadze was critical of what he viewed as Stalin's historical legacy in the Far East – the violation of the 1945 Japanese-Soviet neutrality pact and the 'illegal' seizure of lands from the Japanese. Kunadze also viewed the unilateral renunciation by Khrushchev in 1960 of the 1956 joint declaration as an illegal act. He advocated a new policy with Japan, taking the 1956 formula as a starting point (i.e. returning two of the island groups – Shikotan and the Habomai group), signing a peace treaty, and negotiating from there on the two further islands. This became known as the '2+alpha formula.'

On these points Kozyrev agreed with his subordinate.[169] Other people agreed with this line, including Russia's famous writer Aleksandr Solzhenitsyn, who argued that Russia should sell the islands to Japan for a large amount of money.[170]

Kunadze got the first taste of grass-roots opposition on his visit to the disputed islands in the fall of 1991. He met with groups of the islanders and was assaulted with protests about the rumor of the new Russian government's plan to 'sell out' the islands to the Japanese. The governor of Sakhalin Oblast, Valentin Fedorov, had done much to whip up this frenzy over the 'South Kuril Islands.' The Kuril Island chain was placed under the administrative authority of Sakhalin, and so the islands were technically Fedorov's responsibility. In a series of scathing articles, as well as television and radio interviews, Fedorov unleashed a personal attack on Kunadze, going so far as to question his qualifications as a Russian diplomat and a protector of 'Russia's rights' because of Kunadze's ethnic Georgian background.[171] Fedorov was a crafty, yet entrepreneurial, politician who managed to place himself in the national spotlight, however briefly. At one point Fedorov threatened to separate the Far East from the rest of Russia if territorial concessions were made to the Japanese. "If Moscow cannot defend Russia from national betrayal, then a Far Eastern Republic must save Russia and Moscow itself from a territorial repartition of the Kuril Islands."[172] Fedorov tried to negotiate deals with foreign energy companies to develop Sakhalin's offshore energy resources, without Moscow's approval.[173] When a report appeared in the Japanese press in 1992 on the supposed bid by a company from Hong Kong to develop a resort on one of the disputed islands, both governments had to deal with a hornet's nest of accusations from all sides. It was later speculated that Fedorov planted this false story and even fabricated the lease documents.[174] In spite of the fact that Fedorov was such a professed champion of Far Eastern rights, he was seen by many on Sakhalin as a carpet-bagger from Moscow, and was raising the territorial issue to rally people around a common banner to distract attention from the collapsing local economy.[175] Fedorov was able to gather allies in the Russian parliament, namely Sergei Baburin and Nikolai Pavlov, two nationalist politicians from the Russian Unity (*Rossiiskoe Yedinstvo*) movement in his bid to block any territorial handover to Japan.[176]

The opposition forces in Moscow joined together in denunciation of Yeltsin's Japan policy. On the right were members of the newly formed Security Council, such as Vice-President Aleksandr Rutskoi, Defense Minister Pavel Grachev, and Minister of Internal Affairs Viktor Baranikov. This group's influence over Yeltsin was especially great in his first year because of their crucial support for Yeltsin during the August 1991 coup attempt. On the left some of Yeltsin's erstwhile democratic allies joined in the chorus of opposition to territorial concessions. These included deputies of the Supreme Soviet like Oleg Rumiantsev, Evgenii Ambartsumov, Sergei Stankevich, and Yeltsin's once trusted deputy Ruslan Khasbulatov. In July 1992, two months before Yeltsin's planned Tokyo visit, Rumiantsev organized a series of special hearings in the Supreme Soviet to discuss and debate the territorial issue, and to advise Yeltsin before his trip. In preparation for the hearings Rumiantsev prepared a detailed report refuting all Japanese claims, questioning the constitutionality of any territorial handover, warning that any turnover of territory would be against the national interest and would endanger national security and might open up a Pandora's box of territorial troubles along Russia's periphery from the Baltic to the Pacific. Rumiantsev also argued in his report that the economic value of the mineral sources alone on the disputed islands amounted to approximately $44 billion. Given the tone of this report it is no surprise that the Supreme Soviet came out against territorial concessions, and many lawmakers urged Yeltsin to postpone his visit.[177]

The fishing lobby joined the parliamentary forces in opposing concessions based on the economic value of the waters surrounding the islands. The chairman of the Russian State Committee of Fisheries, Vyacheslav Zilanov, called on the president to protect Russia's territorial waters. A figure of more than $1 billion in the annual economic value of the fisheries in the seas around the disputed islands was cited by several sources.[178] The Russian military was quick to back up the opposition to any territorial compromise. The strategic value of the islands was stated as the reason for the necessity of not only retaining the islands but also maintaining troops there. Several ex-military officers questioned the value of the islands to Russia's crumbling military, including a former KGB general Oleg Kalugin; however, an overwhelming majority of active officers opposed handing over territory.[179] A group of generals led by Chief of the General Staff, General Viktor Dubynin, brought pressure on Yeltsin in 1992 to cancel his upcoming visit to Tokyo. Dubynin gave testimony at the parliamentary hearings and claimed that the loss of the islands would amount to a giant strategic blunder. His battles with the Ministry of Foreign Affairs came to be known as 'Arbat vs. Smolensky' (the locations of the Defense and Foreign Ministries). Dubynin – playing to the domestic audience – grandly stated: "Our Great Russia begins with the crumbs of the Kurils."[180]

In September 1992, bowing to the domestic pressure, Yeltsin cancelled his trip to Japan three days before his planned departure. Although Yeltsin did eventually visit Tokyo in October 1993, he was never able to get past domestic opposition and completely normalize relations with Japan.

The ideational context

The Gorbachev revolution and the collapse of the Soviet Union were the cause of a significant reshaping of perceptions around the world. The Cold War's political divisions that had molded the thoughts and ideas of two generations of the world's citizens had crumbled as suddenly as the Berlin Wall. As is always the case, biases and perceptions were slower to change than the political order. In the case of Japan and Russia this was especially so. Japanese-Russian relations not only were unable to exit from the Cold War mindset but the relationship continued to be hampered by a Second World War-era dispute. At times the rhetoric seemed from an earlier century. Indeed, the way the two governments bandied about the issue of nineteenth-century treaties, one wonders how much the mindset of both nations toward one another had truly changed since that time.

During the 1970s and 1980s the Japanese government had so strongly stressed the nature of the Soviet threat that it was hard for many Japanese to imagine that

Gorbachev's reforms were genuine. The government was just as unsure of the extent and progress of Soviet reforms, and continued to express skepticism even as Western governments and their people were joining in 'Gorby-mania.' Although some at the top level of the Japanese government were willing to think about new policies toward Russia, the majority of the people in Japan remained strongly opposed to any wavering of the obdurate position espoused throughout the 1980s.[181] The Ministry of Foreign Affairs, which had come to dominate Soviet policy in the 1970s and 1980s, was unwilling to try to change the mindset that existed throughout Japan. As the economy of Japan expanded during this time, and the influence of other ministries such as MITI, the Ministry of Finance, and the Construction Ministry grew, the Ministry of Foreign Affairs clung ever more tightly to the one domain where its power was unchallenged. This was Soviet policy, and it was dominated by the handful of men in the 'Soviet School' (later the 'Russian School'). This view of the Soviet threat was backed up by the Japan Defense Agency. But by the late 1980s this interpretation was beginning to be questioned. In fact, the people of Japan were forming different attitudes toward the Soviet Union. But they were not necessarily for the better.

In his book *Japan's Response to the Gorbachev Era*, Princeton scholar Gilbert Rozman makes an in-depth study of the reactions of Japanese officials, academics, and ordinary people to the changes in the Soviet Union. Rozman found that the views of the Japanese toward the Soviet Union were unique; the Japanese took what the West considered fundamental change in the Soviet Union with a more cautious and guarded attitude.[182] The subtitle of Rozman's work, 'A Rising Superpower Views a Declining One,' gives a clear idea of how the Japanese felt at this time. Perceptions and opinions that began to color Japan's attitude toward the Soviet Union in the 1970s had become even more pronounced. Most Japanese had come to feel superior to the Soviets. By the late 1980s, as the economy soared, for the first time in decades ordinary Japanese felt assured and confident about themselves and their country. This self-assuredness manifested itself in Japanese attitudes toward the United States, Europe, and especially toward the Soviet Union. As Gorbachev's *glasnost'* opened the world's eyes to the sorry state of the Soviet economy, attitudes toward the Soviet Union assumed an arrogant posture. Rozman writes how Japanese became "openly disdainful" toward the Soviet Union by the late 1980s.[183]

Interestingly, leaders in the Soviet Union saw in Japan a potential model for economic success. Japan's economic reconstruction after the Second World War was also accomplished under tight central government control. Soviet economic managers were sent to Japan to study Japanese techniques during the 1980s.[184] In the end, it was apparent that the differences were too great for a model of reform to be transplanted onto the Soviet economy. The Soviet economy was in such a bad state that even the most drastic of reforms would be unable to accomplish the gargantuan task of rebuilding what communism had mishandled for 70 years. So low had the Japanese opinion toward the Soviet Union sunk that by 1991, for the first time since the end of the Second World War, the United States had replaced the Soviet Union in opinion polls as the country deemed the greatest security threat to Japan.[185]

The Japanese became even more determined to press their advantage and finally gain control of the 'Northern Territories'. The feeling was that although the Soviets might be in the process of genuine change, and the difficulties may be great for the Soviet people, Japan should not relent in its demands. While the Soviets are down, what better time to take advantage and get the territories back? During the Gorbachev and Yeltsin years there was very little debate in the Japanese press about the substance of Soviet policy. The policy itself was seemingly immune to criticism either from the right or the left (apart from a few dissenting voices). Like the proverb stating that the nail sticking up will be hammered down, those few voices in Japan calling for anything less than a full territorial reversion were hammered down. To be fair, there were a number of Japanese calling for a 'balanced' policy toward the Soviet policy, but the ultimate goal (full territorial reversion) remained unchanged. The collapse of the Soviet Union and the birth of Russia seemed to matter little for most Japanese.

In the Soviet Union, on the other hand, a lively debate sprang up in the press of *glasnost'*. Views ranged from hard-line conservatism, to hard-core criticism of the Soviet Union's Japan policy. Experts and novices alike took up the debate. Veteran Pravda correspondent Igor Latyshev led the hard-line group. The title of his 1994 book 'Who is selling out Russia, and how?' tells enough about his views.[186] Dailies such as *Izvestia, Pravda*, and *Sovietskaya Rossia* backed this view. Far Eastern publications were also unanimously against territorial concessions. On the other end of the spectrum, Georgii Kunadze and another product of the Academy of Sciences, Konstantin Sarkisov, advocated a fundamental review of Japan policy. They did not necessarily call for a return of all the disputed islands, but they did favor the 1956 declaration as a starting point for building new relations.[187] This view was echoed by many others in the Academy of Sciences and in some of the newer academic journals and regular dailies (such as *Nezavisimaya Gazeta* and *Komsomol'skaya Pravda*). Some, including the celebrated author Aleksandr Solzhenitsyn, called for a complete reversion of the islands. Other experts took a middle approach, but admitted that the Soviet Union's Japan policy needed a thorough review.[188] At least one author criticized both Japan and the Soviet Union with equal vigor.[189]

The Soviet people had much more pressing matters on their minds, and Japan barely registered in their consciousness, unless they happened to live in the Far Eastern regions. People in the European regions of the Soviet Union and Russia thought of Japan mainly in the context of the territorial issue, and this did not create a favorable impression. Ironically, it was the Japanese Foreign Ministry that had worked hard to bring this issue to the attention of the Russian people, and it seemed to have backfired. Polls taken in March 1991 reflected the overwhelming sentiment at that time: 71 percent of Russian respondents nationwide were against territorial reversion; in the Far East the number was 85 percent.[190] Some isolated cases of Japanese humanitarian aid (such as to the victims of Chernobyl) produced positive publicity, but these were few and far between. In Japan, the cases that drew the most attention to the Soviet Union, or Russia, were often unflattering. A case in point is the two cancellations of visits by Yeltsin. When the third visit was finally successfully brought off, it was learned only several days later that the Russian navy had been dumping nuclear waste into the Sea of Japan.

It can be said that the few positive expectations that had been raised by the monumental changes in the Soviet Union only caused the disappointment to taste much more bitter when realities and obstacles prevented yet another period of *rapprochement* to result in normalization.

The breakdown

Although optimism did exist in certain circles (particularly Japan) during the 1989–93 period, most experts realized by early 1991 that the *rapprochement* was doomed to fail. Many realized this even before Gorbachev's visit. Although the hope of the early Yeltsin months seemed even more promising, by early 1992 it was again apparent that the differences extended to a level beyond the means of normal diplomacy to rectify.

Indeed, the brief re-warming of relations brought about by the Yeltsin visit in October 1993 fizzled out soon thereafter. The nuclear dumping issue was but the beginning of a new cold period in relations. The December 1993 Duma elections in Russia resulted in a stunning victory for nationalist forces and for the Russian Communist Party. The head of the nationalistic Liberal Democratic Party of Russia, Vladimir Zhirinovsky, was often quoted speaking of Japan and its 'pretensions' in a menacing tone.

Meanwhile, the political situation in Japan had also taken a confused turn. The reformist Prime Minister Morihiro Hosokawa was forced to resign in the spring of 1994 in the wake of yet another corruption scandal, and the weak successor cabinet of Tsutomu Hata soon gave way itself to a patchwork coalition uniting the Japanese LDP with the Socialist Party. Confusion left little room for diplomatic initiatives, while bilateral tension increased. In 1994 in the waters around the disputed islands the number of incidents at sea rose dramatically. Japanese fishing boats and their crews were often stopped by Russian coastal patrol boats in disputed waters and sometimes arrested and taken into Russian custody while their boats (and catches) were impounded. These incidents were known as the 'fishing wars' and they significantly increased the level of public hostility.[191] Negotiations soon smoothed out most of the differences, but Japanese fishermen to this day risk fishing in these disputed waters. A further element of strain came about in 1992 when the central government in Moscow cancelled preferences for newly established firms operating with foreign capital. This would come to harm Japanese businesses operating in the Russian Far East throughout the 1990s.[192] New politics did not necessarily translate into new business opportunities.

Two natural catastrophes further brought the deep distrust between Japan and Russia to the world's attention. An 8.1 magnitude earthquake struck Hokkaido and the disputed islands in October 1994. Close to 60 percent of the homes on the island of Shikotan (population then of just under 6,000) were destroyed. The Japanese government quickly offered aid, medical and rescue crews, and shelter to the islanders. The two sides bickered for precious days about the details, and in the end very little Japanese assistance ever reached the islands. This event was a mirror for the unsuccessful efforts of the Soviet government to deliver aid to earthquake victims in Yokohama in 1923 (see Chapter 1).[193] Then in May 1995 another disastrous earthquake struck the Russian Far East, this time taking the lives of 2,000 people in northern Sakhalin. Again, Russia dithered before finally accepting Japanese emergency assistance.[194]

This lack of cooperation even over matters of life and death was symptomatic of the serious malaise that had afflicted bilateral relations for over a century. Even seismic shifts in the international situation could not alter the fundamentally bad nature of Japanese-Russian relations.

The *rapprochement* of this period was a result of international structural change. Interestingly, the change in the international structure was due primarily to domestic change in the Soviet Union. Unfortunately, the images and perceptions in each country changed very little. But ultimately, and ironically, the growth of democratic government in Russia and its maturation in Japan helped to bring about the breakdown of efforts to normalize relations.

3 Another *rapprochement*

1996–2007: Attempts at normalization

Unlike the earlier periods of *rapprochement* that had always followed relatively lengthy periods of cool relations, the *rapprochement* that began in the late 1990s came on the heels of the improved relationship of the late 1980s and early 1990s. Although some observers might be tempted to draw a direct relationship between these two periods of amicability, they were in fact quite distinct. The early Yeltsin years did draw on the momentum of the Gorbachev years, but the attempts at normalization during the later 1990s were the products of subtle shifts in the domestic politics of both countries and in the international situation in East Asia and elsewhere around the Eurasian periphery in the mid-1990s.

In spite of the differences among the various eras of *rapprochement* during the twentieth century a pattern in Japanese-Russian bilateral relations can be discerned. It was a familiar fluctuating pattern, as had been the case in the periods reviewed earlier in this study.[1] The period 1996–2007, though different in many aspects from earlier periods, should be analyzed with an eye toward these earlier periods. The historical examination of bilateral relations between Moscow and Tokyo demonstrates that international political and systemic change has always preceded or accompanied a Japanese-Russian *rapprochement*. Furthermore, domestic political change has always been prevalent during these periods. A final factor that has often prevented a complete normalization of bilateral relations is the factor of mutual perception, or what has been termed 'ideational factors'. International change and domestic political change have been two of the constant and necessary factors during periods of Japanese-Russian *rapprochement*. Changes in mutual perceptions, however, have been glacial throughout the twentieth century and up to the present.

The turn of 1996–97

Relations between Moscow and Tokyo had assumed a low place on the policymaking agendas of both nations after Boris Yeltsin's 1993 Tokyo visit. Two-way trade did almost reach the $6 billion level in 1995, but this was relatively speaking a minuscule amount for both countries. Domestic political events in both countries between 1993 and 1996 occupied the attentions of senior policymakers. Elections in Japan and Russia coincided with economic problems and social discontent (though of a much different scale in both countries). The disappointment of the failure to normalize relations in 1992–93 was great because of the initial euphoria that accompanied the fall of the Soviet Union in 1991, particularly in Japan. There were still voices on both sides cautioning against a move toward normalization. In a 1995 meeting with US officials in San Francisco, Japan's Deputy Vice Foreign Minister Minoru Tamba had warned that Russia was still a threat to Japan's security. A Japanese military analyst had echoed this warning, noting that once the Soviet Union had recovered from the chaos of the civil war in the 1920s it was able to deploy large forces in the Russian Far East that defeated the Kwantung Army in Northeastern China in 1939. Japan, he went on, "should learn from history the risk of Japan's northern threat recovering."[2] Influential Japanese strategist and former diplomat Hisahiko Okazaki agreed with this assessment, writing in the early 1990s that the outcome of the 'Russian revolution' was not yet clear, and there was no telling whether a Robespierre or a Napoleon would emerge.[3] An editorial in the influential *Asahi Shimbun* in the spring of 1995 stated that Japanese-Russian relations are "as cold as they can get" (*kankei no genjō ha o-samui kagiri)*. In Russia, conservative forces that had effectively quashed any chance for a peace treaty in 1992–93 had increased their power in the Duma after the December 1993 elections. Yeltsin was unable and unwilling to confront these forces and expend political capital on an issue as minor (in the eyes of most Russians) as the territorial dispute with Japan. Hence Yeltsin's reluctance even to accept Japanese humanitarian aid after the earthquakes in 1994 and 1995 that devastated the island of Shikotan and the northern part of Sakhalin. Although Kozyrev remained at his post as Foreign Minister until early 1996, he was increasingly isolated from the Kremlin. Moreover, his deputy, Georgii Kunadze, had been 'exiled' to South Korea to serve as ambassador there.

The year 1996 turned out to be an auspicious year for Japanese-Russian relations. It was the fortieth anniversary of the re-establishment of diplomatic relations between the Soviet Union and Japan. In early 1996 Ryutaro Hashimoto was picked to be Prime Minister by a coalition of the LDP, the Socialist Party, and the reform *Sakigake* party. Hashimoto was the first LDP prime minister since 1993. As MITI minister from mid-1994 to January 1996, Hashimoto had been a strong proponent of expanding Japanese trade and investment with Russia. As president of the LDP from 1995, Hashimoto had made it known that he also favored stepped-up political relations with Russia. Hashimoto had built a strong personal relationship with Russian First Deputy Prime Minister Oleg Soskovets.[4] In March 1996 Hashimoto dispatched Foreign Minister Yukihiko Ikeda to Moscow to meet with new Russian Foreign Minister Yevgenii Primakov. Like Hashimoto, Primakov was an influential player who had just come to office in early 1996. Primakov replaced the unpopular Andrei Kozyrev and was known to have deep and long-standing contacts in Japan. In the late 1980s, Primakov, as Director of IMEMO (part of the Russian Academy of Sciences), had been one of Mikhail Gorbachev's top Asia advisors and had called for better relations with Japan.

In early 1996 contact had been re-established at the highest levels and it was clear to both sides that better relations were a high priority. The next two years saw a never-ending procession of official high-level and mid-level meetings, accompanied by an increase in non-official contacts as well.

The Ikeda-Primakov meeting in March 1996 was a clear signal of the start of serious bilateral engagement. Ikeda conveyed a letter from Hashimoto to Yeltsin, expressing Japanese support for Russian reforms (i.e. Japanese support for Yeltsin in the 1996 presidential election). Primakov pledged that Russia would continue its demilitarization of the disputed territories, pointing out that only 3,500 soldiers remained on the islands, down from 7,000 in 1992. While in Moscow Ikeda co-chaired along with Russian First Deputy Prime Minister Soskovets the first session of a new Russo-Japanese trade commission. Soskovets announced that a tentative agreement had been reached on the re-orientation of a $500 million humanitarian loan that Tokyo had granted as part of the October 1991 $2.5 billion assistance package.[5] Later, the Japanese government announced that it would allow the Japan Export-Import Bank to use the money to invest in Russian industrial and commercial ventures.

Yeltsin and Hashimoto had their first face-to-face meeting as heads of state at an April 1996 summit meeting held in Moscow on nuclear energy safety to mark the tenth anniversary of the Chernobyl disaster. Hashimoto wisely chose to not bring up the territorial dispute on the eve of the Russian presidential election, and after the meeting he declared: "Boris Yeltsin and I established friendly personal relations." The two also affirmed the validity of the 1993 Tokyo declaration and promised to resume negotiations toward a peace treaty (which had been suspended after 1993).[6]

Ikeda and Primakov met again at a gathering of G-7 finance ministers in Lyon in late June 1996. Primakov had earlier made statements suggesting a postponement of territorial talks for "future generations," but in Lyon he backed away from these controversial statements, and soon agreed with Ikeda to resume meetings of a working group to discuss the territorial issue and a possible peace treaty. Apparently, the mixed signals were due to rhetoric emanating from the heated presidential campaign in Russia.[7] After Yeltsin won the election he received a congratulatory phone call from Hashimoto. The two promised to continue political talks aiming at a peace treaty, and Hashimoto expressed interest in an early visit to Russia. Hashimoto declined an invitation to attend Yeltsin's inauguration ceremony as it was coincidentally scheduled to be held on August 9, the fifty-first anniversary of the Soviet Union's attack on Japanese forces in the closing days of the Second World War.

In November 1996 Primakov made his first visit to Tokyo as Foreign Minister and proposed to Ikeda that Japan and Russia jointly develop the economies of the disputed islands. Ikeda replied that in principle Tokyo was not opposed to joint economic development, as long as it did not mean a shelving of the territorial issue, and provided Russia recognized Japan's sovereignty over the islands. The discussion on joint economic development died at this point. Primakov did, however, meet with a wide number of officials and non-governmental specialists in Tokyo and they all assured him of Japan's sincere desire for an improvement in relations. It was announced during Primakov's visit that the Japanese government would join with the French and German governments in helping Russia to build a plant for reprocessing weapons-grade plutonium into fuel for nuclear power plants. The facility would not be operational for some time, but it was yet another positive signal that Japanese leaders realized that bilateral relations were about more than the territorial dispute.[8] Shortly thereafter, an editorial in the conservative Japanese daily the *Sankei Shimbun* endorsed the move toward a *rapprochement* with Russia.[9] The Japanese Foreign Ministry's chief of the Russia division, Kenji Shinoda, had advocated a 'multi-layered approach' (*juusō-teki apurochi*), and the Japanese government was beginning to act along these lines. Formerly, all aspects of the bilateral relationship had been tied to the territorial dispute.[10]

The year 1997 did not start off so auspiciously for Japanese-Russian relations. In early January the Russian oil tanker *Nakhodka* sank off the northwestern coast of Japan. The resulting oil spill threatened not only vital fishing grounds but also water intakes for energy plants supplying electricity to a half million people. The disaster also caused an energy crisis on the Kamchatka Peninsula, to which the oil tanker was traveling when it capsized. The disaster (as well as the 1993 nuclear dumping sensation) highlighted the perilous state of conditions in the Russian Far East and reminded the Japanese people that Russia's social problems were capable of affecting Japan, and that the nation could not simply afford to stand by and watch as the situation in the Russian Far East deteriorated to new lows. The sinking of the *Nakhodka* highlighted the paradoxical situation of the energy situation in the Russian Far East. While multinationals scrambled for energy concessions in Siberia and Sakhalin, these regions were still reliant on outside energy sources.[11]

Also in January 1997, certain officials in Moscow and the Russian Far East were led to believe that the United States was redeploying to Hokkaido part of the force of US Marines stationed on Okinawa. The Japanese government quickly reassured Moscow that no such move was under way. In fact it was later reported that the Japanese had proposed such a transfer, but the United States had rejected the idea.[12] The situation was yet another reminder of how the two governments often misunderstood, misled, or failed to effectively communicate on a variety of issues. For example, in late February 1997 Russia's new First Deputy Prime Minister Viktor Ilyushin was due to visit Tokyo. At the last minute he cancelled the visit, citing pressing domestic matters. Then several days later Ilyushin turned up in Switzerland to support an Olympic bid for St. Petersburg, much to the shock and dismay of Japanese diplomats.[13] The Russian government was unable to give a satisfactory explanation. The episode reinforced Japanese perceptions that Russia was a chaotic and unpredictable place governed by politicians and officials of a similar nature.

Yet, the arrival of Russia's new ambassador to Japan, Aleksandr Panov, was viewed very favorably in Japan, and came at a good time for Moscow. A fluent Japanese speaker and a respected scholar, both sides looked to Panov to help close the communication gap. Panov arrived in late 1996 and the impact was soon felt. When the *Nakhodka* capsized off the coast, Panov arrived at the scene of the beach where clean-up operations were under way, and in television interviews in perfect Japanese he offered his sincere regrets for the mishap.[14] Panov authored numerous articles in the Japanese press and continued to conduct frequent interviews, shoring up Russia's image somewhat.

By the spring of 1997 the pleasant atmosphere of the fall before had returned to the relationship. During a May visit to Tokyo by Russian Defense Minister Igor Rodionov, Japanese leaders agreed to drop their objections to Russia's participation at the upcoming G-7 summit meeting in Denver.[15] They did qualify this support by insisting that Russia be excluded from certain key meetings on economic issues. Japanese Foreign Minister Ikeda visited Moscow the same month and announced that Japan was willing to allow Moscow to defer payment of the $1.5 billion debt for at least three, and possibly six, years.[16]

The Denver G-7 summit of June 1997 was a major turning point in bilateral relations. During the summit Yeltsin and Hashimoto twice met behind closed doors and discussed how to increase Japanese investment in the Russian Far East, and Japan's interest in a natural gas project in Yakutia. Yeltsin pledged to support Japan's bid to become a permanent member of the UN Security Council, and he promised to stop targeting Russian nuclear missiles on Japan. It was clear that the two leaders had found common ground and that they felt very comfortable with one another, often exchanging jokes and stories. The two continued to build on their strong personal relationship, and promised to meet once a year.[17]

In Tokyo on July 24, 1997, in a speech to the Japanese Association of Corporate Executives (*Keizai Dōyuukai*), Hashimoto took a further step toward cementing a *rapprochement* with Russia. In his speech outlining a strategy for Japan's new 'Eurasian Diplomacy' for the 'Silk Road' region, Hashimoto announced that Japan would steer a "new course" with regard to developing relations with Russia. This new course was to be based on three principles: trust, mutual respect for each other's interests, and the building of relations proceeding from a long-term perspective.[18] Hashimoto drafted the speech with the help of top officials at MITI (including Section Chief of the Policy Division Kenji Isayama) and an emphasis was placed on energy cooperation. At the insistence of one of the top officials from the Foreign Ministry (Kazuhiko Tōgō), Hashimoto avoided placing too much emphasis on the territorial dispute. The speech was warmly welcomed in Moscow and was given high marks by Russian officials.[19]

It was revealed much later that in August 1997 the Japanese Foreign Ministry had prepared a secret paper for the Prime Minister, outlining three options for a peace treaty with Russia. One option called for Russia's recognition of Japan's sovereignty and agreement for continued Russian administrative control of the islands until some point in the future. Another option called for Moscow and Tokyo to issue a joint statement pledging to continue talks on the four islands within a limited time frame. The last option envisioned a reissuing of a joint declaration based on the 1956 Joint Declaration, stipulating the return of two of the disputed islands (Shikotan and the Habomai group), and a "clear direction" toward the return of the remaining two islands (Etorofu/Iturup and Kunashir/i). This last option, in fact, was the precursor of the 'two-track' formula that would become quite controversial several years later in Japan.[20] Top Foreign Ministry officials including Vice Minister Shunji Yanai, Minoru Tamba (then Deputy Vice Minister), Mutsuyoshi Nishimura (Director of the European and Oceanic Affairs Bureau), and Yukio Takeuchi (director of the influential Treaty Affairs Bureau) approved this document. It is interesting to note that the document, which was published only in 2002 by the *Asahi Shimbun*, did not appear to have involved Diet member Muneo Suzuki and Kazuhiko Tōgō (who succeeded both Takeuchi and Nishimura in the Foreign Ministry).[21] Suzuki and Tōgō were later harshly criticized for their pursuit of a 'two-track' policy. It appears that this option had widespread support within the Foreign Ministry, and that Suzuki and Tōgō were not alone in their interest in such an approach.

It should be pointed out that in the past eras of *rapprochement* it had often been the Russian or Soviet side that made the first moves toward conciliation. In 1996–97 it was the Japanese that earnestly began cultivating a new relationship with Russia. In a September 1997 cabinet reshuffle Hashimoto appointed Keizo Obuchi as his new foreign minister. Despite Ikeda's hard work, the Russian side was happy to see Obuchi succeed him. Obuchi had been one of the foremost advocates in the Japanese Diet for stepped-up relations with Russia. He was deeply connected with Japanese business circles interested in Siberian development, and earlier in the year had headed a delegation of business leaders to Moscow where he met with then Deputy Prime Minister Anatoly Chubais, one of Yeltsin's closest aides.[22] The stage was set for the first bilateral meeting between Yeltsin and Hashimoto. In Denver, Hashimoto asked Yeltsin and other officials in the Yeltsin entourage (including new First Deputy Prime Minster Boris Nemtsov) if it would be possible to arrange a casual meeting between the Prime Minister and the President, perhaps somewhere in the Russian Far East ("on a weekend, as among friends"). Yeltsin agreed to this.[23] After a meeting in New York in late September, Foreign Ministers Obuchi and Primakov announced that Yeltsin and Hashimoto would meet in the Siberian city Krasnoyarsk on November 1–2, 1997.

The Krasnoyarsk meeting was hailed on both sides a milestone in the bilateral relationship. The decision to hold the summit meeting in Krasnoyarsk was pregnant with meaning. Krasnoyarsk is located approximately halfway between Tokyo and Moscow. The two sides agreed to meet halfway geographically, and as equals in the diplomatic arena. Ronald Reagan and Mikhail Gorbachev met as equals in Reykjavik, Iceland in 1986, halfway between Washington and Moscow. Yeltsin and Hashimoto no doubt were hoping that the Krasnoyarsk meeting would also be seen as an epoch-making event.[24]

Yeltsin and Hashimoto greeted one another like two old friends. Attired in ski parkas, the two boarded a boat and headed up the Yenisei River for a fishing trip. This 'no neck-tie' summit may have seemed more of a photo opportunity for both sides than an arena for concrete discussions. Both leaders, however, quickly demonstrated their desire to get down to business. Hashimoto announced that Japan would officially support Russia's APEC membership, and he unveiled a six-point economic plan (known as the Hashimoto-Yeltsin Plan).[25] The two sides also promised to step up defense contacts. To everyone's surprise, Yeltsin went a step further and declared that Russia and Japan would do their utmost to sign a peace treaty by the year 2000. This stunning announcement not only surprised the Japanese side but apparently it also shocked the Russian Foreign Ministry. Yeltsin had failed to consult with officials from there before the meeting.[26] Unfortunately, the announcement by Yeltsin created a giddy mood of optimism. The idea that the complex issues and problems that had divided Moscow and Tokyo for decades could be erased with one grand statement was far-fetched, as was pointed out at the time by experts in Japan and Russia. Japanese diplomats and officials were cautious, but the public and the press were caught up in a wave of enthusiastic speculation. The Japanese press in particular was given to wild predictions about how a settlement might be reached, and on several occasions both the Japanese and the Russian Foreign Ministries were forced to deny certain press reports.[27]

Ironically, in his memoirs Yeltsin was later to write that the discussion with Hashimoto on the boat resulted in a "deadlock" (*tupik*) over the territorial dispute. How then did he come up with his suggestion at the press conference to sign a peace treaty before 2000 if neither side was ready to make compromises? Yeltsin makes it clear that he felt relations with Japan were important to Russia for one reason only: economics. He goes on to say that the Japanese side was concerned with only one issue: territory.[28] After the Krasnoyarsk summit Hashimoto stated that future negotiations would be based on the 1993 Tokyo Declaration, which states that a resolution of the territorial issue must precede the conclusion of a peace treaty. The Russian side saw things somewhat differently. Some in Russia even speculated that the Japanese had 'tricked' Yeltsin into making certain statements, by taking advantage of the fact that there were no experts present from the Ministry of Foreign Affairs, and that Yeltsin was still somewhat weak from a heart condition. First Deputy Prime Minister Boris Nemtsov attended the talks and was apparently also caught off guard by Yeltsin's sudden declaration. Afterwards he told reporters that the president was known for his impetuous nature, and that his health-related weakness and the chaotic domestic situation may have had something to do with the sudden remarks. Nemtsov pointed out that the Russian constitution binds Yeltsin to maintain Russia's borders, and therefore its control of the Kuril Islands.[29]

Wherein were the two sides to find a common ground for compromise if they could see the differences and yet they continued to talk past them? Misperceptions and misunderstandings were bound to arise as long as the two sides continued to talk past one another, as they seem to have done at Krasnoyarsk. However difficult future negotiations promised to be, Yeltsin's bold statement did manage to produce positive momentum in the relationship. Yeltsin and Hashimoto agreed to refer to one another by their names ('Boris and Ryu'), and they agreed to meet again informally in Japan in the spring.

In spite of the growing political warmth, trade relations were stagnant. In 1996 two-way trade actually dropped by $1 billion from the previous year to $4.9 billion. In 1997 the trade figure barely rose to $5 billion, still one billion dollars less than the all-time high registered in 1989.[30] Late in the summer of 1996 there were a number of incidents involving Russian border guards firing on Japanese fishing vessels operating dangerously close to the waters contiguous to the disputed islands. The local Russian authorities issued stern warnings, but the issue seemed to have been smoothed over before it reached the highest levels. Negotiations between Japanese and Russian diplomats over fishing rights around the waters of the disputed islands were held through 1997. In the end the two sides were unable to reach an agreement before the Krasnoyarsk summit in November.

Economic relations, however, were not all bad. On May 21, 1996 a US-Japanese-European consortium announced that a $10 billion energy agreement known as the Sakhalin-2 project would begin work later that year. The consortium included at the time the American oil companies Marathon and McDermott (these two companies would later withdraw from the consortium), the Anglo-Dutch company Royal Dutch/Shell, Russian Petroleum (Rosneft), and the Japanese companies Mitsubishi and Mitsui. In addition, on June 10, 1996 an international consortium comprised of American, Japanese, and Russian companies announced the signing of a $15 billion agreement to begin work on the Sakhalin-1 energy project. The consortium manages three fields off the Sakhalin coast that are estimated to contain approximately 2.5 billion barrels of crude oil and 15 trillion cubic feet of natural gas. In addition to Exxon-Mobil (30 percent), the Japanese company Sakhalin Oil and Gas Development Co. (SODECO, 30 percent), and the Russian groups Rosneft-Sakhalin and Sakhalinmorneftegaz (40 percent) took part in the initial signing of the agreement.[31] Regional leaders, hoping to use this new momentum to boost trade, met in Fukui in northwestern Japan in late November 1996 and called for expanded trade and cultural relations.[32]

One of the more surprising aspects of the newfound warmth in relations was the dramatic increase in contacts between the defense establishments and the militaries of the two nations. In the 1996 edition of the Japan Defense Agency's White Paper Russia's Far Eastern military forces were downgraded from a "factor of instability." Prior to the 1990s they were deemed a threat, then after 1992 a factor of instability. This phraseology was entirely eliminated in 1996, reflecting new attitudes in the Japanese defense community. In April of 1996 the Director General of the Japan Defense Agency Hideo Usui visited Moscow, the first such visit by a JDA chief. During his visit Usui and Russian Defense Minister Pavel Grachev signed a military cooperation protocol. The agreement calls for advance notice of military exercises, and provides for exchanges of information, training missions, and naval port visits. Usui intimated that Japan might be interested in acquiring advanced weapons systems, including fighter jets, from Russia.[33] In July 1996 a Japanese Maritime Self-Defense Forces (MSDF) destroyer, the *Kurama*, visited the Far Eastern port Vladivostok, the first visit of a Japanese warship to Russian waters since Japan's withdrawal from the Siberian intervention in 1922. In August 1996 it was announced that pilots from Japan's Air Self-Defense Forces (ASDF) would be invited to train in Russia. Articles published later in the *Asahi Shimbun* revealed that two veteran Japanese pilots from the ASDF had been training at the Gromov air base just south of Moscow, learning to fly the Su-27, one of Russia's most advanced fighter jets. The article pointed out that these jets were being sold to China. The ASDF was in fact interested in learning more about these planes, not just because potential enemies might be flying them, but also to carry out research for Japan's own indigenous fighter program.[34] This revelation was surprising because not only was Russia allowing Japanese pilots unfettered access to platforms of the latest technology and to Russian bases but also because the

Japanese ASDF (and the JDA) was studying equipment that would be used by the People's Liberation Air Force in China.

In May of 1997 Russian Defense Minister Igor Rodionov returned the visit his Japanese counterpart had made to Moscow the previous year. His visit to Tokyo was the first visit by a Russian Defense Minister for almost a century. Rodionov and JDA Director General Fumio Kyūma signed a protocol creating bilateral exchanges and joint working groups of defense officials. Rodionov also announced that Russia would be interested in holding joint naval exercises with Japan and the United States. He went on to state that the US-Japan Security Treaty was "necessary" for regional stability, thus ending the Soviet Union and Russia's long opposition to this alliance. One Russian diplomat reportedly expressed interest in the three-way development of a missile defense program.[35] Soon thereafter, the Russian warship *Admiral Vinogradov* visited Tokyo, the first visit to Japan by a Russian naval ship since the visit of Crown Prince Nicholas in 1891.

In the summer of 1997 an event transpired that demonstrated the depth of the Japanese-Russian *rapprochement*. It was on July 22, 1997 that two MSDF destroyers, the *Setogiri* and the *Sawayuki*, quietly departed from the port of Abashiri in northern Hokkaido for a routine mission in the Northern Pacific. Rather than steaming through the Soya Strait (between Hokkaido and Sakhalin) and proceeding southward into the Sea of Japan, they pointed their bows northward and steamed into the Sea of Okhotsk. After passing the 46th parallel on July 23, they turned eastward, steaming just north of the Russian island of Urup in the Kuril chain. The two vessels then passed through the Urup Strait, exiting the Sea of Okhotsk out into the Northern Pacific. These were the first Japanese military ships to pass through these waters since 1945. During the Cold War the Sea of Okhotsk had been a 'nuclear bastion' of the Soviet Union's Pacific Fleet. The Urup Strait is a strategically vital passageway through which Soviet (and now Russian) nuclear submarines passed with great regularity. Formerly Japanese MSDF ships had to proceed into the Northern Pacific through the Tsugaru Strait on the southern side of the island of Hokkaido. Although the waters of the Urup Strait technically lie within international waters, Tokyo undoubtedly notified Moscow, thus seeking approval, before sending the *Setogiri* and the *Sawayuki* through the sensitive strait. The fact that Moscow approved sent a strong, yet quiet, signal. The seemingly innocuous movement of two ships through a strait in the Kuril Islands, though generally unnoticed throughout the world, was a dramatic symbol of the warming of relations between Tokyo and Russia.[36]

1998: A year for negotiations

The first bilateral negotiations to bear fruit in 1998 were the fishing negotiations that had continued off and on for three years. It took 14 rounds of talks for an agreement to be reached that allowed Japanese fishing boats to catch a quota of fish and shellfish in the waters off of the disputed islands, in exchange for cash and fishing equipment to be paid by the Hokkaido Fisheries Association. For some groups in Japan this was seen as the single most important outstanding issue between Japan and Russia. Others also saw this as a sign that Russia was slowly coming around to granting Japan sovereignty, first over the waters, and then over the islands.[37]

In a January Moscow meeting between Russian Deputy Foreign Minister Grigorii Karasin and his Japanese counterpart Minoru Tamba, the two governments agreed to establish a joint commission tasked to explore ways to come up with a draft for a peace treaty. This was similar to the working groups that had met while Gorbachev was President of the Soviet Union. Also, that month a delegation of top Japanese defense officials met with Russia's new Defense Minister Igor Sergeyev in Moscow and again both sides pledged to further bilateral defense cooperation.

In February 1998 Russian First Deputy Prime Minister Nemtsov and Japan's Foreign Minister Obuchi signed the completed fishing agreement during Obuchi's Moscow visit. Obuchi and Primakov also met to chair the first meeting of the newly created joint commission on a peace treaty. Primakov informed Obuchi that Yeltsin would like to visit Japan in April in a follow up to the Krasnoyarsk summit. Most importantly, Obuchi announced that the Japan Export-Import Bank would extend a $1.5 billion untied loan to Russia to help with economic reforms. Kremlin spokesman Sergei Yastrzhembsky indicated that the Russian government planned to use most of the money to construct housing for decommissioned Russian servicemen. The fact that the credit was announced on one of Russia's primary military holidays was yet another signal that ties between the defense establishments of the two countries were growing stronger. A full schedule of meetings between Russian and Japanese military personnel for the upcoming year was announced.[38]

Yeltsin's sudden dismissal of Prime Minister Viktor Chernomyrdin in late March 1998 was cause for concern in Japan that yet another domestic shake-up would force Yeltsin to cancel his visit to Japan. Boris Nemtsov assured the Japanese that Yeltsin would visit as scheduled in spite of the Duma's opposition to Yeltsin's proposed candidacy of Sergei Kirienko to be the next prime minister. In fact, many in Japan saw Chernomyrdin's dismissal as a plus, because he was known to be indifferent toward Japan.[39]

Hashimoto and Yeltsin met once again informally ('no neckties') on April 18–19, 1998 at the Japanese resort of Kawana on the Izu Peninsula southwest of Tokyo. The public enthusiasm created by the Krasnoyarsk summit was tempered somewhat by the meeting in Kawana. Both leaders acknowledged the promise to strive for a peace treaty by the year 2000. At the meeting Hashimoto presented to Yeltsin a new proposal. Although this proposal was not made public, experts and journalists that followed the summit were able to piece together the basic outline of the plan. Hashimoto's proposal called for a demarcation of the border between the two countries indicating Japan's sovereignty over the 'Northern Territories' (the line to be drawn between the islands of Urup and Etorofu/Iturup). Russia would meanwhile maintain administrative control over the islands while the two jointly developed the economy of the islands. Territorial reversion could come at a later date. This was in line with Yeltsin's earlier call to put off a territorial settlement for future generations. It also echoed Foreign Minister Primakov's proposal to jointly develop the islands. Yeltsin promised to study the proposal and get back to Hashimoto. He also added that it looked like a promising proposal and that he was "optimistic" about it. Hashimoto was reportedly upset by Yeltsin's public announcement of the 'secret' proposal, which had caused a spasm of wild speculation in the Japanese press.[40] Beside the ever-present territorial issue the two leaders did discuss concrete economic plans, and they agreed to maintain the engagement of the defense establishments of both countries. Yeltsin also invited Hashimoto to visit Moscow in the fall, and he proposed a Tokyo visit sometime in 1999.

The enthusiasm and optimism that marked the Krasnoyarsk summit were distinctly lacking at Kawana. Hashimoto and Yeltsin were as engaged with one another as they had been in Krasnoyarsk, but the atmosphere between the teams assembled around the leaders was noticeably subdued. Some observers tried to put a positive spin on the Kawana summit.[41] The summit did reaffirm the close relationship between Hashimoto and Yeltsin, and the desire to sign a peace treaty by the year 2000. The summit also seemingly brought the Japanese side closer to the Russian side in terms of putting off a final settlement of the territorial dispute for future generations and agreeing to joint economic development. Even the most optimistic observers, however, were unable to find evidence that the summit had moved the agenda forward. Indeed, the next few days after the summit, officials from both governments stepped back and qualified certain statements made in the press. Foreign Ministry officials in Japan and Russia also were made to reinterpret certain statements made by both leaders.

In a press conference in early May, Foreign Minister Obuchi insisted that any peace treaty must include a settlement of the territorial dispute. This was presumably in response to Yeltsin's announced intention to sign first a peace treaty and then, at a later time, engage in negotiations over the territorial dispute (much as Russia had agreed with China in 1997). Soon thereafter in a radio interview Kremlin spokesman and Yeltsin foreign policy advisor Sergei Yastrzhembsky denied reports that Moscow was planning to return any territory to Japan. In fact, Yastrzhembsky had just returned from the Kuril Islands and Sakhalin where he had been sent by Yeltsin to reassure regional leaders that Russia was not planning for any handover. Yastrzhembsky added that politically Russia was in no position to make concessions. He emphatically declared: "Russia has no land to spare."[42] Newspaper reports in Japan speculated that Hashimoto had actually asked for a demarcation in which three (as opposed to four) islands would be returned to Japan. These reports were heatedly denied by all sides, and added to the increasingly negative rhetoric coming from Moscow and Tokyo.

Undaunted, the two sides pushed ahead with a busy diplomatic schedule. Deputy Foreign Minister Tamba traveled to Moscow to meet with his Russian counterpart Grigorii Karasin to discuss the Kawana proposal. Hashimoto and Yeltsin continued their friendly dialogue at the G-8 summit in Birmingham, England in mid-May 1998. Foreign Ministers Obuchi and Primakov had met just prior to the summit and had promised to push for an accommodation before 2000. Karasin returned Tamba's earlier visit and the two met again in Tokyo in late June 1998. Economic issues were discussed, but the territorial issue dominated the agenda. Bilateral discussions were now returning to a familiar pattern. In 1996–97 much of the talk centered on economics and security, as the Japanese side studiously avoided discussing the territorial issue. Territory once again came to dominate the agenda, as had been the case for the previous four decades.

Russia's new 35-year-old Prime Minister Sergei Kirienko paid the first-ever official visit of a Russian prime minister to Japan on July 13–14, 1998. Unfortunately, his timing could not have been worse. Once again domestic political troubles cast their long shadow on the diplomatic agenda. Just one day prior to Kirienko's visit Hashimoto's LDP suffered a major defeat in the Diet elections for the House of Councilors (the upper house). Hashimoto announced his intention to resign the day the Russian Prime Minister arrived in Tokyo. Kirienko was met by a lame duck Japanese leader (as was the case with Yeltsin on his summer 1993 visit to the G-7 Tokyo summit). Kirienko met with Hashimoto and Foreign Minister Obuchi (soon to be named Hashimoto's successor). The talks centered almost exclusively on economic issues. The two sides hashed out the details of the $1.5 billion loan that Japan had granted earlier in the year to Russia.

New doubt was cast on the future of the relationship because Hashimoto had invested so much personally in building a close relationship with Yeltsin and had striven to drive forward the agenda of the bilateral relationship. Many wondered if the newfound warmth could be rekindled under the new leadership. Negative speculation as to the political future of Yeltsin was also rampant, especially when the Russian economy collapsed under the weight of unpaid short-term loans in August 1998, forcing Kirienko from office.

Optimists continued to insist that relations would not be put off, because Obuchi was named as Hashimoto's successor and Kirienko was replaced by Primakov. The two former foreign ministers, now prime ministers, had been integral parts of the recent *rapprochement* and were interested in seeing the bilateral agenda moved forward. In light of Russia's economic difficulties, however, the Japanese Foreign Ministry indicated that the remainder of the $1.5 billion loan would be withheld until the economic situation in Russia had become settled. The Japanese Foreign Ministry also announced that it was undertaking a thorough policy review of its economic policy toward Russia.

In August 1998 after having been selected as Prime Minister, Obuchi announced he would visit Russia in November. Obuchi made the announcement while meeting with a group of grandchildren of former residents of the disputed islands. He also described the status of the islands as "regrettable" and "abnormal," and said he wanted to see the islands returned to Japan "as soon as possible." Obuchi also presented his new Foreign Minister, Masahiko Kōmura. One of Kōmura's first conversations was with Primakov, who invited the Japanese Foreign Minister to visit Moscow in September. Although the visit was postponed until October when Primakov was named Prime Minister, ex-premier Hashimoto visited Moscow in September. The visit was officially of a personal nature, but Hashimoto did meet with Yeltsin and the two promised to keep the relationship on track.

Keizo Obuchi visited Moscow on November 12–13, 1998. Obuchi was the first Japanese Prime Minister to pay an official state visit to Moscow since Kakuei Tanaka in 1973. Unfortunately, Boris Yeltsin was incapacitated and bedridden much of the time, so Obuchi was only able to visit him on one occasion for 90 minutes.

Obuchi spent a good portion of his time with Primakov, which probably would have been the case anyway, as the political power of Yeltsin in Russia was in dramatic decline. Obuchi expected an official reply to Hashimoto's Kawana proposal to demarcate the islands in favor of Japanese sovereignty. Primakov reiterated Russia's interest first in a peace treaty, then a border demarcation. The Japanese side, however, had been led to believe at Kawana that a peace treaty would include a territorial settlement (in line with the 1993 Tokyo Declaration). It was clear that the two governments were still talking past one another. The two sides did issue the so-called Moscow Declaration, agreeing to establish two subcommittees to discuss the border demarcation issue, and to explore ways to undertake joint economic development of the islands without prejudicing either side's legal claims.[43] In the past Japanese policymakers felt that economic assistance to the islands would imply recognition of Russian sovereignty, therefore Japan's offer to consider this could be seen as a significant concession. In fact, opposition lawmaker Yukio Hatoyama harshly criticized Obuchi, claiming that Japan should continue to use economic assistance to the islands as a wild card.[44] Meanwhile the Russian Foreign Ministry announced that Hashimoto's proposal had been rejected by the Russian government during Obuchi's visit.[45]

Once again, relations seemed to take a step backward. If Kawana was seen as somewhat of a disappointment after Krasnoyarsk, the Moscow summit was a clear signal that normalization was still a distant goal. In December 1998 Aleksandr Panov, Russian Ambassador to Japan, seemed to strike a conclusive blow to the optimists by stating that the Russian government did not expect a peace treaty to be signed until after 2000.[46]

Economic ties also seemed worse in 1998. Two-way trade, $3.86 billion in 1998, sank to new depths last seen only in 1992. These low numbers partly reflect Russia's economic collapse in August 1998. Russia accounted for a mere 0.6 percent of Japan's total trade figures. By 1999 Japanese cumulative investment in Russia was close to $340 million, or 1.3 percent of all foreign investment in Russia. Japan ranked ninth in this category after Sweden.[47] Japanese businesses were still wary of Russia, given the political turmoil and the lack of a legal framework. The climate was particularly difficult for smaller Japanese businesses that operated in the Russian Far East. In a move perhaps designed to jump-start small business ventures between Japan and Russia, Yeltsin appointed Irina Hakamada, a woman of mixed Japanese ancestry, as head of the State Committee for Support and Development of Small Businesses.[48] Nevertheless, small- and medium-sized businesses continued to play only a small role in two-way trade and investment. As part of the Hashimoto-Yeltsin Plan a joint investment fund was founded to seek out potential investment targets in the Russian Far East. The Daiwa Far East and Eastern Siberia Fund was established in 1995 to promote small-scale investment in the regions closest to Japan. Mitsuhiro Ichiyanagi, the Chief Representative of the Vladivostok office, was quite pessimistic in a 1998 interview about the future of economic cooperation. He criticized leaders in Tokyo and the Japanese government for making economic and political decisions based on the big picture, without an adequate understanding of local economic conditions. In 1992–93, when Japanese companies were rushing into the region, some "very stupid decisions" were made in Tokyo, according to Ichiyanagi. Small- and mid-sized Japanese companies, meanwhile, were disillusioned in the late 1990s by prospects in the Russian Far East, apart from the energy projects in Sakhalin and some metal extraction projects in Siberia. The larger Japanese companies were there, Ichiyanagi insisted in 1998, to get what resources they could extract and sell them back to Japan. They spent very little time conducting trade promotion efforts. In 1998 there was even a moratorium on investments from the Daiwa Fund until the political chaos in Russia subsided.[49]

Yet Hashimoto, MITI officials, and politicians from the Diet energy group ('tribe,' or *zoku*) continued to push for large-scale energy cooperation. The Sakhalin-1 consortium announced that spending for the fiscal year 1998 would amount to $200 million, the same amount as in 1997. Drilling for gas and oil in both the Sakhalin-1 and 2 projects began in 1998. Although the economic crisis in Russia caused a halt, both projects were back on track in early 2000.[50] Kirienko's appointment as Prime Minister in the spring of 1998 was seen as a big plus in energy circles, because at the time he had been serving as Russia's Fuel and Energy Minister. Furthermore, Kirienko had a close relationship with Anatoly Chubais, chairman of United Energy Systems, former Deputy Prime Minister, and a close Yeltsin confidant. The two also maintained close ties with First Deputy Prime Minister Boris Nemtsov, who was also active in promoting Japanese-Russian energy cooperation.

Many of the large Japanese trading companies (*shōsha*) had enthusiastically gone into Russia in the early 1990s. In unison with MITI, they had pressed for market reform in Russia (in line with the other G-7 countries). In addition, MITI and some Japanese trading companies began paying special attention to Russian industrial policy and had become somewhat involved on an advisory basis. By 1998 many of these trading companies were experiencing heartburn in Russia. In spite of the stepped-up publicity given to projects such as a Toyota-Mitsubishi joint-venture assembly plant in Moscow and Far Eastern energy cooperation plans, Japanese direct investment in Russia still accounted for barely 1 percent of Japan's total overseas investment. In 1998 Russia accounted for less than 1 percent in Japan's trade turnover, and Japan for less than 5 percent in Russia's foreign trade. Top leaders in the Japanese government were still strongly pushing for stepped-up trade and investment in Russia, but private firms were no longer inclined to listen. Even MITI was no longer enthusiastic. Once Hashimoto stepped down in the summer of 1998, MITI officials who had been encouraging investment in Russia lost any remaining influence over Russia policy. By 1998 many of the Japanese trading companies had become more interested in Central Asia than in Russia. They recognized the opportunities in that region when US and European companies had begun investing. Japanese trading companies felt better going into the region now that Western companies were present. In particular, Japanese companies were attracted to opportunities in Kyrgyzstan and Uzbekistan. Hashimoto's 'Silk Road' diplomacy had another aspect to it besides China and Russia, and Japanese firms hoped that profits could be realized elsewhere on the Eurasian periphery.[51]

Recognizing the necessity of long-term economic cooperation with Russia the Japanese government extended a hand in areas other than resource extraction. In the spring of 1998 Japan promised to deliver a low-level liquid radioactive waste facility to prevent waste from being dumped into the Sea of Japan, and to help dispose of decommissioned nuclear submarines in the Russian Far East. The facility cost roughly $300 million and was included as part of the $1.5 billion loan announced earlier in the year.[52] In December 1998, Anatoly Chubais, chairman of United Energy Systems, Russia's major power supplier, visited Tokyo to discuss Japanese assistance in modernizing Russian energy generation facilities. Chubais also reportedly discussed an ambitious plan to link the Russian island of Sakhalin and Japan's northernmost island Hokkaido with a gas pipeline.[53]

There was no respite in defense contacts in 1998, in spite of the political impasse. In January a delegation of top Japanese defense officials traveled to Moscow and met with Russian Defense Minister Igor Sergeyev. The two parties discussed security issues in the Asia-Pacific region. In addition, Sergeyev promised to further scale back Russian forces stationed on the disputed islands, noting that the number of troops stationed there was now only 1,300. On May 31, Admiral Kazuya Natsukawa, chairman of the Japanese Defense Agency's Joint Staff Council, began a lengthy trip to Russia, visiting Moscow and Vladivostok. It was the first time Japan's highest-ranking soldier had visited Russia. Natsukawa met with his Russian counterpart Anatoly Kvashnin and Defense Minister Sergeyev. Later that summer the two navies undertook joint search and rescue operations – also a first – off of the Russian coast near Vladivostok. Three ships from the MSDF then visited that Russian port city. The exercises did not go off without a hitch, as the Japanese remained wary about Russian observers on MSDF ships. "Both of us still have Cold War wariness ... [and] we don't want to display our skills too much to each other," one MSDF source was quoted as saying.[54]

At the end of the year defense relations ended up on a positive note with the visit to Tokyo by Russian Chief of Staff General Anatoly Kvashnin. Kvashnin visited Yokosuka naval base near Tokyo, which is adjacent to the largest overseas American naval base. Kvashnin and Japanese Chief of Joint Chiefs of Staff Admiral Natsukawa agreed to continue Russian-Japanese naval exercises and make regular exchanges of port calls by the two countries' navies.

1999: The pace slows

If 1997 was a high point for establishing the *rapprochement*, and 1998 a period of negotiation, 1999 could be called a year of reflection for the two sides. Mid-level diplomatic activity continued apace, as did defense contacts, but no summit meeting was convened in 1999.

Reports early in the year surfaced about a Russian proposal to turn the southern Kuril region into a special economic zone linking it with the Japanese island of Hokkaido. According to the scheme, proposed by Prime Minister Primakov, Japanese firms would be permitted to lease land for periods up to ninety-nine years. Analysts in Moscow correctly surmised that this proposal would meet not only with a lukewarm reaction from the Japanese, but also with regional opposition in the Russian Far East.[55]

Russia's new Foreign Minister Igor Ivanov (a loyal Primakov deputy) visited Tokyo for three days in mid-February 1999. On his way to Japan Ivanov stopped off in Sakhalin and the Kuril Islands to assess the deteriorating economic situation there and to reassure the local populace of Moscow's forthcoming financial support (such visits have become a tradition for Russian leaders visiting Japan). In Tokyo, Ivanov hinted that the 2000 peace treaty deadline was unattainable based on the present circumstances. To the surprise of many, Ivanov admonished both Tokyo and Washington for the lack of transparency in the US-Japan defense cooperation guidelines, and for a US-Japanese proposal to jointly build a regional missile defense system.[56] A March visit to Tokyo by Russian First Deputy Prime Minister Yuri Maslyukov was no more successful than Ivanov's visit. Maslyukov had hoped to secure Japanese support for the release of funds from an IMF loan and for the remaining tranche of the $1.5 billion loan that Japan had held up since the 1998 economic collapse. His attempts "failed utterly," according to Russian newspaper reports.[57]

More controversy was stirred up in August when Japanese Diet member Muneo Suzuki, Deputy Chief Cabinet Secretary of the LDP, suggested that Japan could accept an interim peace treaty if it were to specify Japan's sovereignty over the disputed islands. He contradicted one of the Foreign Ministry's top Russia hands, Minoru Tamba, who had announced only a few days prior to Suzuki's statement that Japan would in no way accept an interim treaty. Suzuki's statement apparently also met with opposition within Prime Minister Obuchi's cabinet, and Suzuki was forced several days later to recant his statement.[58]

In a gesture seen as conciliatory by many, the Japanese government announced that it would open a consulate in Yuzhno-Sakhalinsk, the capital of Sakhalin Oblast. The consulate was formally opened in 1999. Because the Soviet government had never signed the 1952 San Francisco Treaty, many Japanese had insisted that Russia had no right over Sakhalin (the San Francisco Treaty did not specify to whom Japan would cede the Kuril Islands or southern Sakhalin). The Japan Communist Party and the Socialist Party had for many years claimed that the Soviet/Russian occupation of these territories was illegal and demanded the return of the entire Kuril Archipelago and Sakhalin. Thus, Japan's recognition that Sakhalin was not part of the disputed territorial zone was implicit in the opening of the consulate. Apparently, the decision was taken over the objections of the powerful Treaty Bureau within the Japanese Foreign Ministry.[59]

In another gesture that did receive notice in Russia, at the end of August 1999 Japan agreed to 'unfreeze' the remaining portion of the $1.5 billion loan granted 18 months earlier. The announcement came during the Tokyo visit of First Deputy Prime Minister Viktor Khristenko. The obstacles to the loan disbursement were removed when Russia reached agreements with the IMF, the World Bank, and the Paris Club of debtors.[60] During Khristenko's visit the issue of Yeltsin's planned visit to Tokyo was brought up. Khristenko said that Yeltsin would indeed come before the end of the year. No date, however, had been set, and with the upcoming Duma elections in the fall and the continued political shake-up in Moscow (Yeltsin had just appointed his third prime minister for the year – a relative unknown by the name of Vladimir Putin) it seemed less and less likely that any visit would take place.

In September, Japan's new ambassador arrived in Moscow. Minoru Tamba, a career diplomat from the Foreign Ministry's 'Russia School', was viewed in Japan as a wise choice. The upcoming diplomatic agenda was full, and the 2000 deadline for a peace treaty was drawing near. Although Tamba was seen as one of the primary architects of Hashimoto's new Russia policy in 1996–97, he was also known as a tough negotiator. During the Cold War Tamba had been one the Foreign Ministry's most severe critics of the Soviet Union (in a ministry full of hard-line anti-Soviet critics). This may have been due partly to the fact that Tamba was born on Sakhalin and his family was forced to leave at the end of the Second World War while he was still a child. Tamba appeared to be the right combination of a strong negotiator, but one with enough vision to recognize the long-term strategic importance of Japan's relationship with Russia. The arrival of Tamba in Moscow was given fairly extensive coverage in the Russian press. Several of the larger dailies published interviews with him. Much was made of the fact that Tamba was born on Sakhalin, speaks Russian well, and has spent much of his career at the Foreign Ministry covering Russia.[61]

Upon his arrival, however, Tamba was forced to deal with a tense hostage situation in Kyrgyzstan, when four Japanese geologists were kidnapped in August by Tajik and Uzbek bandits in a remote valley along the Kyrgyz-Tajik border. Since Japan had no embassy in Kyrgyzstan, the staff at the embassy in Moscow was forced to handle this delicate issue. A good number of Japanese diplomats in Moscow were sent to monitor the situation and to help the Kyrgyz government with negotiations. The Russian government also cooperated with both the Japanese and the Kyrgyz governments during the negotiating process. Upon resolution of the crisis Japanese Prime Minister Keizo Obuchi extended a personal note of thanks to Russian President Yeltsin and Prime Minister Vladimir Putin.[62]

By this time Russia was again waist-deep in the Chechen morass. The Japanese government's reaction to the renewal of the war was muted, compared to the other members of the G-7 group. In early November 1999, amid a rising clamor in the West to cut off aid to Russia, the Japanese Foreign Ministry announced that it considered Chechnya to be an internal Russian matter.[63] Later, perhaps with the scrutiny of Western governments and the Japanese media in mind (from the beginning the press in Japan had been critical of Russia's actions in Chechnya), the Japanese government announced that it would extend $1 million in humanitarian aid to Chechen refugees in the North Caucasus. Nevertheless, the Japanese Ministry of Foreign Affairs has been careful not to allow the war in Chechnya to upset the relationship with Moscow. At the same time that Michel Camdessus, director of the IMF, began hinting that his organization would consider cutting off aid to Russia because of the campaign in Chechnya, the Japanese government announced in November 1999 that the Japan Export-Import Bank would release $375 million in credit to the Russian government. This credit was linked to the February 1998 loan package of $1.5 billion. Aleksandr Livshits, the Russian president's envoy to the G-8 Group, praised the Japanese government, saying, "Japan is the only country which is keeping its credit line to Russia open and running, and we highly assess that."[64] Interestingly, the announcement made little impact in Moscow, where all attention was focused on the war and the upcoming Duma elections. Only one paper, *Izvestia*, had a lengthy article reporting the credit extension, and no mention was made on television. Russian leaders were in no position to give relations with Japan much consideration at this point, especially when it might mean surrendering territory, which would be anathema in the midst of a 'secessionist' war in the Caucasus. In spite of the Japanese government's cautious stance and its attempts to show support for the Russian government, it soon became apparent that Moscow had no plans to reward Tokyo.[65]

The visit of former Japanese Prime Minister Hashimoto to Moscow in mid-November 1999 garnered more attention in the Moscow press than the credit extension package. Hashimoto was attending a *kendō* tournament (a traditional Japanese fencing sport). The visit was widely viewed as an attempt by Tokyo to probe the Russian leadership on the status of the territorial issue, and to firm up the dates for a pending Yeltsin visit to Japan. Following the Japanese Ministry of Foreign Affair's line, Hashimoto said that the Chechen issue was Russia's internal matter and expressed his understanding of the difficulties that this problem has created for the president of Russia. The Russian Duma backhandedly responded to Hashimoto's goodwill visit by drafting a statement declaring that territorial concessions to Japan are impermissible and reminding the president of "his constitutional duty to take steps to protect the sovereignty, independence and territorial integrity of the Russian Federation." The resolution, which was overwhelmingly supported, also said, "Any treaty implying the loss or restriction of the sovereignty of the Russian Federation over the South Kuril Islands has no prospects of being ratified by the State Duma."[66]

A week later Yeltsin, whose health had been fragile all fall, attended an Organization for Security and Cooperation in Europe (OSCE) summit in Istanbul, Turkey. There he announced that he would not be visiting Japan in 1999. Later the Kremlin announced that he would visit instead sometime in the spring; few in Japan were convinced even this spring visit would come off. As relations stagnated once again most Japanese press reports assumed a negative tone. On the eve of the Duma elections in Russia, editorials in the *Sankei Shimbun* and the *Asahi Shimbun* warned of a nationalist backlash that could irreparably damage relations between Moscow and Tokyo.[67]

Economic indicators were mixed again. Two-way trade in 1999 hovered at the low figure of $4.2 billion, but Japan's assistance packages for Russia were starting to be implemented, while energy cooperation promised even bigger investment figures in the future. Energy projects continued to draw the big headlines. The Exxon (later Exxon-Mobil) Corporation announced in May 1999 that its subsidiary Exxon Japan Pipeline Ltd. had signed an agreement with Japan Sakhalin Pipeline FC Co. Ltd. on a feasibility study for the building of a gas pipeline connecting Sakhalin and Japan. Ironically, however, the first bilateral energy cooperation program to get off the ground involved the transfer of energy from Japan to Russia. The deteriorating economic situation on the four disputed islands was cause for the Japanese government to send 400 tons of fuel oil in March 1999 to the beleaguered residents of Shikotan and Kunashiri.[68] In April the bilateral sub-commission on economic relations met in Tokyo and participants discussed six large-scale projects. Included were four new gas pipelines in the Russian Far East: on Sakhalin Island, on the Kamchatka Peninsula, in Khabarovsk Territory, and in the Yakutia Republic, as well as completion of the construction of the Bureya hydro-power station on the Amur River and creation of production facilities for polymetallic ores in the Primorye (Maritime) Territory.[69] Later, three Japanese firms, Mitsui, Itochu, and Sumitomo, announced that they would be supplying pipelines for the Russian-Turkish-Italian Trans-Black Sea gas pipeline project known as 'Blue Stream.'

Nuclear energy and environmental concerns also continued to play a role in the relationship. In the spring of 1999 Japan promised Russia $200 million in financial assistance to help dismantle 50 aging nuclear submarines forming part of the Russian Pacific Fleet. Japanese Foreign Minister Masahiko Kōmura informed Russia of these plans during his May 1999 visit to Moscow, and Obuchi made the public announcement in June at the Cologne G-8 summit meeting.[70] Japanese citizens had become increasingly aware of the implications of the 'rot' in the infrastructure of the Russian Far East. The sinking of the *Nakhodka* in 1997 was but one sign. In fact, the sinking of the *Nakhodka* was a comparatively minor detail compared to the threat of nuclear contamination. This specter was raised in 1993 with the nuclear dumping controversy. It had emerged again in 1998 when the Russian government brought treason charges against a naval officer, Grigorii Pasko, who had allegedly passed state secrets on nuclear waste disposal to the Japanese television station NHK. Pasko's revelations brought to light a series of accidents and near disasters in the Russian Far East that threatened a nuclear catastrophe "worse than Chernobyl."[71] What was no doubt insulting to many Japanese, in light of the growing nuclear waste problem in the Russian Far East, was a quiet proposal in early 1998 by officials from the Russian Ministry of Atomic Energy to the Japanese nuclear energy industry for the renting out of waste storage facilities on one of the disputed (Kuril) islands. At the time the Japanese nuclear energy industry was experiencing a series of crises, including the shutdown of its Rokkasho plutonium separation plant project, causing spent fuel to build up at several reactor sites. The fact that the initiative never made a large splash in the press means that it probably died a quick and ignominious death.[72] The Japanese government, fearing environmental effects, became yet more eager to cooperate with Russia and the United States in large-scale nuclear clean-up projects.[73] The offer in the spring of 1999 was an example of the Japanese desire to help with the shattered Russian nuclear infrastructure.

Defense contacts, meanwhile, continued in earnest in 1999. Russian Pacific fleet commander Mikhail Zakharenko visited Japan in April. The visit, which took place in the framework of military exchanges between the two countries, was the first official visit of the Pacific fleet commander to Japan in the history of bilateral relations. The two navies announced that they would begin cooperating in joint maritime safety and security operations. The announcement was made in the wake of the March 1999 incident wherein ships from North Korea were suspected to have violated Japan's maritime territorial zone. The two sides also promised to extend cooperation to include anti-smuggling enforcement and operations against drug trafficking.[74]

In the fall of 1999, the head of the Japanese Defense Agency Hosei Norota became the first chief of the JDA to visit the Russian Far East. His five-day trip included visits to Moscow and Vladivostok. In Moscow he met with Russian Defense Minister Sergeyev. The two signed a memorandum to further promote defense-related dialogue and exchanges between their countries. In Vladivostok the two sides agreed to establish a military hot line to warn each other of potential threats at sea. Shortly after Norota's visit the Russian guided missile cruiser *Admiral Panteleyev* visited Yokosuka naval base outside of Tokyo, yet another first-time event, affirming the growing defense cooperation.

2000–01: A flurry of activity

Suspense over whether Boris Yeltsin would visit Japan in 2000 ended on December 31, 1999 when Yeltsin announced his resignation in a televised speech in Moscow. Yeltsin's resignation was not completely unexpected, and he was due to step down anyway in the summer of 2000 after the presidential election. Yeltsin's resignation assured the succession of Vladimir Putin, the newcomer who had rocketed from obscurity in the summer of 1999 to command astronomical public support in four short months. Certain Japanese officials and Russia experts expressed hope that a strong leader in Russia was what the relationship needed for serious peace negotiations to get under way before the end of 2000. A strong leader like Putin with widespread public support, they argued, might be the only one able to push through a territorial settlement, returning the islands to Japan. Such thinking was, however, in the words of Russian Foreign Minister Igor Ivanov, "illusional." Putin's popularity was based on his actions of strength in Chechnya and his promise to protect Russia and to restore order. One of the first things Putin did upon assuming office was to place phone calls to Washington, Beijing, Paris, London, and Berlin. No phone call was made to Japan until January 28. This was a particularly inauspicious start for Japan at the beginning of the year 2000.

Putin announced that he wished to make economic relations the centerpiece of his foreign policy. This should have been a good sign for Japan. Soon, however, it became apparent that Japan was not necessarily high on Putin's agenda. In late February Deputy Foreign Minister Ivan Ivanov (no relation to Foreign Minister Igor Ivanov) gave an interview to the daily *Nezavisimaya Gazeta*, in which he stressed the importance of building a strong economic foundation for Russia's new diplomacy. And yet, throughout the lengthy interview, Ivanov made no reference to Japan. Ivanov mentioned the members of ASEAN and the nations of Latin America, but Japan was suspiciously excluded. That same week Sergey Karaganov, a former foreign policy advisor to Yevgenii Primakov, and a man with his finger on the pulse of the Kremlin, published an article covering the same theme in the weekly *Moskovskie Novosti*. In his lengthy piece mention of Japan was also absent.[75]

To its credit, the Japanese leadership did not give up hope, and continued to push for a full agenda with Moscow. So anxious were the Japanese to get placed on Putin's agenda, they invited him to visit Japan twice in 2000 – once after his May 5 inauguration, and again at the July G-8 summit meeting in Okinawa. Prime Minister Obuchi even offered to visit Russia before the July summit. Obuchi had visited Russia in November 1998, so diplomatic protocol called for the Russian head of state to visit Japan next. Obuchi was apparently willing to overlook this.

Japanese officials undauntedly pushed on, and a whole series of top-level meetings were planned for the year. Both high- and mid-level governmental contacts between Japan and Russia saw increasing activity at the beginning of the year. Russian Deputy Prime Minster Mikhail Kasyanov (soon to be named Prime Minister) visited Japan in mid-January 2000, along with the head of the Central Bank of Russia Viktor Gerashchenko. Foreign Minister Igor Ivanov was the next high-ranking official to visit Japan, arriving in mid-February. Although he was quick to downplay the chances for a peace treaty, Ivanov conveyed Putin's desire to oversee cordial relations and he made a bold argument for Japan and Russia to be more deeply involved in negotiations on the Korean Peninsula.[76]

What received even more notice in the press of both nations was the mid-February 2000 visit to Moscow and Vladivostok of the Chief of Staff of Japan's Maritime Self-Defense Forces, Admiral Hosei Fujita. It was the first visit of a Japanese naval chief to Russia, and built upon the bilateral contacts between the two nations' military forces. The two sides discussed holding joint Russian-Japanese naval exercises as part of a program to prevent accidents, as well as conduct search and rescue operations. During Fujita's visit to Vladivostok it was announced that Japan would extend another $120 million in financial assistance to help clean up nuclear waste in Russia's Pacific Fleet ports.[77] Putin did reciprocate somewhat by overseeing the ratification in the Duma of a Japanese-Russian agreement on the encouragement and protection of Japanese capital investments in Russia. At the end of March 2000, Russian First Deputy Finance Minister Aleksei Kudrin announced that Russia expected to receive a loan of $100 million from Japan as part of a coal sector adjustment loan from the World Bank.

At the beginning of April 2000 Prime Minister Obuchi collapsed at his official residence in Tokyo and fell into a coma. He soon thereafter passed away. A small group of LDP leaders met behind closed doors to decide on a strategy before it was announced to the public that Obuchi had fallen ill. In a session eerily reminiscent of the days of the Communist Party dictatorship in the Soviet Union, the LDP leadership decided in secret that Yoshiro Mori would be the new prime minister. The announcement drew tremendous criticism in Japan. The Russian press, however, emphasized the fact that Mori had special ties to Russia through his father, a former mayor of a small town on the northwest Japanese coast. The elder Mori had developed close links with a small town near Lake Baikal in the Russian Far East. One of the elder Mori's dying wishes was to have a portion of his remains buried there. The younger Mori has visited the area several times, and maintained a warm relationship with local politicians there. The press in both countries grabbed onto this story and gave it much play.[78] It was hoped that the new blood at the top of both governments would serve to reinvigorate relations. The two leaders had to first rebuild the level of trust that had existed under Hashimoto and Yeltsin, and to a lesser extent under Obuchi.

The first meeting between Mori and Putin took place in St. Petersburg at the end of April 2000. Obuchi had planned to visit Russia in the spring and Mori followed through with his plans. It was an unofficial visit that Mori took on his way to Europe. It was, however, his first visit abroad. There was significance in this because normally Japanese prime ministers reserve their first foreign visits for the United States. During the two-day visit Putin and Mori discussed Japanese financial credits for Russia, Japan's potential participation along with the United States in the development of a new anti-ballistic missile system, and Putin's promised visit to Japan later in the year. Additionally, the two leaders discussed geopolitical issues, and how the two nations could address strategic problems together. Putin was obviously quite interested in such issues.[79] Mori studiously avoided discussion of the territorial issue. Mori and Putin agreed to call one another by their respective first names, but this did little to patch up what was again becoming a frosty relationship and did little to improve prospects of reaching a peace treaty by the end of the year. While Russian leaders were willing to bide their time, among Japanese leaders there was a growing sense of urgency. According to reports in the Russian press, a number of Japanese diplomats and politicians were ready to freeze relations if no progress were made on the territorial issue by the end of the year. Mikio Aoki, Secretary-General of the LDP in Japan, had to publicly repudiate a statement by Japanese Ambassador Tamba who had stated that there would be no peace treaty by the end of 2000.[80] Concerned about these leaks and the increasingly negative atmosphere, the Japanese government attempted to regroup.

During the summer of 2000 Putin and Mori met two more times in Japan. The first meeting took place at the G-8 summit on Okinawa in late July. Vladimir Putin emerged as the star of the Okinawa summit. Fresh from a visit to North Korea, he captivated the leaders of the other G-7 nations. German Chancellor Gerhard Schroeder described his report on the situation in North Korea as "brilliant." Because of Putin's crowded schedule, Mori was only able to meet with him for about an hour. They talked about the agenda for the upcoming September summit in Tokyo, though neither side was prepared to enter into a lengthy discussion on bilateral matters. This actually may have been the best time to catch Putin in a generous mood, for these were the halcyon days of Putin's first political honeymoon (his second being in the last two years of his presidency as oil prices climbed and the state coffers expanded). Things changed dramatically for Putin in August 2000. A terrorist bombing in central Moscow, the sinking of the nuclear attack submarine *Kursk*, and the Ostankino television tower fire fell upon Russia and Putin in rapid succession during that month. His response to the latter two tragedies left many Russians questioning his crisis management skills. This was to have a profound influence on Putin's behavior in Tokyo during the September summit.

Several days after the Putin-Mori meeting on Okinawa, Hiromu Nonaka, the new Secretary-General of the LDP, shocked the Japanese public and the Ministry of Foreign Affairs by stating that the territorial dispute should not stand in the way of a peace treaty between Japan and Russia. This has been Russia's position all along. Foreign Minister Yohei Kōnō was quick to dismiss this idea, and Mori did as well.[81] But this was apparently no mere slip of the tongue. Nonaka's remarks were not necessarily directed toward Moscow. In fact, Nonaka seemed to be embroiled in an inter-governmental spat involving both the LDP and the Ministry of Foreign Affairs. Several LDP leaders (including deputy chief cabinet minister Muneo Suzuki, who acted as one of former Prime Minister Obuchi's special envoys to Russia), and a few members of the Foreign Ministry, unhappy with the direction of Japan's Russia policy, had been floating a trial balloon. The old guard of the Foreign Ministry's Russia School was quick to quash this experiment, with the support of the Prime Minister.[82] All of the major Japanese dailies quickly joined in the criticism of Nonaka's statement.[83]

Before the September Tokyo summit meeting Putin visited Sakhalin Island. This was his first trip to the Far East since becoming president in May. The Sakhalin administration of Governor Igor Farkhutdinov has been adamant about not handing over any territory to the Japanese. Putin's visit was designed to demonstrate to all Russians that he had no intention of buckling to the Japanese. This should obviously have been seen as a signal to Tokyo. In a similar vein, the head of the Russian State Fishery Committee Yuri Sinel'nik published a lengthy piece in the *Nezavisimaya Gazeta*. Sinel'nik claimed that the waters of the seas around the southern Kuril Islands could yield up to $1 billion annually in fishing revenues.[84] Furthermore, Putin's visit to Sakhalin came on September 2, the anniversary of the end of the war with Japan (V-J Day). The message from Russia was clear: no to territorial concessions.

The Mori-Putin summit was held as planned in Tokyo in early September 2000. The atmospherics were perfect. Putin dined with the Emperor, lunched with business executives of the *Keidanren*, and even had time to tumble on the judo mat. Not had a Russian leader been so warmly welcomed in Japan since Soviet Premier Mikhail Gorbachev visited in 1991. Putin declared that relations were "at their highest level since the end of World War II."[85] Despite the mutual admiration and the pageantry (which included Mrs. Putin dressing in a Japanese kimono at the formal state dinner), the results of his talks with Mori were mediocre at best, which came as no surprise to most observers. The two leaders signed 15 documents encompassing trade, investment, security, the environment, and military cooperation, and they reaffirmed the Tokyo (1993) and the Krasnoyarsk (1997) Declarations.[86]

Nevertheless, both sides came away from the summit rather disappointed. The Japanese expected some sort of new initiative to emerge, and had rolled out the red carpet for Putin. Mori went all the way to the airport to meet Putin, and the Emperor came to greet the Russian President at the door of his residence in Akasaka Palace, a gesture rarely extended to foreign guests. Hoping to renew Japanese interest in Russian investment, Putin had announced in Sakhalin that Russian Minister of Trade and Economic Development German Gref (known as a Westernizer, or liberal in Russian political circles) would oversee the major production-sharing agreements (PSAs) the Russian government had signed with Western energy majors in the mid-1990s. Although this was seen as a boost to investment protection in the energy field, both Mori and business executives of the *Keidanren* pointed out to Putin that until the poor investment climate in Russia and the Russian Far East was rectified, no large-scale Japanese investment could be expected.[87]

During the Tokyo summit Putin reportedly affirmed the validity of the 1956 joint declaration, thus in principle agreeing to cede to Japan control of the two southernmost and smallest of the disputed islands in exchange for a peace treaty. The status of the two remaining islands remained unclear. After the summit Russian Deputy Foreign Minister Aleksandr Losyukov explained that Japan and Russia have differing interpretations of the 1956 agreement, based presumably on whether two islands were the starting or ending point of discussions.[88] Apart from the disagreements on interpretations and starting points, Japanese officials expressed guarded optimism about Putin and his ability to deliver. In talks with this author in Tokyo after the summit, several Japanese government officials and advisors expressed appreciation that between the two visits to Japan Putin seemed to have done his homework, demonstrating a good grasp of the historical details and legal technicalities of the territorial dispute. They remained confident that Putin could possibly be the leader to make territorial concessions. They stressed, however, that rapid progress was essential, and that a policy of inaction by the Russian government could damage relations to an even greater extent.[89]

In spite of the relatively sanguine mood at the governmental level, public opinion in both countries was anything but positive. The summit was seen to have no benefit to either side. Russian editorials published prior to the summit warned Putin that he should not give in. The respected centrist daily *Kommersant'* (August 30) exhorted Putin not to "sell out the Motherland." Two other centrist papers, *Vremya* and *Nezavisimaya Gazeta*, published pieces on the fifty-fifth anniversary of the Japanese surrender (September 2), asserting essentially that, "to the victors the spoils." Several Japanese editorials on September 6 decried the meager results of the summit. The *Sankei Shimbun* warned that anti-Russian sentiment in Japan was bound to grow given a continued stalemate over the territorial issue. The *Yomiuri Shimbun* argued that no more large-scale economic assistance should be extended to Russia until a peace treaty was signed. Even the normally placid *Nihon Keizai Shimbun* scolded Putin for being too concerned about short-term domestic political considerations. Putin himself pointed to the general positive trend in Russo-Japanese relations over the past several years. But this seemed to be putting a brave face on what was a fruitless summit. In the end, as usual both leaders agreed to continue discussions in the future.

It seemed clear to most at this point that there would be no peace treaty in 2000. A meeting of the foreign ministers in Moscow in November produced no results, though Igor Ivanov and Kōnō Yohei promised to "step up the pace" before the end of the year.[90] Mori and Putin met one more time at the APEC summit in Brunei on November 15, 2000. Mori expressed his desire to meet with Putin in Siberia late in the year or early in 2001. The two also briefly spoke about a peace treaty and Mori was happy to report that the talks were "very good and useful." Nevertheless, after the meeting Russian Deputy Foreign Minister Aleksandr Losyukov announced that there would be no accord in 2000. The Japanese vowed to push on and Mori announced that he would be ready to meet with Putin in the city of Irkutsk, near Lake

Baikal in Eastern Siberia. It appeared that the meeting would happen sometime in early 2001.[91] Mori sent Muneo Suzuki to Moscow to firm up a date and Suzuki met with Putin's trusted lieutenant Sergei Ivanov, chairman of Russia's National Security Council (and soon to be named Minister of Defense). Suzuki and Ivanov had developed a working relationship and would meet or correspond regularly throughout the 2000–01 period.[92]

By early 2001 the Russian government was still not being clear about the exact date of the Irkutsk summit. The Kremlin suggested late February, and then a week later changed its mind and said late March. Tired of the perpetual merry-go-round Japanese Foreign Minister Kōnō criticized the Russian government and demanded a firm date in a January phone call with Russian Foreign Minister Igor Ivanov.[93]

By the time both sides were ready to meet in Irkutsk in late March 2001, Mori's political standing at home was in serious question. His ratings had been low since his Byzantine-like appointment a year before, and with each policy gaffe and each slip of the tongue he only dug himself a deeper hole. By the March Irkutsk summit it was once again apparent that Russia's leader was meeting with a lame duck Japanese prime minister.

The Irkutsk summit was held on March 25, 2001. At the meeting – the sixth meeting between the two leaders – Mori broached the idea of a 'two-track' policy that envisioned two sets of talks: one focusing on the 1956 agreement (and the handover of Shikotan and the Habomai group), and the other on the sovereignty of the two additional islands (Kunashir/i and Etorofu/Iturup). This was a substantial shift in Japanese policy. Up to this point the Foreign Ministry and the LDP had always insisted on the return of all four islands at the same time. Attempts to separate the islands had been consistently refuted by the Japanese (especially the MOFA) since 1956. This shift in policy brought the Japanese closer to the Russian perspective and could be seen as a dramatic compromise. The argument for this new tactic was outlined in a series of articles published in the *Sankei Shimbun* by a high-ranking Japanese diplomat under the pen name of 'Mr. X.' In these articles Mr. X assured readers that the Foreign Ministry was not giving up on the two northern islands; it was simply trying a new approach to gain positive momentum so that all four islands might one day be returned.[94] Valerii Zaitsev, a Japan expert at IMEMO in Moscow, suggested that once Russians got over the initial shock of initially handing over two islands, then it might not be as difficult agreeing to the return of the latter two islands. He cited as an example NATO expansion in 1997, when the West decided to include Poland, Hungary and the Czech Republic in the alliance. The Kremlin was adamantly opposed and the Russian people joined in opposition. Nevertheless, expansion occurred and Russia has learned to live with it. Zaitsev pointed out that psychologically the Russian people are now less opposed to further NATO expansion because a precedent has been established. The handover of Shikotan and the Habomais (in line with the 1956 joint declaration), he added, would be the most difficult for Putin to deliver. But once emotions were spent on this issue, Kunashir and Iturup could probably be delivered.[95]

Mori's new tactic at Irkutsk was the result of serious infighting within the Japanese government. Behind the idea was LDP deputy chief cabinet secretary

Suzuki. Muneo Suzuki had allies within the Foreign Ministry, and most importantly he was close to Prime Minister Mori, through his relationship with LDP Secretary-General Hiromu Nonaka.[96] Soon thereafter the discord spilled out into the public arena and caused a shake-up in personnel at the Foreign Ministry. The bitter inter-governmental struggle did not end with Irkutsk. More would come of this later.

The reaction in Japan to the Irkutsk summit was bitter. Articles in the press expressed serious concern at Japan's 'retreat' on the issue. This was especially the case when it later was revealed that Putin had told Mori that Russia could return no more than two islands to Japan.[97] The Russian press and the public were in general pleased with the results of the Irkutsk summit. Several articles in the Russian press and subsequent interviews with a number of Japan experts in Moscow bear out this assessment.[98]

Mori was forced to step down at the end of April 2001. His successor was Jun'ichiro Koizumi, also of the LDP. Koizumi at first announced that he could be flexible on the territorial issue, suggesting tacit support of the results of the Irkutsk meeting. But his new Foreign Minister, Makiko Tanaka, immediately let it be known that she was opposed to the 'two-track' policy outlined by Mori. Soon Koizumi also called publicly for the return of all four islands simultaneously.[99] This new line was bolstered by the return to Tokyo of a Foreign Ministry official, former Section Chief of the Russia Desk, Jirō Kodera. In early 2001 Kodera had voiced his opposition to the 'two-track' policy and was subsequently sent to a new posting in London after the Irkutsk summit. Tanaka recalled Kodera in May.[100] Tanaka also said in a Diet address that she felt that there had been no true achievement in Japanese-Russian relations since the 1973 Moscow summit, which coincidentally (or not) was attended by her father, former Prime Minister Kakuei Tanaka. With one statement she had brushed over the accomplishments of Japanese politicians and diplomats in 1993, 1997, 1998, and 2001. Suzuki publicly blasted Tanaka's comments, saying that they invited a return to the days of the Cold War.[101]

Fuel was added to the growing fire when ex-Prime Minister Mori insisted in an interview that the Russian government had agreed to return the two islands specified in the 1956 joint declaration. The Russian government issued a sharp denial.[102] Then, two days later the governor of the Far Eastern Khabarovsk region, Viktor Ishayev, suggested that the Russian government give up the four islands to Japan. This was a dramatic gesture coming from a regional governor, who had in the past been against territorial reversion.[103]

In the summer of 2000 yet another controversial issue came to dominate bilateral relations. This time the issue revolved around a Russian-South Korean agreement to allow Korean fishing vessels the right to fish the same waters around the disputed territories that the Japanese had painstakingly labored so long to acquire the rights to fish in. The Japanese government strongly protested to the Russian and South Korean governments, insisting that the agreement was an infringement on Japan's sovereignty over the islands. The *Yomiuri Shimbun* stated in a series of editorials that South Korea was "flouting" Japan's sovereignty, that the agreement was an "intolerable incursion," and that the fish "were not Russia's to give away." This point of view was echoed all summer long in a series of editorials in all the major Japanese dailies.[104] South Korean boats began fishing the waters in July and continued until November 2001, as had been agreed upon. Japan sent envoys to both Moscow and Seoul in an attempt to have the agreement abrogated, but they were unsuccessful. Foreign Minister Tanaka futilely protested to Foreign Minister Ivanov in a July 2001 letter. Prime Minister Koizumi failed to persuade Putin when they met at the G-8 summit in Genoa later that month. Insult was added to injury when it was announced in August that North Korean and Ukrainian boats would also be granted fishing rights in the disputed waters.[105] Finally, in October 2001, perhaps in response to continued Japanese protests, the Russian government agreed on a tentative accord banning third country fishing in the disputed waters in 2002.[106]

Meanwhile, Prime Minister Koizumi seemed to back track from his hard-line position when he met with Putin at the October 2001 APEC summit in Shanghai. Koizumi acknowledged that the 'two-track' negotiating strategy that had originated under Mori could be effective.[107] This change reflected the still unsettled situation in the Japanese Foreign Ministry. It also was a reflection of the influence that Muneo Suzuki still exerted over Japan's Russia policy. But by the end of 2001 the situation, which had seemed somewhat promising as recently as the spring, was as muddled as ever. By this time the political situation in Russia had become somewhat settled, and the Russian economy was actually showing signs of life. But a war had erupted between Foreign Minister Tanaka and Suzuki (the 'Makiko vs. Muneo' war), and now the political situation in Japan was no longer conducive to creative diplomacy.

Two-way trade continued to be sluggish in 2000–01. Two-way trade in 2000 was merely $5 billion, an insignificant increase from 1999. As has been was pointed out, however, Tokyo was slowly beginning to free up the credits it had promised two years earlier. A Japanese consortium comprising the firms Mitsui, Sumitomo, and Itochu under the guarantees of MITI and the Japan Bank for International Cooperation (the Japan Export-Import Bank) received a contract worth $400 million from the Russian-Turkish-Italian Trans-Black Sea gas pipeline project known as 'Blue Stream' to provide 300,000 tons of steel pipe.[108] Russian energy giant Gazprom also received a $20 million credit from Marubeni Corporation.[109] In April 2001 Russian Atomic Minister Yevgenii Adamov visited Tokyo to take part in the thirty-third conference of the Atomic Industrial Forum. Adamov met officials from the Japanese Foreign Ministry, MITI, and the heads of some of the leading trading companies, including from Itochu, Sumitomo and Marubeni. The two sides were interested in the expansion of bilateral contacts in the atomic energy sector. In May 2001 a delegation of the Japan Association for the Promotion of the Economic Development of the Sakhalin Region, led by Eifu Yamada, traveled to Sakhalin to discuss oil and gas projects for the development of Sakhalin Island shelf and the construction of a pipeline from Sakhalin to Hokkaido. These projects, as mentioned, involved the Mitsui, Mitsubishi, Marubeni, Itochu, and SODECO companies. In a July visit to Tokyo, Russian Communications Minister Leonid Reiman announced the signing of a major contract for the purchase of Japanese satellite equipment by a Russian state-owned company, *Kosmicheskaya Sviaz*

(Space Telecom). Japan's NEC Corporation, Mitsui Corporation, and Sumitomo Shoji trading company agreed to supply the Russian company with satellite electronics for a total sum of $102 million.[110]

The head of the Russian energy monopoly Unified Energy Systems (UES), Anatoly Chubais, accompanied Putin to Tokyo in September 2000. Chubais hoped to find investment for the construction of a proposed Sakhalin-Hokkaido 'Energy Bridge.' UES and the Sakhalin regional government had signed an agreement to coordinate the construction of a 4,000-megawatt gas-fired power station, in an effort to export electricity to Japan by the year 2012. Chubais was hoping the money would come from Japan. Although analysts were skeptical, the Sakhalin Energy Consortium (Sakhalin-2) considered assisting the project, by selling some of its production to UES. Japan's Mitsui and Mitsubishi Corporations are participants in the Sakhalin-2 consortium. During the first year this project produced 1 million tons of crude oil. In June 2001 the first shipments of oil from Sakhalin began reaching Japan.[111] Mitsubishi increased its stake in the Sakhalin-2 project to 20 percent by buying part of the stake of Royal Dutch/Shell (although their position was weakened substantially in 2007). Exxon-Mobil also expressed interest in constructing a gas pipeline linking the Sakhalin fields to Japan. It remains to be seen whether Japan's LNG needs will meet the minimum requirements necessary to make such a project profitable.[112]

A *Keidanren* delegation visited Khabarovsk in the early fall of 2000 to discuss the upgrade and modernization of the Russian port of Vanino. The initial interest was in improving the facilities necessary to export coal (several Japanese firms are involved in coal mining in the Russian Far East). The Japanese delegation reportedly also was interested in stepping up energy export facilities, since the Sakhalin projects were starting to produce. In addition, the Japanese Ministry of Transportation sent representatives with the *Keidanren* group to discuss upgrading and modernizing the trans-Siberian Railroad. The trans-Siberian offers the potential to ship goods to Europe faster then the maritime routes.[113] In line with the trilateral cooperation (American, Japanese, and Russian) in nuclear clean-up operations the Japanese government invested $40 million in a liquid waste processing plant at the Bolshoi Kamen submarine base in Primorskii Krai. This plant became operational by the end of 2001.[114]

In June 2001 a large delegation of Japanese business executives, led by *Keidanren* President Takashi Imai, visited Russia. Unlike the group in 2000, this was the first delegation from the *Keidanren* to visit Russia with official Japanese government sanction since the end of the Cold War. The large mission divided into several groups and visited industrial centers in the Russian Far East, Western Siberia, and near Moscow. The group was reportedly excited about certain opportunities in the Far East in the tourism and consumer goods area. The Sakhalin energy projects were also touted as a major success story.[115]

In spite of the promise, Japanese firms continued to be wary of Russia and for good reason. Several high profile joint ventures in the Russian Far East ended badly for Japanese businesses, and for Russia's image in Japan. A medium-sized Japanese construction company had helped to build a new airport terminal in

Vladivostok. When the project was completed the Russian partners simply said, 'Thank you, now go home.' No money was returned to the Japanese.[116] Another Japanese trading company from Niigata, Tairiku Bōeki, similarly lost out after it constructed a resort hotel in Yuzhno-Sakhalinsk. The firm put up $20 million to build the Santa Resort, a four-star, 89-room hotel, with restaurants, shops, gym, sauna, and casino. Once the complex had been completed, the Russian 'partners' seized the property and pushed their Japanese 'partners' out.[117] Similar cases were reported and after Russia's economy collapsed in 1998, Japanese investment appreciably dropped off.[118]

Several multinational corporations are involved in the large-scale energy projects in Siberia and the Russian Far East. These companies have much deeper pockets than smaller firms that specialize in goods and services. When Putin came into power he put legislation through the Duma on production sharing agreements (PSAs) that was specific to the Sakhalin energy projects. Companies such as Exxon-Mobil, Mitsubishi, Mitsui, and Shell felt confident that the Russian government was backing them, though this confidence would later be shaken. Smaller Japanese companies in the Russian Far East had little clout and were forced to deal with the notoriously corrupt local governments.

Defense contacts continued at a torrid pace in 2000–01. Concerned over the rising number of smuggling cases and illegal incursions, the two sides pushed for cooperation between their respective border services. In March 2000, 15 Russian border guards, serving in the Pacific Border District, arrived in Japan at the invitation of the Yeltsin-Obuchi Center and met with Japanese officers. At virtually the same time Russian and Japanese law enforcers met in Tokyo for a two-day conference on prevention of arm, drug and marine products smuggling. In April 2000 the Russian coast guard patrol vessel *Yuzhno-Sakhalinsk* visited the Japanese port of Yokohama. This was the first visit of a Russian border guard vessel to Japan. In September 2000 Russian and Japanese naval ships took part in joint exercises off of Kamchatka. Taking part in the exercises were two Russian anti-submarine ships of the Pacific Fleet and the Japanese destroyers *Hiei* and *Hamagiri*. In late December 2000 Colonel-General Konstantin Totsky, Director of Russia's Federal Borderguard Service (FBS), visited his Japanese counterpart Shige Arai, Head of the Maritime Safety Agency (MSA) in Tokyo in an effort to increase cooperation in combating illegal cross-border violations in the Sea of Japan region.

Russian leaders, however, were becoming more critical of Japan's stepped-up role within the framework of the US-Japan Security Treaty. Particular attention was given to Japan's role in the development of a regional ballistic missile defense system (BMD). Russian leaders saw in this program the beginning of the technical means to deploy a nationwide national missile defense system in the United States, which would essentially negate the 1972 Anti Ballistic Missile (ABM) Treaty. For the first time, the Russian Foreign Ministry issued public statements warning the Japanese not to become involved in such a program.[119] Previously, Russian diplomats had privately urged Japanese counterparts to disavow participation in any such program. The decision to publicly denounce Japanese participation in a TMD program was a dramatic gesture. At their St. Petersburg in April 2000 meeting Putin reportedly raised the issue with Mori.[120] Russian Defense Minister Igor Sergeyev again raised the issue when he visited Tokyo in December 2000.

Defense relations were further damaged when an MSDF officer, Shigehiro Hagisaki, was arrested in Tokyo in September 2000 on suspicion of having divulged classified information to a Russian naval attaché serving at the Russian embassy in Tokyo, Captain Viktor Bogatenkov. Russian officials denied Russian involvement.[121] Later Hagisaki confessed and implicated Bogatenkov. By this time Bogatenkov had already returned to Moscow. This affair hurt the standing of the JDA in Japan's Russia-policy arena. The voice of the JDA had assumed more and more importance in Russian policy since 1996. The Hagisaki affair was a serious blow.[122]

Nevertheless, defense contacts continued in 2001. Lower level officers and officials began exchanging visits and, on occasion, information. Thirty Japanese officers visited Khabarovsk and Vladivostok in January meeting with the Commander of Russia's Far Eastern Military District (FEMD), Colonel-General Yuri Yakubov. This was part of an increasing effort to move defense contacts beyond the higher levels to contacts at a level more conducive to actual joint operations.[123] The Chief of Staff of the Japanese ASDF visited Moscow in January 2001. Admiral of the Fleet Vladimir Kuroyedov, Commander-in-Chief of the Russian Navy, visited Japan in April, as did Chief of the Main Ground Forces Department of the Russian General Staff Yuri Bukreyev. During his visit Bukreyev broached the idea of the joint training of small army units.[124]

But there were yet more setbacks. In the spring of 2001 Russian warplanes allegedly violated Japanese air space on two different occasions for the first time since the 1980s.[125] In addition, Russian officials informed Japanese officials that the annual visit between the defense ministers would be 'difficult' to schedule in 2001. Many speculated that this was in response to the diplomatic row over fishing rights that had erupted in the summer of 2001. This left defense officials in Japan wondering whether the positive momentum that had been building for five years would be destroyed.[126] Defense relations had come to reflect the vicissitudes of the political relationship.

2002: The bottom falls out

There were no great expectations in either Japan or Russia going into 2002. But neither side was prepared for the damage that a political scandal in Japan wrought on the relationship. Since 1998 the influence of one man on Japan's Russia policy had grown inordinately powerful. Diet member Muneo Suzuki was an inveterate backroom negotiator who had great influence in the LDP and in the Diet. His abrasive style, however, had earned him many enemies in the Diet, even within his own party. His influence in the Foreign Ministry, which had begun to grow in 1997–98, had reached new heights by the year 2001.

In January 2002 ex-Prime Minister Mori visited Moscow in a private capacity but was granted a meeting with Putin. Mori and Putin agreed that Koizumi and Putin should meet as early as possible and continue to pursue peace treaty negotiations. In a sign that the 'two-track' policy was still an option for Japan, Mori spoke of the need to return Shikotan and the Habomai group, but on the status of the Etorofu/Iturup and Kunashir/i, Mori spoke merely of the recognition of sovereignty.[127] Suzuki accompanied Mori on his trip. An article in the Russian press pointed out Suzuki's still active influence on Japan's Russia policy.[128] Japanese observers continued to insist on the danger of the 'two-track' approach, saying that all Russia wanted was to 'eat and run' (*kuinige suru* – take two islands and leave the table).[129]

Suzuki received some help from the top in his war with Foreign Minister Makiko Tanaka when Prime Minister Koizumi relieved Tanaka of her post in late January 2002. Koizumi's decision was unrelated to Russia policy. Tanaka's constant warring with Diet members, like Suzuki, and her diplomatic blunders forced the hand of the Prime Minister. The new Foreign Minister, Noriko Kawaguchi, was forced to jump right into Russia policy. Russian Foreign Minister Igor Ivanov arrived in Tokyo on an official visit only hours after Kawaguchi was asked to replace Tanaka. The two discussed the territorial dispute and the peace treaty, but there was little substance to the rushed meetings. As one press report said, "the cast of characters continues to change but the results remain largely the same."[130]

Suzuki's 'victory' over Tanaka was fleeting. In mid-February 2002 in Diet budget committee deliberations a member of the Japanese Communist Party, Kensho Sasaki, accused Suzuki of profiting from the use of ODA (Overseas Development Assistance) funds dispersed to fund the construction of a lodging facility on the disputed island of Kunashiri. Sasaki pointed out that Russian residents of the island refer to the facility as the 'Muneo House.' To bolster his claims, Sasaki pointed out that the contract to build the facility was won by a Hokkaido firm managed by an executive of Suzuki's political booster group in his constituency in the city of Nemuro. The facility in question, known as the 'Friendship House' (*Yuukō no Ie*), was built in 1999 at a cost of about $3.5 million and was used to accommodate Japanese groups visiting the islands.[131]

Subsequent investigations and Diet deliberations revealed that the bidding process to construct the facility had indeed been rigged. The Foreign Ministry itself released several reports showing that two construction firms from Suzuki's home district (Watanabe Kensetsu Kogyo Co. of Nemuro, and Inukai Komuten Co. of Nakashibetsu) had received tacit assurances before the bidding process began that they would be awarded the contracts. In addition, it was revealed that these two firms, as well as five subcontracting firms, had made political donations to Suzuki between 1995 and 2000 to the amount of approximately a half million dollars. Suzuki denied that he had pressured anyone to award the bid to firms in his district, though he did say that he strongly supported their bids as he felt that districts where former islanders lived should be rewarded.[132]

In late February Diet deliberations, lawmakers harshly criticized Suzuki's undue influence over the Foreign Ministry. In the ensuing days Foreign Minister Kawaguchi asked both Kazuhiko Tōgō and Mutsuyoshi Nishimura, two officials said to have strong ties to Suzuki, to resign. Tōgō, of course, was one of the Foreign Ministry's top Russia experts. Nishimura, also from the Ministry's Russia School, had preceded Tōgō as head of the Bureau of European and Oceanic Affairs. They were both posted as ambassadors in Western Europe at the time they were recalled. Others within the Russia School with links to Suzuki were either demoted or moved to unrelated posts elsewhere in the ministry. One Japanese journalist termed it a "massacre" of the Russia School.[133]

The affair became even more sordid when it was announced that Suzuki had also probably benefited from contracts awarded to build schools in Tanzania and a power plant in Kenya. It was also reported that on a trip to the disputed territories Suzuki had punched and kicked a Foreign Ministry official who had dared to contradict him. The final nail in the coffin came when it was revealed that in a conversation several years earlier with a Foreign Ministry official Suzuki had accused Japan of being too obsessed with the Northern Territories and that their return would be of "no benefit" to Japan.[134] Soon thereafter the LDP expelled Suzuki from its ranks. Calls for his resignation went unheeded for a while, but the subsequent arrest of his secretary on suspicion of rigging the aforementioned bids put Suzuki's political and legal career in serious question. Suzuki was finally arrested on June 19, 2002 on suspicion of receiving illegal contributions in return for political favors. Meanwhile the Japanese government announced that it was halting all aid projects in the Northern Territories.

Russia's ambassador in Tokyo, Aleksandr Panov, publicly lamented the firing of a good portion of the Foreign Ministry's Russia School. He also defended Suzuki as a friend of Russia. In an editorial in *Izvestia* (March 10) veteran Japan correspondent Vasilii Golovnin said that after the backlash over Suzuki has died down, Japanese and Russian diplomatic sessions will revert to the days of one side saying, 'give back the islands,' and the other saying, 'no we won't.' Not long thereafter the Russian government announced that it had no intention of carrying out negotiations based on a 'two-track' policy. As if on cue lawmakers in both the Japanese Diet and the Russian Duma denounced the 'two-track' policy and vowed to protect the sovereignty of their respective countries. Russian Deputy Foreign Minister Aleksandr Losyukov, normally a mild-mannered diplomat, said he was "pessimistic about the current state of Russia-Japan relations." Bilateral talks appeared to be dead in the water.[135]

The tremendous Japanese push to normalize relations with Moscow and sign a peace treaty that had begun in 1996–97 ended in ignominy on the floor of the Diet amid allegations of corruption and undue political influence. What had once been grandiose visions of geopolitical maneuvering died in the crossfire of political mudslinging.[136]

2003–04: Energy cooperation and new diplomatic forays

Prime Minister Jun'ichiro Koizumi made it clear from almost the beginning of his time in office in 2001 that relations with Russia would rate far down his list of diplomatic priorities. This was understandable given the rise of China, the continued imbroglio on the Korean Peninsula, and the need for Japanese leaders to firm up the bilateral partnership with the United States. Koizumi had met with Putin in

July of 2002 on the sidelines of the Kananaskis G-8 summit in Canada. There the two leaders promised to try to further develop the relationship, and Putin invited Koizumi to visit Russia in early 2003. Nevertheless, it was evident that the damage from the political scandal in Japan would take some time to repair. There was one factor, however, that kept the Japanese government interested in maintaining at least some semblance of cordiality with the Russian government: the potential for an energy bonanza in Russia's Far East.

In early January 2003 Koizumi traveled to Moscow and to Khabarovsk in the Russian Far East on a four-day visit. Koizumi was the first Japanese leader to ever visit the Russian Far East. The discussions between Koizumi and Putin in Moscow were focused first and foremost on the potential for energy development and investment cooperation. Although Koizumi did mention the territorial dispute during talks with Putin, he otherwise studiously avoided the sensitive issue, lest he offend his host. The two leaders signed an "action plan" calling for long-term bilateral cooperation in trade, energy development, and other areas. The so-called "Six Pillars of the Japan-Russia Action Plan" were: (1) peace treaty negotiations; (2) deepening of the political dialogue; (3) cooperation in trade and economic areas; (4) development of relations in defense and security; (5) advancements in cultural and interpersonal exchange; and (6) cooperation in the international arena.[137] The two leaders issued a statement expressing their determination to expand bilateral ties, while reiterating the need to resolve the territorial dispute by signing a peace treaty "as early as possible."[138]

The Japanese delegation in Russia included executives from energy firms and other high-profile trading companies. In Khabarovsk, the Japanese met with local leaders and lobbied stridently for the construction of an oil pipeline going all the way to the Pacific, thus bypassing China. Initially, in late 2002 the Japanese had announced that they would be willing to invest up to $5 billion in such a project. A Japanese Foreign Ministry official was quoted as saying that the Japanese offer was a "powerful counter" against a Chinese proposal to bring the pipeline into Northeastern China.[139] The Japanese government also let it be known that it would be willing to purchase up to one million barrels of oil per day from Russia.[140] The Russian government, meanwhile, seemed content to sit back and let the Chinese and the Japanese attempt to outbid one another.

By early 2003, both the Japanese government and Japanese business began making serious overtures to the Russian government for a stake in the development and transportation of oil and gas resources from Eastern Siberia. Of particular interest was, of course, the proposed oil pipeline linking the Siberian oil depot Angarsk, northwest of Lake Baikal, with the Pacific coast. China's interest in this project has been well documented.[141] As oil prices climbed from 2003 (partially because of the war in Iraq, but also because of growing Chinese consumption), Japan's interest in the pipeline project grew in proportion. Japanese investment had been limited to the Sakhalin shelf, but now the Japanese became interested again in substantial investment deep in eastern Siberia, a dream that has existed since the early twentieth century, and which Prime Minister Tanaka held such great hope for in the 1970s.[142] One Japanese official noted optimistically that with the development of Eastern

Siberian oil resources, Japan could lessen its dependence on Middle Eastern oil (which comprises 85 percent of Japan's oil imports) by 15 percent.[143]

Originally, in October 2002 Russia and China agreed for an oil pipeline to be constructed from Angarsk to the Chinese city of Daqing in northeastern China, well short of the Pacific coast. The initial estimates for the cost of the project ran to about $2.5 billion. China was unable, however, to secure a firm pledge from Vladimir Putin during his state visit to Beijing in early December 2002.[144] Japan jumped squarely into the picture with Prime Minister Koizumi's Russia visit in January 2003 (although the Japanese began active lobbying for a Pacific route in the autumn of 2002). Eventually, the Japanese government announced that it was prepared to offer up to $8 billion for the so-called Pacific route, which would include $1 billion in investment in the infrastructure of the beleaguered Russian Far East.[145] The Chinese reacted testily to the Japanese overture at the eleventh hour, to which the Russian government responded by announcing that a branch line could easily be built to China.[146] Over the next several months the announcements coming from Moscow varied about what was to be done, normally according to who gave the commentary. But it was clear that Putin himself was reluctant to put all of his eggs in one basket, by building a pipeline directly into China where all the energy would be consumed. The Russian government was wary of the experience of depending on one consumer, especially given the case of the Blue Stream gas pipeline running across the Black Sea from Russia to Turkey (of which Gazprom was the lead actor). As Turkish demand slumped, the Turkish government asked for lower and lower prices. The concern was that China, as the sole consumer, could set prices for oil imports.[147] The Chinese government had already demonstrated to the Russians in early talks in 1999 (concerning the possibility of gas deliveries to China from the massive gas field at Kovykta) that it would be a tough bargainer in energy negotiations.[148] The Pacific route offered access to all of the markets of the Asia-Pacific region, including China, Japan, South Korea, and the United States. In May 2003 the Russian government designated the so-called Pacific route (East Siberia-Pacific Ocean, or ESPO – *VSTO* in Russian) as the priority, but quickly added that a branch to China would also be built.[149]

Through the summer of 2003 numerous Japanese government and business delegations visited Russia as part of an intense lobbying campaign to see the Pacific route realized. This included a trip by Japanese Foreign Minister Kawaguchi to Vladivostok in late June. Kawaguchi offered up a figure of $7.5 billion in Japanese investment.[150] Meanwhile, during the summer of 2003 the private Russian oil company Yukos was negotiating independently with the Chinese government about pipeline options from the Russian Far East to Daqing. Different actors in the Russian government and energy establishment seemed to have different opinions about the pipeline route, which they were not afraid to state publicly. This has caused much angst among the Japanese and the Chinese leadership. As one analyst noted, the Russian government seems to have "been weighing up the moves by Japan and China . . . in its basic approach of playing both ends against the middle."[151]

The lobbying campaign for the pipeline route was more than just between the

Chinese and Japanese governments. There was also a large Russian domestic component at this point, involving the privately owned Russian oil company Yukos, and the state-owned pipeline firm Transneft. The CEO of Yukos, Mikhail Khodorkovsky, had political ambitions that ran counter to Putin. He also favored the China pipeline, arguing that it made the most sense economically. Transneft wanted to build a pipeline to the Pacific. Transneft, in the words of one Russian scholar, is a major "agent of the Kremlin's policy of control," in the energy sector.[152] Khodorkovsky wanted big change in the energy sector (commercialization along Western lines), and he wanted change at Transneft. What he was proposing was contravening what the Kremlin was planning in terms of a state-controlled energy complex in the Russian Far East. In 2003 the Russian Far East was reportedly selected by the Kremlin as the "testing ground" for this new state energy strategy.[153] In May 2003, the same month that the Russian government tentatively proposed a Pacific-route (ESPO) pipeline, Yukos concluded an arrangement with the Chinese government, wherein the company proposed a pipeline supplying roughly one half million barrels of oil a day to northeastern China. Yukos executives felt that this could be completed at a cost of $2 billion within two years.[154] This was but one of many battles Khodorkovsky was fighting with the Kremlin at the time, and when he evinced interest in entering politics, Putin and his entourage made a political power move against him. Khodorkovsky was arrested and imprisoned in October 2003 (for tax evasion), and the China route lost a big ally.[155]

Evidently, the Putin versus Khodorkovsky battle was not the only domestic dispute concerning the competing pipeline routes. The Russian Ministry of Foreign Affairs was not convinced that the ESPO pipeline made sense economically, and pushed for an acceleration of work on the China route to Daqing. The Russian Ministry of Natural Resources backed the ESPO pipeline, insisting that it would be a boon for the further economic development of the Russian Far East and the Baikal-Amur Mainline (BAM) railroad. The Russian Ministry of Energy backed the Pacific route, as well, for strategic reasons. Energy Minister Igor Yusufov met with Japanese officials in 2002 and suggested that since Russian was committed to a strategic partnership with the United States, petroleum resources in the Russian Far East should be directed to the Pacific. Not surprisingly, the governors of the two far eastern regions, which a Pacific pipeline would pass through, also publicly declared support for this option.[156]

Putin and Koizumi again met at the APEC summit in Bangkok in October 2003, and although the two discussed trade and investment issues, North Korea was clearly highest on the agenda at that time. In December 2003 Russian Prime Minister Mikhail Kasyanov traveled to Japan, meeting with Koizumi, Foreign Minister Kawaguchi, and top Japanese officials to discuss trade, investment, and energy issues. Prior to his Tokyo visit, Kasyanov had to suffer an uncomfortable visit to Beijing, where he was forced to answer tense questions about prospects for a route to Daqing.[157]

A big decision concerning the East Siberian pipeline was made by Transneft President Semyon Vainshtok in February 2004, which bode well for the ESPO pipeline and Japan. The pipeline was originally slated to begin in Angarsk northwest of Lake Baikal, and would proceed south of the lake, along the Mongolian and then Chinese borders. Instead, Vainshtok announced that the pipeline would begin in Taishet, even further to the north, and would pass in a loop to the north of the lake, whence it would run parallel with the BAM railroad, and then proceed in a southeasterly direction to the town of Skovorodino, where it could easily be routed to either China or the Pacific coast. Transneft also proposed a three-stage construction plan. The first stage would include the simultaneous construction of pipeline segments at the beginning (Taishet) and ending points, including a terminal at Perevoznaya Bay near Nakhodka on the Pacific.[158] This was clearly a victory for advocates of the ESPO pipeline. Also in February 2004 Energy Minister for Russia Igor Yusufov announced that his government wanted Transneft to build a new $7 billion crude oil link to Perevoznaya so that sales could be opened to all buyers. According to Yusufov, the decision was based on "pure economic terms." But Yusufov went on to add, "The northern route [Pacific route] is a strategically important and a priority project."[159] By spring 2004 the Russian government seemed to have come to a decision that it preferred the longer pipeline skirting Northeastern China and ending up at Perevoznaya, from whence oil could be shipped to Japan, South Korea, China/Taiwan, and across the Pacific to the United States and elsewhere.

Meanwhile, on the political front the Russian government indicated that it was prepared to come to some sort of compromise with the Japanese. In November 2004 in a televised interview on a Sunday morning television talk show in Moscow, Russian Foreign Minister Sergei Lavrov suggested that the Russian government would be prepared to fulfill its obligation under the 1956 Joint Soviet-Japanese Declaration (hence returning two islands after the signing of a peace treaty). In what could be seen as a possible trial balloon for the Japanese, Lavrov stated, "Russia hopes for [the] complete normalization of relations with Japan. To that end, signing a peace treaty is important and the territorial dispute must be resolved."[160] Lavrov's statement was reiterated by Putin at a cabinet meeting on the following day. The announcement came on the heels of Putin's successful visit to China in October 2004. In Beijing Putin had reached an agreement with Chinese leader Hu Jintao, which effectively put an end to the Sino-Russian border demarcation issues that had divided the two nations for decades. In doing so Putin had agreed to the partial transfer of the last three disputed islands in the Amur and Ussuri Rivers (against local opposition).[161] The Russian government undoubtedly had Putin's pending Japan visit in 2005 in mind, and made legal references to bolster their line of thinking. Both Lavrov and Putin stressed the necessity of meeting all obligations that Russia had inherited from the Soviet Union.[162]

The Japanese side has demonstrated some flexibility on this issue, even agreeing to recognize Russian administrative control for years to come, in exchange for Russian recognition of Japan's ultimate sovereignty over the islands. But the Japanese government has been very firm in its insistence on the return of all four – not just two – islands. At a press conference in his residence on November 16, 2004 Koizumi said that as long as the sovereignty of all four islands is not clearly stated, then Japan has no intention of signing a peace treaty. But he did suggest that discussions could go forward.[163] As one veteran journalist commented, "In Tokyo, you can count the diplomats and politicians capable of simply speaking of the possibility of a compromise on the islands on the fingers of one hand."[164]

While the Japanese government was negative but guarded, in Russia the political opposition was swift to rise against any talk of compromise. Dmitry Rogozin, leader of the Duma faction Rodina (Motherland), stated that Lavrov and the Ministry of Foreign Affairs have no right from a legal standpoint to make such offers to the Japanese. He continued, saying that a peace treaty with Japan is "far from necessary for us and that Russia will not allow itself to be dictated to about sovereignty by an aggressor state from World War II." Sergei Ponomarev, a deputy of the local Sakhalin Oblast Duma, announced the region's intention to block any efforts to cede territory to the Japanese, and he suggested that the local legislature would have no problem enlisting the aid of the Communist Party and the Liberal Democrats.[165] Grass-roots opposition in the Russian Far East was also heated and visibly demonstrated on Sakhalin and in Vladivostok. One article in *Pravda* quoted local islanders as saying that they would conduct guerilla operations on the islands against either the Russian or Japanese government in the event of a territorial handover.[166]

Nevertheless, the Russian government had just recently concluded a settlement of the long-standing territorial dispute with China, despite strong local opposition in the Far East, therefore any proposal by the Russian government – no matter how casually thrown out – would have to be taken seriously by the Japanese. Additionally, Putin's recent re-election as president and his firm control of the Duma suggested to some that he may have had the clout to push through an agreement with Japan. Similarly, in Japan Prime Minister Koizumi had recently retained his post as party president and had amassed a strong political base. Koizumi also made it clear that he desired to go down in history as the Japanese Prime Minister who signed the peace treaty with Russia. Additionally, observers were suggesting that both leaders were eager to shore up the respective geopolitical positions of their nations in the face of a rising China.[167]

The brief excitement created by Putin's mid-November 2004 announcement that the Russian government would be prepared to honor the 1956 Soviet-Japanese declaration evaporated as quickly as it appeared. At their meeting less than a week later in Santiago, Chile at the APEC summit, Koizumi and Putin were unable to come to any sort of agreement; they were also unable to finalize the dates of the Putin visit to Japan in 2005. In Santiago, both Koizumi and Putin said that it was "necessary" to sign a peace treaty and to resolve the territorial dispute. Koizumi went so far as to say that a *rapprochement* would "serve the strategic interests of both nations." An editorial in the conservative *Yomiuri Shimbun* the next day elucidated these very points in relation to the development of Japanese-Russian relations, alluding to China's rise.[168] But there appeared to be differing opinions on the means, and both leaders admitted that the two sides were still far apart and no date for Putin's visit was set. Specialists from both governments speculated that the failure to set a date for the visit suggested that Putin had held this up as a means of exerting pressure on the Japanese government to reach some sort of compromise agreement, whether agreeing to a peace treaty without any exchange of territory, or the return of just two islands.[169]

As in earlier periods of Japanese-Russian discussions, however, the two sides failed to clearly communicate their intentions and to properly understand one another. A good example of this was the proposed visit by Putin to Japan in 2005. The Japanese government initially wanted Putin to visit on the 150th anniversary of the establishment of diplomatic relations: February 7, 2005. Yet February 7 is also known in Japan as "Northern Territories Day," when all of Japan's major dailies carry editorials demanding the return of the disputed islands. Government ministry buildings in Tokyo hang banners declaring, "The Northern Territories are OUR territories," or "Give us back the Northern Territories!" Right-wing Uyoku sound trucks come out in force to blare patriotic messages in front of the Russian Embassy in Tokyo. Asking Putin to be in Tokyo on this particular day was sending the wrong message.

Another example of misunderstanding was the apparent illusion in Japanese government circles that Putin would be willing to give back the islands because he had recently won his second term. The most common analogy given is that only a strong leader like Nixon could have visited China. But visiting a country is a far cry from ceding territory back to a neighbor. The same analogy was applied to Yeltsin after he won re-election in 1996, but he did not make any compromise over the disputed islands.

The year 2004[170] did end on a positive note for Japanese-Russian relations when the Russian government announced on the last day that it had given the final approval for the go-ahead for the construction of the Taishet-Perevoznaya pipeline to the Pacific (ESPO).[171] As expected, the state-owned company Transneft was given the green light for the project. The goal was for the pipeline to pump 80 million tons of crude oil annually, but the cost estimates continued to balloon, up to $15 billion by this time. Upon the announcement of a Pacific route the Japanese government was oddly quiet. Not much was made of the decision in the Japanese press. This was probably due in part to the territorial impasse, but also due to the need to temper enthusiasm about a costly project that was still far from being commenced. Both the Japanese and the Russian governments were probably also wary of angering China with a grand public pronouncement on the heels of what was undoubtedly a great disappointment for Beijing. In a nod to addressing China's energy concerns, officials in Russia promptly indicated that eventually a pipeline spur would be opened running to Daqing in northeastern China.[172]

2004–05: Russia's last offer

Vladimir Putin's offer to negotiate with the Japanese at the end of 2004 on the basis of the 1956 declaration along with the decision to favor the ESPO pipeline marked the high point after the Suzuki scandal for the potential of normalization between Tokyo and Moscow. Unfortunately, Japanese politicians seemed unable to rise to the occasion. Koizumi dithered enough over the offer to create doubt about whether Japan was genuinely sincere about compromise. But this did not stop the two governments from further engagement in the energy sector, and it did not stop companies from the two nations from exporting to one another.

Two-way trade, which totaled just under $6 billion in 2003, jumped almost 50 percent to total $8.8 billion in 2004. Japanese exports to Russia via third countries boost the figure another $3–4 billion, some experts estimate. Meanwhile, by 2005 cumulative Japanese investment in Russia reached $700 million; this figure goes up $3 billion if the Sakhalin investment figures are included.[173] At the end of 2004 Toyota Motor Corporation announced that they would establish an assembly plant near St. Petersburg and start assembling vehicles in 2006. In 2003 sales of Toyota models tripled from the year before to 25,000 units, and the manufacturer hoped to cash in even more.[174] Military exchanges continued throughout 2004–05, but not quite at the frantic pace of the previous few years. In February General Hajime Masaki, Chairman of the Joint Staff Council of the Japanese Self-Defense Forces, visited his Russian counterpart Yuri Baluyevsky in Moscow in May, where the two discussed, among other things, Japanese-Russian cooperation in the Six-Party talks on the Korean Peninsula.[175] More importantly, cooperation in nuclear clean-up operations in the Russian Far East got under way by early 2005, near Vladivostok and on Kamchatka. Japan has pledged up to $180 million toward the dismantlement of 40 or so decommissioned nuclear submarines at Russian bases in the region.[176]

Russian industry leaders also viewed the Japanese energy market with hungry eyes. At an investment meeting in Niigata in Japan, Alexei Mastepanov, an advisor to Gazprom chairman Alexei Miller, announced the firm's new "Eastern Program." Gazprom management (in line with the Kremlin's strategy of a far eastern energy complex) hopes to significantly boost gas exports to Northeast Asian partner nations, including Japan. But Mastepanov and others in Russia warned that such a program will require tens of billions of dollars of both public and private investment. Gazprom's merger in 2005 with the state-owned oil giant Rosneft (which has a stake in five of the six major Sakhalin energy projects) was one of the first steps in the establishment of a state-controlled export monopoly for Russian oil and gas to the nations of East Asia and the Pacific Rim. Russian leaders hoped to capture 15 percent of the Asian gas import market share by 2020.[177]

Although energy became the *sine qua non* for Japanese-Russian bilateral cooperation by 2005, the territorial issue could not help from rearing its ugly head. In a January foreign ministerial meeting between Sergei Lavrov and Nobutake Machimura, the Russian Foreign Minister reportedly told his Japanese counterpart that two islands were the final offer.[178] Again, the date for the Putin visit to Tokyo was left undecided. Meanwhile, the upper chamber of the Japanese Diet passed a non-binding resolution on February 7 (on the 150th anniversary of the establishment of diplomatic relations between Moscow and Tokyo) calling for Russia to return the Northern Territories lost in the war. The interpretation in Russia was that Japan was calling for all of the Kuril Islands. The resolution was somewhat vaguely worded to appeal to all Japanese parties in order to get a unanimous vote, as had been the practice in the past. For the Japanese it was nothing out of the ordinary. But the Russian press took the statement literally and it received extensive coverage.[179] Soon thereafter the Russian ambassador to Tokyo, Aleksandr Losyukov, gave an interview in which he stated that Japanese-Russian relations were in their worst state.[180] As the summer of 2005 approached, Japan's political relations with Russia seemed scarcely better than they had been during the Cold War.

There was even a growing concern in both capitals about the viability of the pipeline project. An April 2005 interview with Kremlin Chief of Staff Dmitry Medvedev raised eyebrows in both nations. Medvedev named Siberia and the Russian Far East as two of Russia's most troubled areas in terms of the potential for fragmentation. He stated that the construction of an oil pipeline through the region could boost the local economy. But, at the same time, Medvedev pointed out that Russia's weakness there is not merely economic but also demographic. Medvedev warned that Russia needed to address the panoply of problems in those regions; otherwise, "cold and emptiness will reign there." While supporting the construction of the pipeline, he said that Russia should not rush to build it merely for the sake of building it. He gave the example of the Baikal-Amur Mainline (BAM) railroad as a wasted project with no real viability. Medvedev seemed to indicate that Russia could not afford to chase grandiose projects, as had often been the case in the Soviet Union.[181]

Leaders in Japan continued to remain quiet on the Siberian pipeline. Following the December 31, 2004 announcement by the Russian government that the pipeline would be constructed to Perevoznaya, there was little reaction from the Japanese side. Russians who had lobbied on behalf of the Japanese also received little in the way of acknowledgement from those in Japan they had supported. One Russian official, who claims to have personally lobbied President Vladimir Putin on behalf of the Pacific route, says he was stonewalled when he subsequently visited Tokyo in January. "We are still studying the feasibility [of the pipeline project]," he was tersely told by one Japanese official.[182]

Putin was reportedly displeased with the cool attitude of the Japanese, and many of Tokyo's allies in Moscow (what few there are) expressed similar irritation. Meanwhile, the decision by the United States and Japan (at Tokyo's behest, no doubt) to include a statement about the resolution of the Northern Territories dispute as a "common strategic objective" in the February 2005 2+2 meetings (the US secretaries of state and defense with Japanese counterparts) further angered Moscow.[183]

Medvedev's interview raised new doubts about the status of the Pacific pipeline, whose economic feasibility has been doubted for years by both Russian and Western experts. The Russian government meantime began increasing shipments (by rail) of oil to China and gave preliminary indications that it was prepared to allow Chinese participation in equity holdings of Russian energy firms, something that had eluded the Chinese for years.[184] The China National Petroleum Corporation (CNPC) also began earnest investigations into gas deliveries from the Exxon-controlled Sakhalin-1 energy project. The potential costs of a Sakhalin-China gas pipeline were estimated as high as $14 billion, but the Chinese did not seem put off.[185] Japanese observers speculated that the Russian government was allowing Chinese involvement in the construction of a gas pipeline as some sort of compensation for having been disappointed in the Russian decision to choose the ESPO pipeline.[186] But it seems to have been more a case of the Russian government making it clear that it would find the necessary capital to finance energy projects, whatever Japan's attitude may have been. When Russian Industry and Energy Minister Viktor Khristenko visited Tokyo in April 2005, he issued several 'vague hints' that the ESPO pipeline decision was not a final one and that further review of the practicalities was necessary. His Japanese counterpart, METI Minister Shoichi Nakagawa, made it clear that any pipeline that included a route to China would fail to receive Japanese funding. Khristenko promptly softened his language by adding, "There is no problem of choice between the 'Japanese' and 'Chinese' routes of the pipeline because a decision was already made on the Taishet-Perevoznaya (ESPO) route."[187] But, in fact, Putin had been saying for months that Russia seeks multiple outlets for both its oil and gas, and that no one nation would be the exclusive recipient.[188] In October Deputy Foreign Minister Alexander Yakovenko stated that, "Russia is not making the building of the oil pipeline contingent on its relations with Japan. We have sufficient resources to build this oil pipeline independently."[189]

It was about this time in early 2005 that a number of reports emanated from Russia concerning the economic value of the disputed islands. In February the Russian Ministry of Natural Resources published a report listing the minerals and natural resources that could be found in the disputed Southern Kuril Islands/ Northern Territories. Besides natural gas, gold, iron, rhenium, titanium, and other precious metals, the report suggested that there could be as much as 364 million tons of oil under the islands and the seabed of the surrounding waters.[190] Meanwhile, during a July 2005 inspection of Russian troops on the disputed islands, Russian Defense Minister Sergei Ivanov announced that Russia had no intention to make concessions to the Japanese.[191] In a September interview Putin reiterated Ivanov's statement. In October 2005 the Russian government announced a plan to develop the Kuril Islands through 2015, by investing more than $670 million in the infrastructure of the islands.[192] In November on the eve of Putin's Tokyo visit, a report by a respected defense expert in *Novaya Gazeta* described the strategic importance of the Kuril Islands for Russia's nuclear arm, particularly Russian nuclear submarines.[193] This was typical background noise in preparation for Putin's upcoming visit to Japan.

Putin's long-anticipated Japan visit finally came about on November 20–21, 2005, when he paid an official state visit to Tokyo, his first visit to Japan in over five years. Putin's two-day trip disappointed those expecting some sort of breakthrough on the territorial issue. But the trip did have significance in several ways. Putin was escorted to Tokyo by a large delegation of business executives (over 100 people), primarily drawn from the energy industry. The ESPO pipeline was obviously a major point of discussion among business and government executives from both sides. This marked by far the largest Russian business presence at a bilateral summit. Normally, it is the Japanese side that brings large business delegations to Russia. In Tokyo Putin and Koizumi signed 12 different documents on economic cooperation, joint anti-terror efforts, tourism, and information technology.[194] As for the ESPO pipeline, Putin was careful to make any concrete announcement about whether the ESPO pipeline would include a spur to China. Koizumi again decided against harping on the territorial dispute, probably a sound move. But observers and editorials in Japan lamented the government's punch-less policy vis-à-vis Moscow. At least one editorial criticized Japan's new investment policies in Russia, suggesting that Tokyo was sending Moscow the mistaken signal that the territorial issue would be permanently shelved.[195]

Where Japan once seemed to possess the carrots in the relationship, Russia now seemed to have the upper hand economically, due to the energy equation. Tokyo could once threaten to withhold economic assistance and investment if Moscow skirted the territorial issue. At one time, particularly in the early 1990s and again after the 1998 economic crisis in Russia, Russia was desperate for Japanese investment, particularly in the Russian Far East. But Russia was now cash rich thanks to the historically high price for oil and gas. Additionally, Russia now had other suitors for investment and economic cooperation, as half a dozen nations besides Japan continue to invest in energy projects in the Russian Far East and Siberia, particularly on Sakhalin. There had been a fundamental shift in the bilateral relationship that was markedly apparent during Putin's visit to Japan in late 2005. Japanese businesses are moving more rapidly into Russia, no matter how the government in Tokyo feels about the situation. As mentioned, one of Japan's flagship corporations, Toyota Motors, is building an assembly plant near St. Petersburg. Large trading houses, including Mitsubishi and Mitsui, are deeply involved in energy projects from the Caspian to the Pacific. It could be argued that the time in which Japan could realistically expect to recoup all four islands had passed by the end of 2005 and that the territorial dispute can no longer hold up the development of the Russian Far East. If so, then the 2005 Putin visit marked a watershed in Japanese-Russian relations. At the beginning of the year, Putin seemed ready to hand over at least two islands (which admittedly comprised only 7 percent of the disputed territory), but the Japanese were unprepared. By the end of 2005, perhaps the last window of opportunity for the next several years or decades for a territorial settlement was slammed shut with Putin's visit.

2006: Energy machinations and political fallout

Energy matters again dominated the relationship in 2006. For Japanese leaders the importance of the ESPO pipeline increased ever more so in their eyes when the Ministry of Energy, Trade, and Industry (METI) announced that Japanese imports of Middle Eastern oil in 2005 had increased its share of total oil imports to over 90 percent.[196] Meanwhile, the Kremlin was finalizing its plans for the establishment of a state-controlled far eastern energy complex.

In this vein, Putin traveled to Yakutia in January 2006, and while touring energy facilities he outlined his plan for a unified energy system in the region.[197] In February Industry and Energy Minister Viktor Khristenko said in an interview that the Russian government hoped to increase its energy exports to East Asia to 30 percent by the year 2020, up from the current level of 3 percent.[198] In April, Khristenko's Ministry of Industry and Energy issued a detailed outline of this energy complex, stressing the strategic importance of the Russian Far East for Russia's development in the twenty-first century. This report pointed out that the Russian Far East sits over 84 percent of Russia's coal deposits, 40–50 percent of Russia's oil deposits, 14 percent of Russia's natural gas deposits, and that the region accounts for 76 percent of Russia's hydro-energy generation. The report again stressed the importance of the Asia-Pacific export market, but toned down the goals for Asian exports to 15–20 percent of Russia's total energy exports by 2015–20.[199] A new amendment to an existing law on subsoil use was announced, which places ceilings on foreign participation in oil and gas development projects.[200] Putin also insisted on the construction of the ESPO along the northern route, skirting Lake Baikal scores of kilometers, so that there would be no adverse environmental impact given an accident.[201]

Politically, Koizumi continued to focus on other issues within Japan and elsewhere in Asia (particularly China), and Russia seemed low on the priority list. In what many saw as a gesture toward Moscow, however, Koizumi skipped the February 7 Northern Territories Day celebrations in Tokyo. In January Japan Defense Agency Director-General Fukushiro Nukaga traveled to Moscow to meet with Russian Defense Minister Sergei Ivanov, continuing the parade of defense visits.[202] In March Japanese leaders awaited the results of Putin's trip to China. In China, Putin was similarly non-committal about Russian intentions for the ESPO pipeline. Some good news did emerge in the spring when it was announced that construction on the ESPO pipeline had begun in April in Taishet.

Putin and Koizumi met at the G-8 summit in St. Petersburg in July 2006. At their meeting Putin threw cold water on the Japanese hopes for a single ESPO pipeline, and refused to give a state guarantee that the pipeline would even be built. Putin insisted on the "commercial nature" of the project, and stated that the Russian government is in no way "obliged" to be involved.[203] This was an incredible statement considering the plans unveiled by Putin's own government earlier in the year for a state-controlled far eastern energy complex, and Putin's intervention to re-route the pipeline plans far to the north to avoid the Lake Baikal watershed.

In August 2006 another round of the 'fishing wars' broke out. On August 16, in disputed territorial waters north of Hokkaido and south of the Habomai group of islets, a Japanese fisherman was killed by a shot fired from a Russian Coast Guard patrol boat. Three other crewmembers of the Japanese fishing vessel *Kisshin Maru No. 31*, including the captain, were arrested on the spot by Russian authorities and taken into custody on Kunashir.

While this incident may have been swept under the carpet in early 2006, the timing could not have been worse, given the fiftieth anniversary of the restoration of diplomatic relations between Moscow and Tokyo in the fall. Although there have been a number of incidents in the disputed waters over the past decade or so (30 Japanese fishing boats have been seized by Russian authorities since 1994, and seven Japanese have been wounded by gunfire in those incidents), this is the first fatality since 1956. In a cruel irony, this was the year relations were restored. The Japanese daily *Sankei Shimbun* remarked in an August 17 editorial, "The Russian side has left a stain" on Japanese-Russian relations in the fiftieth year of their bilateral relationship.

Russian authorities insisted that the *Kisshin Maru* resisted calls to stop and be boarded, and tried to escape when warning shots were fired from the Russian boat. The Japanese victim was supposedly struck in the head by random fire. Russian authorities made it very clear that the boat was in forbidden waters and taking part in illegal fishing for crabs. The Russian press repeatedly referred to the Japanese fishermen as "poachers." Some Japanese fishing boats from the Hokkaido port of Nemuro have the right to fish for flatfish, octopus, pollack, and to harvest seaweed, but crabs – which can fetch high dollars on the commercial market – are strictly forbidden. The Russian coast guard reportedly found large numbers of crabs and at least 25 crab traps on board the *Kisshin Maru*. While some Japanese observers concede that it might have been operating illegally, the Japanese public had a hard time fathoming how the Russian authorities could have delivered such a stern and malicious response to illegal fishing and taken a life.[204]

The Japanese Foreign Ministry dispatched diplomats to Moscow and to the region to win the release of the two other crewmen, which occurred within days. But the captain was released only in October and forced to pay a fine. The incident only added to the long-standing image of Russia among Japanese as a dangerous place where the law is applied arbitrarily. This image exists among Japanese diplomats who feel stonewalled at every meeting with Russian counterparts, Japanese businessmen who have had assets seized by organized crime in Russia, and Japan tourists who have been randomly attacked in Moscow and St. Petersburg, as well as in the Russian Far East.

The incident did spur one LDP member to say in Diet debate in August that had Japan accepted the two-island deal Putin had been offering, the incident may never have happened. This prompted a wave of criticism from the usual suspects on the right of the government and in the press.[205]

More ominous for the Japanese side were the movements being made within the Russian government to help Gazprom secure a majority stake in the Sakhalin-2 project at the expense of Shell, Mitsubishi, and Mitsui. On September 18 a Russian high court ordered the temporary suspension of operations at the Sakhalin-2 project. The order followed a complaint filed by the Russian Ministry of Natural Resources, which claimed that the project was violating environmental regulations.[206] The Russian government then told the project leaders that they had one month to correct the problems.[207] The Sakhalin-2 energy consortium (known simply as Sakhalin Energy) was at the time 55 percent controlled by Royal Dutch-Shell, while the Japanese trading companies Mitsui and Mitsubishi controlled 25 percent and 20 percent, respectively. Evidently, the Kremlin was peeved by the continuing and massive cost overruns, thus depriving the Russian government of profits at a time of soaring energy prices.

The Russian Ministry of Natural Resources also asked Sakhalin Energy to review its planned gas pipeline routes, suggesting that certain portions be rerouted to avoid sensitive areas such as primary-growth forests, grey whale breeding waters, and salmon spawning areas. At the same time, Gazprom was hovering in the background. Gazprom had been eager for years to get in on the Sakhalin action to meet growing energy demands across East Asia, and in Russia itself. Additionally, President Putin's plan for a state-controlled far eastern energy complex necessitated a strong Russian presence in the Sakhalin projects. For months Gazprom had been in negotiations to acquire a 25 percent stake in Sakhalin-2 in an asset swap with Shell.[208] Shell was initially hesitant to take Gazprom management up on their offer, but the pressure from the Russian government was intense enough that Shell finally decided, along with its Japanese partners, to come to an agreement.

In December 2006 the heads of Shell, Mitsubishi, and Mitsui traveled to Moscow to sign an agreement with Gazprom CEO Alexei Miller in the Kremlin. A beaming Putin also attended the ceremony and announced with the signing that the environmental transgressions had been "resolved." The four firms signed a deal giving Gazprom a 50 percent-plus-one-share stake in the project in exchange for $7.45 billion in cash. Subsequently, the Shell share of the project decreased to 27.5 percent, while Mitsui and Mitsubishi fell to 12.5 percent and 10 percent, respectively. Gazprom Deputy Chief Executive Alexander Medvedev trumpeted, "Gazprom's entry into Sakhalin-2 is a powerful impetus for implementation of this large-scale development in the area of energy export to Asia Pacific and North America."[209]

As this crisis was breaking, in Japan a new leader had emerged. Shinzo Abe, formerly the Chief Cabinet Secretary to Koizumi and staunch LDP conservative, had replaced Koizumi in September. Abe was seen in many circles as a tough-minded nationalist. Faced with one of his first foreign policy issues, he responded in strong fashion, in line with his reputation. "A significant hold-up in this [Sakhalin-2] project, which is a symbol of Japanese-Russian cooperation," Abe said, "will have . . . negative repercussions on the whole of our relations with Russia."[210] Practically every major daily in Japan called for the Japanese government to respond strongly to Russia's "high-handed" behavior. In the end, the Japanese could do nothing but accept the agreement, as unfair as it may have been.

Almost immediately the other Sakhalin energy projects began feeling pressure, including the Exxon-Mobil controlled Sakhalin-1 project. In August 2006 the Russian government announced that it wanted to review the three major Production Sharing Agreements (PSAs) that it signed with Western and Japanese energy firms in the early 1990s.[211] These three PSAs were negotiated when the price of oil was around $15 per barrel in the mid-1990s. At the time, the Russian government was eager to attract investment, and agreed to terms that were far less than they would be able to negotiate today. Under the PSAs, the Russian government can only see profits once the projects themselves begin to recoup their massive cost outlays. Costs for Sakhalin-2 had risen from $10 billion to $20 billion.

Exxon-Mobil was told that it could not begin regular shipments of oil to the Asia-Pacific region from the Sakhalin-1 project until mid-November, when terminal inspections had been completed. The company was also under fire for a proposed gas pipeline to China from Sakhalin. The Russian government insisted that gas extracted in Russia be sold first domestically. In China, Exxon could get closer to market rates for gas, while rates in Russia are below market value. Gazprom, however, had its own plans for a gas pipeline to China through the Altai highlands near the Kazakh-Russian-Mongolian border. Any Exxon-Mobil pipeline to China would be in direct competition with Gazprom's strategy for supplying gas to China. And once Gazprom gained a controlling stake in the massive Kovykta gas fields north of Lake Baikal from BP in June 2007, the way was set for Russian domination of all the major eastern energy projects, with an export monopoly over natural gas and oil.[212]

Abe and Putin did have the occasion to meet on the sidelines of the APEC summit in Hanoi in November 2006. Although the two did not engage in detailed discussions, the territorial issue did come up in their talks. Konstantin Kosachev, chairman of the State Duma international affairs committee, attended the discussions, and afterwards announced that no compromise was in sight.[213]

Japanese Foreign Minister Taro Aso, evidently frustrated at the state of relations with Russia (which the Russian ambassador to Tokyo Alexander Losyukov termed "catastrophic") suggested at a meeting of the Diet Committee on Foreign Affairs a 50–50 division of the disputed islands. The three southern islands (the Habomai group, Kunashir/i, and Shikotan) would be returned to Japan, and the larger northernmost island (Etorofu/Iturup) would be divided about a third of the way up, thus marking the border between the two nations. His proposal was rejected not only by his Russian counterpart Lavrov, but by almost every member of his own government.[214] Had Aso made this proposal in early 2005, he might have met with a better reaction in Russia. But the fact that he made the proposal a full two years after Putin's offer (and given the fact that Aso had no support even within his own government) doomed any effort on his part to failure. The controversial year that saw shooting incidents and government-backed corporate raids ended with a whimper. The final window of opportunity during the hopeful decade since 1996 had seemingly slammed shut.

2007: Status quo frozen

There was little to recommend optimism for Japanese-Russian relations as the year 2007 – the 100th anniversary of the first of the secret Russo-Japanese treaties in the early twentieth century – began. The death of Russia's first president Boris Yeltsin in April may have literally and figuratively marked the end of the decade of promise begun by Yeltsin and Hashimoto in 1996.

Russian Prime Minister Mikhail Fradkov and Russian Industry and Energy Minister Khristenko visited Tokyo in late February to discuss ways of increasing cooperation in the conventional and nuclear energy fields, and inevitably the territorial issue came up. Fradkov warned his hosts not to let the territorial dispute hinder economic relations. The Russian delegation made clear their interest in Japanese technology, and solicited Japanese participation in a number of projects in the information and telecommunication industries and in the transport sector (particularly airline and railroad technology) of Russia. Fradkov and Khristenko also assured the Japanese that gas deliveries from the Sakhalin-2 project would be delivered in 2008, as had been originally planned before the Gazprom takeover.[215]

Japanese Foreign Minister Aso was next in line for a visit and he arrived in Moscow in April and met with Foreign Minister Lavrov. Lavrov did say that Russia was prepared to discuss the territorial issue, but his statement may have been pro-forma. The two did discuss cooperation in the context of the North Korean issue, as well as avenues of further economic cooperation.[216] The next important meeting took place at the Heiligendamm G-8 summit in Germany when Abe and Putin were able to sit down for an amiable discussion. In a press conference prior to the summit Putin announced with typical steely resolve that Russia was amenable to a settlement, but under the current circumstances it would be "difficult to find new measures [leading to a settlement]." At Heiligendamm, however, Abe offered to increase Japanese investment in the Russian Far East and the Kuril Islands, undoubtedly hoping to secure a favored position in Putin's eyes. Putin promised to work toward some sort of deal toward a peace treaty, but he cautioned against moving too fast. Something prior to the G-8 summit that should have given the Japanese government a clear signal that territorial compromise is perhaps years or even decades ahead was the visit to the disputed islands in early June by Russian Foreign Minister Lavrov. There Lavrov reiterated Russia's firm position vis-à-vis the islands.[217]

Progress on the ESPO pipeline was steady, and by the beginning of 2007 more than 500 kilometers of the first stage had been laid down by Transneft. Transneft management aimed to have more than 12,000 kilometers constructed by the end of 2007, and so the progress was good. But what kept many experts wondering was where the oil would come from to supply the pipeline. The aim of the ESPO pipeline is to eventually ship 1.6 million barrels per day to Skovorodino, from whence 1 million barrels (80 million metric tons per year) could be passed on to the Pacific coast, while 600,000 barrels (30 million metric tons per year) could be siphoned down to Daqing. Estimates, however, for oil production in East Siberian fields in the year 2015 suggest that extraction will amount to only 800,000 barrels per day.[218] This predicted shortfall has been the source for much of the discussion of Sino-Japanese competition over the ESPO pipeline. China offered to pay for the spur to Daqing and agreed to start construction from Daqing to the north in 2008.[219] Japan offered to pay between $7 and $8 billion for the Pacific route. Meanwhile, in July 2007 Transneft CEO Semyon Vainshtok claimed that his firm had already invested $6 billion in the first stage of construction. He also stated that East Siberian sources of oil would suffice to fill the pipeline, and that 1,000 kilometers of the ESPO pipeline had already been built.[220]

At a June 2007 government meeting in Moscow the Russian Ministry of Industry and Energy presented a detailed blueprint for the far eastern energy complex (Program to Develop a Unified System of Gas Production, Transportation, and Supply in Eastern Siberia and the Far East). The ministry estimated the costs for the far-reaching program to amount to over $90 billion by the year 2030, most of this involving a massive expansion of gas infrastructure, including both pipelines to China and elsewhere in Asia, and liquefaction plants.[221] But in a more sober note (and for the Japanese perhaps more somber than sober), it was announced in July that the completion of the ESPO pipeline might be delayed until 2015, in order to allow for the development of the Eastern Siberian oil fields.[222] But indications were strong that Moscow would push for increased gas and oil exports (by pipeline and rail) to China in the years up to 2010.

It was on this sober note that Japanese-Russian economic and political relations seemed once again pushed to the back burner.

4 The international context

1996–2007: System in transformation

The international transformation that resulted from the dissolution of the Soviet Union and the end of the Cold War is still in the process of being settled. The events of September 11, 2001 may have put an end to the era that for ten years had been known simply as the 'post-Cold War' period, but for the decade of the 1990s the ramifications of the end of the US-Soviet confrontation were still unknown. Furthermore, given the unsettled situation in the Middle East and China's rise in Asia early in the twenty-first century, it is still uncertain how the international system will play out over the next few decades, and how the Japanese-Russian relationship will be defined by international events. The influence of international and systemic change over the bilateral relationship has been indisputable during the past century.

A combination of certain crucial events in the mid-1990s became a major catalyst for a change of thinking in policymaking circles in Japan and Russia. The *rapprochement* between Moscow and Tokyo in the late 1980s and early 1990s was largely in response to domestic policy changes in the Soviet Union and the actual collapse of the Soviet Union. This *rapprochement* was peripherally related to international events that happened subsequent to the change of regime in Moscow. The *rapprochement* of 1996–2007, in contrast, was a direct response to the changes in the international system that came about as a result of the dissolution of the Soviet Union, as well as domestic change in each country. During the latter half of the 1990s events transpiring in Europe, Central Asia, and Northeast Asia kept the attention of leaders in Moscow and Tokyo. In Europe, NATO expansion was a major factor driving the foreign policymaking process in Moscow and, to a much lesser extent in Tokyo. The birth of the eight new nations in Central Asia and the Caucasus also played an important role in the thought process of leaders both in Japan and Russia. And in East Asia, China's emergence and the role of the United States have been major factors in the reorientation of foreign policy for Moscow and Tokyo. Additionally, though generally unnoticed, the situation on the Korean Peninsula again began to play an important role in the bilateral relationship. Korea is one of only two countries that are neighbors of both Japan and Russia. One hundred years ago it was the central focus of Japanese-Russian relations. The Japanese occupation of Korea (1910–45) and the Cold War kept Korea from occupying the central role in the bilateral relationship for a long time. But in the 1990s Korea reemerged as a major catalyst in the bilateral relationship.

The rise of China

China's central role in the Japanese-Russian bilateral relationship is clear to even the most casual of observers. Before the Second World War China was a spoil to be fought over. Japan and the Soviet Union (or Russia) had disputed railroad and port concessions and territory in Manchuria and Inner Mongolia since the 1890s. During the Cold War, China's role was of a more subtle nature. The People's Republic of China after 1949 was a strong, independent nation. Neither Japan nor the Soviet Union sought new territory at the expense of China; neither did the two nations dispute economic concessions. Instead, they both sought the good graces of China in the context of larger strategic issues, primarily involving the United States. From the end of the Second World War Japan had desired mainland China as a trading partner; the Soviet Union had desired China as an ally to be used as a bulwark against Japan and the United States. Neither side viewed China as a major strategic threat, until the 1970s when the Nixon administration's China approach was viewed with alarm in Moscow. But China's threat to the Soviet Union was only viable with American backing. The Soviet Union maintained (as Russia does today) a disproportionate superiority in nuclear weaponry, and the sharpness of the Chinese threat to the Soviet Far East was dulled by this reality. The Sino-Japanese *rapprochement* of the 1970s was aimed primarily at the Soviet Union, but also, in Japan's case, toward the United States, whose sudden change in policies toward China resulted in a grave 'shock' to leaders in Tokyo. Therefore China was seen as a potential partner for both Japan and the Soviet Union during the Cold War. In both cases, at different times, China *did* become a partner for the two nations.

During the 1990s, however, China's role took on an entirely new form. China's economy had undergone a tremendous transformation beginning in the 1970s. After the death of Mao Zedong in 1976 and the accession to the top of the Chinese Communist Party of Deng Xiaoping in late 1978, China began the longest sustained period of high economic growth by any country in modern history. The Chinese economy averaged a GDP growth rate of roughly 8 percent for three decades (China's GDP growth rate in 2006 topped 11 percent). China's political and military power increased proportionately with the economy. By the mid-1990s, with the collapse of the Soviet Union, China was clearly becoming the strongest military power in East Asia, apart from the United States. Politically, China had emerged as a power broker, not just in East Asia but also throughout Eurasia. Since 1971 China has had a permanent seat on the UN Security Council. Its policies are felt in Central Asia and the Middle East. In the early years of the twenty-first century the Chinese government began carrying out diplomatic charm offenses in Africa and Latin America. In the early- to mid-1990s China was involved in negotiating disputes in Cambodia and on the Korean Peninsula, whereas Japan and Russia had only minor roles in settling these disputes. The United States, whose prestige and standing in Northeast Asia were perhaps somewhat diminished in the eyes of some, was still the major power broker in that region in the 1990s and into the 2000s. Leaders in the United States began to view China as the emerging key power in the Asia-Pacific region in the twenty-first century. Japan and Russia were concerned about being overlooked in Northeast Asia. China's economic, military, and political rise threatened to entirely eclipse Japan and Russia, not only in Northeast Asia but also in the entire Asia-Pacific region, and in Central Asia. The idea of a Chinese 'threat' gained credence in both nations (as well as in the United States). This was a big change for the leaders and the peoples of both nations, because an independent Chinese threat had never been an issue for Japan or Russia during the entire twentieth century.

A number of analysts feel that the 'China Factor' was perhaps the single most important issue driving Japan's new policy initiatives toward Russia beginning in 1996–97.[1] China's rapid economic development and emergence as an economic competitor to Japan caught many in Tokyo by surprise. As recently as the early 1990s, experts around the world spoke of Japan as the emerging Asian (and global) power that would stamp its mark on the twenty-first century. No longer does one hear stories of Japan leading the 'V-formation' of Asian geese flying to economic greatness. People now understand that China, not Japan, will be the new leader in Asia in the twenty-first century. Japan, meanwhile, often feels left out of the global and regional political agenda. As mentioned, China has a permanent seat on the UN Security Council, something Japan covets greatly.[2] Economic retrenchment, continued domestic political shake-ups, and a performance during the 1991 Gulf War that was ridiculed around the world, served as reminders to the Japanese leadership in the 1990s that Japan is still a political lightweight on the world stage. Japan recognizes that China will play a major – if not *the* major – role in matters concerning Asia's political and economic agenda.

Fear about China's rise has been present in Japan since the 1980s, but this fear truly gained credence as Japan's economic troubles began in the early 1990s.[3] Although few Japanese officials ever publicly claim that China is a potential threat, discussions with Japanese officials on the subject of Japanese-Russian relations invariably turn to China. A series of events beginning in 1994–95 began to test the limits of Sino-Japanese friendship, and forced Japanese leaders to sit up and take notice of the new strategic environment in East Asia.

In 1994 Japan's Prime Minister Morihiro Hosokawa began criticizing not only Beijing's nuclear testing but also the dramatic growth of China's defense establishment. In 1995 the first set of offshore missile tests conducted near Taiwan by the Chinese People's Liberation Army (PLA) drew the attention of Japanese political and military leaders. Continued nuclear testing in China through 1995 caused the Japanese government to announce that it would suspend grant assistance to China, and it threatened to suspend the larger and more important Overseas Development Assistance program if China continued testing. During subsequent bilateral negotiations on a five-year concessionary loan program (1996–2001) the Chinese government requested roughly $15 billion. The Japanese government countered with a three-year package of close to $6 billion.[4] That same year (1995) in a speech to the Diet, Japan's Foreign Minister Yohei Kōnō mentioned China as a potential source of instability in the Asia-Pacific region. The next year Prime Minister Hashimoto took Kōnō's statement one step further and expressed concern that China's policies "might be heading in the wrong direction."[5]

Perhaps the biggest wake-up call for Japan to the potential for strategic instability was the March 1996 Taiwan Strait missile crisis. Suddenly, being drawn into an East Asian conflict between two giants no longer seemed like such a remote possibility. US troops stationed in Japan had been put on alert during the crisis, and the aircraft carrier USS. *Independence*, home ported in Yokosuka, Japan, was dispatched to Taiwanese waters. The Taiwan crisis had touched a sensitive spot for many Japanese politicians who were sympathetic to Taiwan, and it was complicated by the Sino-Japanese contention over the Senkaku/Diaoyutai Islands in the East China Sea north of Taiwan.[6] The 1996 Japanese Defense Agency White Paper made note of China's potential for creating instability. This was the first time China had been specified as a potential threat to Japan's security. During the Cold War, the Soviet Union had occupied the central position in Japan's Defense White Paper. Now that the 'Northern threat' had gone away, the Defense Agency had begun closely monitoring Chinese actions.

> The situation [with regard to China] must be watched with caution in terms of the promotion of nuclear weapons and the modernization of the navy and air forces, expansion of naval activity, and heightened tension in the Taiwan Strait as seen in the military drill near Taiwan.[7]

In the spring of 1996 tension in Sino-Japanese relations was becoming clear. Not coincidentally, this is when Japanese leaders began pursuing a new relationship with Moscow.

Friction between China and the United States in the late 1990s further put Japanese leaders in an awkward position, as they found that they might one day be forced to choose between Japan's closest ally and its increasingly powerful neighbor. As dangerous as a US-Chinese conflict may be for Japan, US-Chinese collusion is also a source of Japanese concern. Many analysts in Japan feel that any warming of relations between Washington and Beijing could be in direct conflict with the interests of Tokyo. They see the triangular relationship between Beijing, Tokyo, and Washington in terms of a 'zero-sum' game. The analysts point out that during the twentieth century whenever the United States has become close with either China or Japan, relations with the other party have suffered. They fear that any subsequent warming of ties between the US and China will inevitably spell bad news for Tokyo.[8] The highly publicized June 1998 visit by US President Bill Clinton to China is a case in point. The Japanese press was awash with stories about 'Japan-passing' (as opposed to the 'Japan-bashing' of the 1980s). This refers to the United States 'passing' over Japan to consult with China on various issues (such as a Korean Peninsula settlement). Clinton literally did just this on his China trip, spending almost a week in China, while bypassing Japan. An editorial in the *Sankei Shimbun* suggested that Clinton's itinerary represented a loss of confidence in Japan by the United States. To add insult to injury, both the United States and China issued calls for Japan to get its economic house in order.[9] Prime Minister Hashimoto later reminisced that he had been extremely unsettled with the Japan bashing in the United States in the early years of the Clinton administration, and when Clinton later bypassed Japan on his trip he felt that the United States was acting 'two-faced' and simply using Japan as a pawn.[10]

Hardest for many Japanese to swallow has been China's economic rise, even though Japanese companies have profited handsomely from China's spectacular growth (two-way trade by 2006 had reached $200 billion annually).[11] A commission brought together by the LDP in 1997 warned that China's economy would overtake Japan's economy by 2010.[12] Japanese leaders watch with a certain angst China making bold policy initiatives in the Middle East, Central Asia, and in Siberia – all key energy producing regions. For example, in 1997 China announced a $9.5 billion investment deal with the government of Kazakhstan to help develop the potentially rich Ozen oil field and to construct a 3,000-kilometer pipeline into Western China. China and Russia have also reached tentative deals to construct pipelines linking Western Siberian oil and gas fields to a Chinese grid.[13] In a crisis, Japan could find itself deprived of Middle Eastern energy sources, only to find that the spigot for alternative sources is controlled by Beijing. Japan's reliance on sea lines of communication (SLOCs) puts it at the potential mercy of China in times of crisis, as a large portion of the vital Persian Gulf-Japan SLOC skirts China. Given this vulnerability China's growing naval expenditures worry Japanese leaders. In addition, China's growing consumption of oil could potentially divert sources from flowing into Japan, even in times of peace, or drive world prices skyward. This has been cause for alarm in some circles in Japan.[14] A report issued by the *Keidanren* in 1994 not only warned about China's growing energy consumption, but also of the potential for dire environmental side-effects.[15]

Competition between Tokyo and Beijing over the Russian ESPO pipeline heated up in early 2003 (as alluded to in Chapter 3). Starting with a visit to Moscow and Khabarovsk by Japanese Prime Minister Koizumi in January 2003, Japanese negotiators began offering incentives to get Moscow to change the planned route of an oil pipeline from the Siberian oil field of Angarsk.[16] In the late 1990s an agreement was made with the Chinese to build the pipeline to the Chinese city of Daqing. Business and political leaders in Japan made known their desire to have the pipeline bypass China and terminate at Nakhodka or Perevoznaya, just south of Vladivostok, from whence Russian oil could be shipped to Japan and elsewhere in the Asia-Pacific region. A pipeline to Nakhodka would cost at least two to three times as much as one to Daqing would. Japan has offered to finance the deal through the receipt of oil shipments. To make the Russians take this offer even more seriously, the Japan National Oil Corporation offered to invest more than $1 billion to help with infrastructure development in the Russian Far East.[17] After Koizumi's visit to the Russian Far East, Foreign Minister Kawaguchi also visited Russia on another occasion in 2003 to promote this idea, as did numerous other officials. Japan's effort was somewhat clumsy (economically it does not yet make sense), but it did interest President Putin enough for him to personally recommend in 2003 a re-examination of the existing agreement with China.[18] Chinese leaders were apparently taken aback by Japan's aggressive strategy, and the Chinese media let it be known that Sino-Russian and Sino-Japanese relations would suffer a big blow were Russia to agree to a Pacific route.[19] Although the Kremlin went back and forth on this issue numerous times during 2003–05, the Russian government began construction of the ESPO pipeline in 2006, and hopes to feed both China and Japan with the resources that are yet to be extracted. A series of newspaper articles in the Japanese daily *Yomiuri Shimbun* in January 2007 highlighted Japan's insecurity vis-à-vis China in the scramble for energy in the Russian Far East.[20]

Energy dependence has been a thorn in Tokyo's side since the beginning of the modernization drive in Japan in the late nineteenth century. One need only look to Japan's expansionist drives into Southeast Asia in the 1930s and 1940s to understand this. Japan has also been forced recently to reconsider expansion plans for its domestic nuclear energy industry, which was developed to a high degree in the wake of the oil crises of the 1970s. Reactor breakdowns, waste disposal issues, and a growing public discontent with the industry could account for further reliance on oil and gas imports. Taking all of this into consideration, Japan has launched its own energy initiatives in Central Asia, and in Siberia and the Russian Far East. Hashimoto's speech to the *Keizai Dōyuukai* in July 1997, urging a new Russia policy with a heavy emphasis on energy development, came on the heels of the announcement by Beijing and Moscow that they would jointly develop pipelines linking the two countries. It was no coincidence that the push for normalization with Moscow came during the mid-1990s, as ties with China were fraying.

The role of history also came to cast a shadow on Sino-Japanese relations. During the late 1960s and early 1970s Mao Zedong and the Chinese leadership had kept the issue of Japan's Second World War guilt under the carpet so as to not alienate the Japanese leadership, and to line up Japan in its anti-Soviet campaign. In fact, Chinese leaders had neither forgotten nor forgiven Japan for its wartime transgressions in China. The issue of war guilt entered into the bilateral dialogue in 1978, ironically the same year Japan and China signed a Peace and Friendship Treaty.[21] In the 1980s and 1990s Japanese politicians drew flack from around East Asia for the increasing frequency of remarks (or 'blunders') about Japan's role in the Second World War. Ryutaro Hashimoto, chairman of the Japan War-Bereaved Families Association, (not yet Prime Minister at this time) was quoted as saying in 1994 that it was a "matter of definition whether Japan committed aggression against Asian nations during the Second World War."[22] In 1996 Prime Minister Hashimoto also visited Yasukuni Shrine, the site where Japan's war dead, including convicted war criminals, are memorialized. Japanese leaders (through Koizumi and his successor Shinzō Abe) have continued making regular visits to Yasukuni Shrine, in defiance of nations across Asia that were victimized by Japanese aggression. China has been the most critical nation about Japan's refusal to atone for the past wrongs it inflicted on its neighbors during the war. In his 1998 state visit to Japan, Chinese President Jiang Zemin repeatedly brought up the issue of a war apology, embarrassing his hosts, who were seemingly tongue-tied in response. The two sides argued bitterly over the language of a joint declaration that was meant to clarify Japan's remorse over the events of 1931–45 in China. In the end, the Japanese government refused to issue an apology for the war, and with Jiang's departure the issue was left unresolved.[23]

Chinese leaders began questioning Japan's military 'build-up' in response to US calls in the mid-1990s for Japan to take on a larger share of the regional security burden. An April 1996 report in the *Renmin Ribao* voiced concern that the

US-Japan security relationship was making a "qualitative change" in its primary function from a defensive to an "offensive" alliance.[24] Another article in early 1996 (before the April US-Japan security declaration) suggested that improvements in the Japanese Self-Defense Forces (SDF), particularly the Maritime SDF, give Japan the capability to project military power around the region. It added that, "[China] cannot preclude the possibility that Japan might try to resolve disputes over economic and maritime interests in Asia by force. As the Japanese SDF gets more involved in Asian security, Japan will steadily expand its sphere of influence through military means."[25] Chinese leaders made it known that they were not happy about Japan's proposal to help the United States develop a regional theater missile defense system (TMD). Chinese leaders recognize that such a system could render China's nuclear arm obsolete. Also of paramount concern to Chinese leaders is Japan's increased role in the framework of the US-Japan security treaty. In 1995 Japan revised its National Defense Program Outline (NDPO), and in 1996 issued a joint security declaration with the United States that included a call for new bilateral defense guidelines, in which a strong emphasis was to be placed on Japanese participation in regional crises. China's ire was raised in 1997 when the LDP's Chief Cabinet Secretary Seiroku Kajiyama announced in a televised speech that if a military confrontation were to erupt over Taiwan, the new defense guidelines would enable Japan to assist the US forces engaging Chinese forces. Prime Minister Hashimoto visited China one month after Kajiyama's statement, and received a wave of indignant protests. Hashimoto calmly insisted that the defense guidelines have no geographical definition (purposefully leaving the issue of Japanese support vague). Ironically, Hashimoto reportedly was concerned about growing US-Chinese defense ties and hoped to bolster Sino-Japanese defense contacts with his trip.[26]

In the 1990s Chinese leaders also took a great interest in the Japanese-Russian *rapprochement*. Premier Li Peng, in response to the Hashimoto-Yeltsin summit at Krasnoyarsk, said, "It seems Japan's diplomacy has become diversified." Wary of growing Japanese-Russian defense contacts, Chinese Defense Minister Chi Haotian was dispatched to Tokyo in early 1998 to boost Sino-Japanese defense contacts. The Chinese leadership was also concerned about Hashimoto's diplomatic initiatives toward the United States and Russia, and hoped to pull Japan closer to the Chinese camp, or at least keep it from getting too close to Russia. A Japanese report quoted a foreign diplomat in Beijing as saying that, "It seems that China is not just restraining Japan, but also Prime Minister Hashimoto himself, who has gotten closer to Russia."[27]

Leaders in Japan were growing concerned about improving Sino-Russian ties, particularly the fact that Russia was selling to China advanced, high-technology weapons systems at seemingly bargain prices. Hashimoto voiced this concern to foreign leaders on more than one occasion.[28] As mentioned, at this time Japanese pilots were in Russia training on the same type of fighters that China was purchasing from Russia (see Chapter 3). Japanese leaders did not want to see the re-emergence of a strong Sino-Russian alliance aimed at the US-Japan security treaty. As relations between the United States and China, and the United States and Russia, worsened during the late 1990s (particularly after NATO expansion, the Taiwan Strait crisis, and the bombing of Yugoslavia), Japanese leaders felt somewhat at the mercy of larger forces, less predictable than they had been during the Cold War.[29]

Various groups in Japan in the 1990s came to view improved relations with Russia as critical for a variety of reasons. One prominent Japanese academic, Shigeki Hakamada, describes Russia's role in Sino-Japanese relations as that of a potential bridge between Tokyo and Beijing. As Sino-Russian and Japanese-Russian relations improve, he argues, Russia might be able to convince leaders in Beijing to support Japan's bid for a permanent seat on the UN Security Council.[30] Leverage with Russia has not always had the desired effect. In 1998 Prime Minister Hashimoto unsuccessfully proposed to China a four-way Northeast Asian security summit involving the United States and Russia. When this failed the Japanese government looked to utilize its support for Russian APEC membership to get Russian leaders to convince China to agree to such a summit. Moscow was unable to convince Beijing on this occasion.[31] Others in Japan view improved relations with Russia through a geopolitical lens. A move by Tokyo to cement a Japanese-Russian *rapprochement* could be seen as a hedging policy to balance against an increasingly powerful China, and against an improvement in US-Chinese relations, according to one Japanese analyst. Japan's alliance with the US reassures Tokyo, but US policy could change gradually, or even dramatically, with the accession of a new administration. The level of the US military commitment to the Asia-Pacific region, though stabilized at present, could conceivably be drawn down further. This partly explains the interest of Japanese leaders in the 1990s in expanding political, economic, *and* defense contacts with Moscow.[32]

Officials in Tokyo, especially in the Foreign Ministry, are hard-pressed to admit the role of a Chinese 'threat' in the formulation of Russia policy. Foreign Ministry officials usually bring China into any discussion on Japanese-Russian relations, though it is couched in non-confrontational terms. Minoru Tamba (former Ambassador to Russia) insists that the 'China factor' in Japanese-Russian relations is much more important for Russia than for Japan. Mutsuyoshi Nishimura (former Director of the Foreign Ministry's European and Oceanic Affairs Bureau) feels that basing Japanese-Russian relations on the idea of a China 'threat' would be foolhardy. Besides drawing the ire of China, it also presupposes knowing how China will act in the future. One should be careful about China, he added, because it is still in a transitional phase from a party dictatorship to some other form of governance. He also suggested that it is impossible to estimate what China will be like in even five to ten years. Kenji Shinoda (also of the Russia School in the Ministry of Foreign Affairs) stated that Japanese-Russian energy cooperation could help to ease China's growing energy demand. Japanese companies, he pointed out, can profit by helping to construct pipelines linking China and Russia.[33]

Other agencies and ministries in Japan are less sure of the potential for co-operation with China in the future. A secret report written in 1995 by the Japan Defense Agency was much more explicit on the implications of the rise of China. It predicted that in 20 years time, Korea would probably be reunified, while the Taiwan issue would remain unresolved. The United States military presence in

East Asia, the report stated, would likely be scaled down because of a withdrawal from Korea, while China's economic and military expansion would continue. The report urges greater defense cooperation with the United States and Russia in response to such trends.[34] However Japanese officials may differ in their policy statements, the 'China factor' undoubtedly weighs heavily on their minds. A secret internal document issued by the Foreign Ministry in 1997 (but only made public in 2002) clearly indicated the link with China in bilateral relations with Russia, declaring that the *rapprochement* was part of Japan's "China strategy."[35] Once he had resigned his post as Prime Minister, Hashimoto was much less guarded in his opinion about the role of China in the *rapprochement* with Russia. In an interview with the journalist Yoichi Funabashi in August 1998, Hashimoto stated, "We have to make Russia an Asian player. We must make Russia Japan's ally. We don't want China and India fighting for supremacy in Asia in the twenty-first century. That's why we need to keep Russia as a balancer."[36]

In Russia the concept of a *rapprochement* with Japan is deeply rooted in China's growing power. Moscow clearly desires a strong relationship with Beijing. Although relations with China were stepped up in 1993, leaders in the Kremlin began pursuing a 'partnership' with China in earnest in 1996 when Evgenii Primakov became Foreign Minister. After four Jiang-Yeltsin summits in 1996–98 the two sides hailed the new 'strategic partnership.' The 2004 treaty signifying the resolution of the Sino-Russian border dispute was truly a milestone for Sino-Russian strategic relations. Yet Russian leaders (especially Putin) have been careful to balance relations with China by improving relations with Japan (and with other nations in Asia). Even while China was weak before the Second World War, Russian/Soviet leaders always wanted secure relations with Japan while dealing with China.[37] Primakov was able to maintain good relations with both Beijing and Tokyo, when he served as Foreign Minister and then as Prime Minister for three years from 1996 to 1999. By the end of Yeltsin's term in 1999, when Primakov was no longer in office and was in fact in the opposition camp, this balance became somewhat skewed toward China (dangerously so, according to some in Russia at the time). Accordingly, Vladimir Putin initially put some distance between himself and Chinese leaders in order to rectify this imbalance.[38]

One Russian author writes, "We should in no way idealize our relations with China, nor should we simplify them."[39] In other words, Russia's relationship with China is very complicated and cannot be oversimplified, as many observers in the United States are often wont to do.[40] A variety of factors weigh in Russia's decision-making process vis-à-vis China. Russian leaders are now forced to consider regional political forces in the Russian Far East, most of whom are extremely concerned about the growing economic and political clout of China. China presents a wide array of both opportunities and dangers for Russia. The country is potentially Russia's greatest trade market. Russian weaponry right now dominates two-way trade, but energy exports to China are fast becoming a leading source of Russian export income. At the same time, fears of Chinese political and economic domination of the Russian Far East are never far below the surface. Chinese immigration (legal and illegal) into the Russian Far East has been a sore topic for Moscow and Beijing.

In a 1998 visit to Moscow Chinese Premier Li Peng had to continually address the issue of Chinese immigrants, not just to the Kremlin, but to the Russian press as well.[41] In July 2000 in the border city of Blagoveshchensk Vladimir Putin warned that if economic and social conditions in the Russian Far East are not improved, residents of the region might be speaking Chinese, Japanese, or Korean in future generations.[42] Partially in reaction to fears in the Russian Far East about illegal Chinese immigration, the Russian government passed a law in early 2007 limiting the number of foreign workers operating in Russian markets to 40 percent.[43]

Beginning in 1991 leaders in Moscow and Beijing signed a series of agreements that resolved outstanding territorial issues (including over 700 islands in the Argun, Amur, and Ussuri Rivers) that had been the cause of border clashes in 1969.[44] Three of the larger disputed islands in the Ussuri and Argun Rivers were left in Russian hands, but were eventually divided amicably between the two governments at the end of 2004, thus resolving a fluid border situation which had kept the two nations at odds. Opposition to the border agreements was strongly voiced in the Russian Far East. Governors from Khabarovskii Krai, Primorskii Krai, and Sakhalinskaya Oblast vigorously denounced the agreements (the same group of governors who denounce Japan's 'pretensions' on the southern Kuril Islands). Local officials and experts, concerned about Chinese demographic encroachment, have contributed to the denunciation, arguing that Chinese traders in the region are devious, their goods shoddy, and the cultural differences too great to be bridged. Immigration should be halted, they argue. China, they argue, is not the key to the reconstruction of Russia's Far East, it is a threat.[45]

Russian defense officials, while eager to sell sophisticated weaponry to China, are perhaps most concerned about China's growing power. Russian military officers try to put a good face on the 'strategic partnership.' Bilateral meetings between defense officials are conducted regularly, and the jovial atmosphere conveys images of the Sino-Soviet alliance of the 1950s. Many high-ranking Chinese defense and government officials were trained in the Soviet Union and speak passable Russian. In fact, Chinese President Jiang Zemin addressed the Russian Duma in Russian during his 2001 trip to Moscow. Meanwhile, Russian defense manufacturers that sell weaponry to the Chinese military insist that the weapons are never top of the line. Nevertheless, the element of the 'China threat' is never far from the surface. Occasional slips happen, such as when Russian Defense Minister Igor Rodionov in a speech in late 1996 referred to China as a "potential threat." Russian air force and naval officers grumble that Chinese officers are manning weapons systems in better repair than their own. New systems are being sold overseas because the Russian government cannot afford them. Even with the booming oil prices of the period 2003–07 available funds go into badly needed infrastructure projects and the so-called stabilization fund, whose purpose is still less than clear. Though not as beleaguered as during the 1990s, the Russian military is still in poor shape and the best weapons systems are being exported. Intelligence leaks to the Chinese are numerous and many Russian scientists are reportedly relocating to China to help establish a Chinese weapons manufacturing base.[46]

Some leaders in Moscow, however, feel that relations with China are a key to preventing a unipolar American domination of the world. Evgenii Primakov was one of the foremost advocates of this view. Primakov recognized that potential problems with China exist, as evidenced by his desire to shore up relations with Japan. But, NATO's eastward expansion in 1997 to include former Soviet satellite states, and the 1999 bombing of Yugoslavia (against Chinese and Russian protest), further pushed Russia and China together on the strategic front. Near the end of his presidential administration in 1999 Yeltsin was in fact seen to be leaning perhaps too heavily toward China.[47] Worsening relations between Moscow and Washington following the 2003 Iraq war strengthened the arguments of those in Moscow who look to balance against the United States with the help of China. Nevertheless, Russian leaders have been continuously concerned about the future of Russia's Far Eastern regions. Duma member Aleksei Arbatov wrote, "China is potentially the most serious direct external problem for Russia's security."[48] Russian leaders face a serious dilemma. While the role of China in Japanese-Russian relations is termed by Japanese policymakers as the 'China factor,' Russian leaders are more apt to refer to it as the 'China dilemma.'[49] Thus, the nature of Russia's relations with Japan consists partially of a strategic balancing factor, involving not just China but also the United States. Vladimir Putin seems to have grasped this almost immediately upon assuming office early in the year 2000. The stepped-up summitry between Moscow and Tokyo was noticed in Beijing, and Chinese leaders began to feel somewhat slighted. In a nod to China, Putin did oversee the signing of a Sino-Russian Treaty of Peace and Friendship in the summer of 2001, and the formation of the Shanghai Cooperation Organization (SCO) in 2000. Additionally, the 2004 Sino-Russian border agreement was an epochal agreement. Putin, however, has been careful to keep a very keen balance in his Asia policy, not favoring either China or Japan, and increasing contacts with both North and South Korea, and India as well.[50] Some leaders in both China and Japan, however, feel that Putin is too focused on Europe and the West.[51]

Analysts in Russia recognize that Japan is a potential partner whose help may one day be needed to balance against China. Perhaps in answer to those who say that China is the key to the future of the Russian Far East, Aleksei Bogaturov of the Academy of Sciences argues that the key to the political stability and economic health of the region lies just south of the Kuril Islands – Japan.[52] Local officials in the Far East concur with this assessment and add that they also desire a strong American and Korean economic presence in the region, as well.[53] The feeling in Moscow is no different, though officials are careful not to directly refer to China. In a talk with a group of Asia experts at IMEMO in 2001, Deputy Foreign Minister Aleksandr Losyukov said, "In Russia there is the recognition that today we [Japan and Russia] need one another like never before." Former Deputy Speaker of the Duma Vladimir Lukin argues that Russia needs strong relations with Japan to deal not only with China but also the United States. Mikhail Galuzin, Former Director of the Russian Foreign Ministry's Second Asia Department (Japan and Korea) simply says that it is "strategically important" for Russia to resolve its differences with Japan. Vladimir Pushkov, chief of the Foreign Ministry's Japan desk, concurred that a Japanese-Russian *rapprochement* is necessary for strategic reasons.

In geopolitical terms it is clear to both sides that they need each other, he said, and the fact that leaders on both sides recognize this is a sign of clear progress in bilateral relations. But, he added, in Russia the primary need for good relations with Japan is economic, not geopolitical.[54] A majority of academics and Japan experts in Moscow agree that in the future China will be a 'problem' for Russia, and that both the United States and Japan need to be involved in the Russian Far East.[55] Given these trends and the statements from officials and experts there seems little doubt that Japan has come to play a major role in Russia's 'China dilemma.'

For both Japan and Russia, China's rise presents a host of new issues. As Sino-Russian and Sino-Japanese relations improve, leaders in Moscow and Tokyo will come to understand that the third leg of the triangle (Japanese-Russian relations) offers tangible political, economic, and strategic benefits in their respective relationships with China and the United States. Furthermore, some day leaders in Japan and Russia may find that the Kuril Islands territorial dispute pales in comparison to what might be future confrontations in Northeast Asia. Gilbert Rozman, a top US scholar on Northeast Asian issues, has pointed out that a Japanese-Russian accommodation "does not mean that ... [these] ties are targeted against China. Rather it indicates that in the shadow of a strong, rising power its principal neighbors find it advantageous to consolidate their own ties, maximizing their ability to work together."[56]

The role of the United States

As with China, the role of the United States in Japanese-Russian relations during the 1990s and early in the twenty-first century took on a new form. Prior to the Second World War, the United States earned the enmity of Japan and the Soviet Union/Russia in Northeast Asia, thus serving to help bring the two together on different occasions (see Chapter 1). During the Cold War Washington played what many claim was an obstructionist role, preventing Moscow and Tokyo from normalizing relations. Beginning in the 1990s, however, US leaders actually encouraged a Japanese-Russian *rapprochement*. In addition, certain US actions in the Asia-Pacific region, in Europe, in the Middle East, and in Central Asia have given cause to officials in Japan and Russia to reassess their bilateral relationship.

As the sole remaining superpower in the world after the dissolution of the Soviet Union in 1991, the United States was able to carry out an agenda around the world much to its liking. Three regions that drew particular attention in Washington were East Asia, Europe, and the new states of Eurasia. In line with the stated aim of the Clinton administration to spread the ideals of liberal democracy and free trade throughout the world, US officials saw the newborn democracies of Eastern Europe (including Russia) and the newly independent states of Central Asia as the ideal candidates for economic and political reform. Certain East Asian countries were held up as models. Nations such as South Korea, the Philippines, Taiwan, and Thailand had thrown off the chains of dictatorship and were experiencing dizzying economic successes (though not necessarily in this order). The idea was to pass on this bit of wisdom to motivate the newly emerged democracies and 'republics' around the Eurasian periphery. The United States strongly encouraged Japan to help with the political education and economic reconstruction of these new states. On several occasions President Bill Clinton personally delivered this message to Japanese Prime Minister Hashimoto.[57]

In East Asia the United States was resolved to remain politically active after the Cold War. US trade with the nations of East Asia reached new heights in the mid-1990s. In 1994 the US engaged in more than $435 billion of trans-Pacific trade. In 1995 trade with East Asia accounted for 37 percent of the total US foreign trade figure (compared to 18 percent with Europe).[58] Meanwhile, US officials began to downsize the US military presence in the region. In 1990 there were 135,000 troops stationed in the Asia-Pacific region east of Hawaii. Following a withdrawal from bases in the Philippines in 1992 (following a Philippine government decision) this number shrank to 100,000. The 1994 Korean nuclear crisis convinced US policymakers to hold firm at the level of 100,000 troops. The 1996 Taiwan Strait missile crisis gave further motivation to US leaders to remain militarily engaged in the region.

Japanese leaders, however, were growing nervous about the US commitment to the region by the mid-1990s. Not only was the United States demonstrating strategic ambiguity, but US leaders were also seen by many Japanese to be acting unilaterally, and often not in Japan's interests.[59] US-Japanese relations were in a state of unease when the Taiwan Strait crisis erupted. Trade friction had come to dominate the US-Japanese relationship since the mid-1980s, threatening to damage security ties. Up to this time the US-Japan security alliance had seemed to be on permanent automatic pilot. Now that the rationale for the alliance (the Soviet threat) had disappeared, defense planners in both nations began to redefine their relationship. With no clear threat to steer security relations, and increasing acrimony in trade relations, it was suggested by some that the alliance had gone "adrift."[60] US strategic planners had been trying for years to get Japan to play a more active role in regional and global affairs. In the mid-1980s, the Japanese had agreed to patrol the sea lanes of communication (SLOCs) in a rough triangle between Japan, the Philippines and Guam. In the event of a Far Eastern contingency directly affecting the security of Japan, however, there was no basic plan on how US forces would be supported by the Japanese. As mentioned, Japan's slow response in agreeing to back up US and coalition forces during the Persian Gulf War in 1990–91 angered many US politicians. Many in the United States began to question what they saw as a security 'free-ride' for the Japanese. Meanwhile Japan was perceived by many in the United States to be a primary long-term economic threat. Many US leaders bemoaned the loss of American economic competitiveness, while Japan-bashing became fashionable among American blue-collar workers, politicians, and academics alike. US-Japanese relations seem to have reached a low point in the mid-1990s, highlighted by the Clinton administration's intense pressure on Japan to negotiate 'results-oriented' trade agreements.

In Japan the rape of a young schoolgirl by three American soldiers in Okinawa in September of 1995 brought to the surface latent grass-roots opposition to the US military presence in Okinawa (where fully 60 percent of US forces in Japan were stationed). Many in Japan began to wonder whether they needed a 'mercenary force' to protect their interests. There was concern expressed in some circles in Japan that the Clinton administration was backing away from its support for Japanese claims to the 'Northern Territories.'[61] Such conflicting signals, coupled with President Clinton's 'Japan passing' trip to China in 1998, left Japanese leaders concerned about the direction of the US-Japan security relationship.[62] It was in this context that Japanese officials began exploring new avenues of political and security cooperation with Russia.

Although Japanese leaders were unsure about US strategic intentions in East Asia, officials in the United States have made it very clear to Japanese leaders how they felt about Japanese-Russian relations. The first Bush administration had offered to mediate the territorial dispute, only to be politely told by Japanese officials that it was a bilateral issue. In spite of the offer to act as a mediator, the US maintained its position of support for Japanese territorial claims. In 1995 US Ambassador to Russia Thomas Pickering aroused anger in Russia when he announced that the disputed islands should be returned to Japan.[63] President Clinton repeatedly urged top Japanese and Russian officials to improve relations. Prior to the 1997 Denver G-7 summit meeting, President Clinton directly asked Prime Minister Hashimoto for Japanese support of Russia (including agreeing to a regular Russian seat at the G-7 meetings) to help ease Russian concerns as NATO was expanding. The United States was asking Japan to think strategically, and Hashimoto was receptive. "Japan had to improve relations with Russia because US-Russian relations in the mid-1990s warmed so quickly that a tough [Japanese] stance toward Russia no longer reinforced US-Japan security ties," wrote Michael Green, "indeed, it only increased the dangers of abandonment."[64]

Beginning in the mid-1990s US officials participated in a series of annual 'Track Two' meetings with mid-level Japanese and Russian political and military officials. This group was known as the Trilateral Forum on North Pacific Security, and all participants noted the positive atmosphere that these unprecedented gatherings created. Among the Japanese and Russian officials regularly attending were Minoru Tamba, Kazuhiko Tōgō, Aleksandr Panov, and Aleksandr Losyukov. Officials from the Clinton administration included the State Department's top Russia hand, Strobe Talbott, and its top Asia hand, Winston Lord. Also attending were Pentagon officials, and individuals who would become key players in the second Bush administration, such as Paul Wolfowitz (Deputy Secretary of Defense) and Richard Armitage (Deputy Secretary of State).[65] The United States had clearly become interested in a Japanese-Russian *rapprochement*. The second Bush administration's approach to Russia in 2001–02 as a partner in the anti-terror coalition was also a further signal that Washington wanted Moscow as a partner, and that Japan should help Russia when and where it could. The United States has been criticized by some analysts for its lack of initiative in the Japanese-Russian relationship, but actually US leaders have gone out of their way to bolster ties between Moscow and Tokyo.[66] Japanese leaders were understandably anxious to shore up relations with Russia, not only due to US urging, but also because of the perception in Japan that that the United States was displaying a new strategic ambiguity. At least one

Russian analyst, Mikhail Krupyanko, detected a strategy within Japan to use the Russia card against the United States, not China. Krupyanko reasoned that economic friction with the United States, as well as differences over strategy, were driving Japan closer to Russia.[67]

The Japanese government under Prime Minister Shinzō Abe became concerned in 2007 about emerging differences between Japan and the United States over various issues. Concerns about America's 'strategic drift' in East Asia were still prevalent among the leadership in Japan, especially given the US preoccupation with the Middle East.[68] The decision by the US government to reach a joint agreement in 2007 with Pyongyang without consulting Tokyo was seen as yet another case of 'Japan-passing.' The Japanese government was concerned not only about North Korea's growing strategic capabilities, but also about the much more personal issue of abduction of Japanese citizens by the DPRK.[69] Additionally, the proposed 2007 resolution brought forward by the US Congress concerning Japan's war guilt (specifically its role in forcing prostitution upon foreign nationals across Asia during the Second World War) further brought tensions into the relationship between Tokyo and Washington. Although Moscow and Tokyo see eye to eye on the North Korea issue, it is unlikely that Moscow would in any way support Japanese protestations about war guilt from the 1930s and 1940s, especially given the nature of the territorial dispute (also stemming from the Second World War).

Russian views of the United States are every bit as complicated as Russian views of China. Russia sees in the United States both a potential threat and a partner. But where Russians see a demographic, economic, and security threat emanating from China, the perception of the United States as a threat is couched purely in geopolitical terms. NATO expansion, NATO intervention in Yugoslavia, the US unilateral withdrawal from the ABM Treaty, US intervention in the Middle East, and the deployment of US troops to three of the former Soviet republics (Georgia, Kyrgyzstan, and formerly Uzbekistan) have been landmark events, which in the West are heralded as signposts of the Cold War's demise. For many Cold Warriors and conservatives in Russia, however, they are terrifying signs of a new American dominance. Even so-called 'liberal Westernizers' in Russia were concerned about the US dominance and the American penchant to meddle in Russia's traditional sphere of influence. Among these were two Duma members of Yabloko, a liberal Russian political faction, Aleksei Arbatov and Vladimir Lukin.[70] Russia's overtures to China are a direct result of the power and actions of the United States. Gilbert Rozman writes that in the Sino-Russian relationship "geopolitics are at the core and opposition to the emerging world order is the *raison d'être*."[71] Russian officials and various experts have claimed that the Sino-Russian 'strategic partnership' is not directed at any 'third country' (i.e. the United States).[72] But in fact, in his memoirs, Boris Yeltsin states explicitly that the 'strategic partnership' was aimed at the growth in American power. "We cannot allow all of the buttons of the remote control to be controlled by one hand."[73]

Leaders in Moscow, of course, recognized the danger of leaning too close to China. Former Russian Defense Minister Pavel Grachev stated in 1997 that, "if NATO goes East, we shall go East too [meaning both toward China *and* Japan]."[74]

Strengthening relations with Tokyo gives Moscow leverage with Beijing, which in turn gives Moscow leverage with Washington. In the emerging quadrilateral relationship Moscow is looking for equal leverage in all directions.[75] Significantly, Russian views of the US-Japan security treaty have changed in large part due to China's rise. During the Cold War the US-Japanese alliance was regularly denounced in Moscow. After the collapse of the Soviet Union there was a complete reversal in the Russian position. Today the US-Japanese alliance is seen as less of an issue in the context of the unipolar dominance of the United States. In his visit to Tokyo in 1997 Russian Defense Minister Igor Rodionov stated that the US-Japan security treaty was "necessary" for regional stability. Nevertheless, Russian officials have expressed concern about the proposal of a joint US-Japanese effort to develop a ballistic missile defense system. There is also concern about the ambiguous regional definition given to Japan's security responsibility outlined in the US-Japan defense guidelines. Russian leaders have joined Chinese leaders in expressing opposition to a missile defense system, and asking for a clarification of Japan's regional responsibilities.

Under Putin the Kremlin has declared strong opposition to US missile defense plans for Europe, but there has been nary a word of protest about an Asian-based system. Russian leaders recognize that a regional theater missile defense system in Asia, though potentially a springboard to a larger strategic system, is not as great a threat to Russia as it is to China. It is a matter of speculation how far Russia would go to support China against the United States on other strategic issues such as Taiwan, arms sales, and nuclear proliferation.[76] With the creation of the Shanghai Cooperation Organization it appeared that China and Russia were ready to collaborate to oppose US interests in Central Asia, but after September 11, 2001 Russia indicated that it did not completely oppose a stepped-up US political and military presence in the region (at least for the time being).[77] Subsequent SCO summits have brought Beijing and Moscow closer together on certain strategic issues in opposition to the United States, but there is still much wariness in Moscow about China's intentions in Central Asia.[78] Meanwhile Moscow hopes to use the Collective Security Treaty Organization (CSTO – which includes Armenia, Kazakhstan, Kyrgyzstan, Russia, Tajikistan, and Uzbekistan) as the premier security organization in Central Asia.

Many consider it unlikely that Russia would go so far as to damage relations with the West in order to appease China's aspirations in East Asia. Ironically, during the Cold War, it was China that tacitly supported the US-Japan security treaty, while the Soviet Union was in permanent opposition. Today, the specter of a Japan rearming under the protection of the American nuclear umbrella does not sit well with Chinese leaders. As Harry Gelman wrote, China and Russia are "in the process of trading traditional positions" on the issue of the US-Japan security treaty.[79] Russia's leaders recognized that maintaining an equal distance from each of the three other powers in Northeast Asia afforded the Kremlin the opportunity to maneuver more freely.

By the beginning of the twenty-first century Japan continued to lean heavily toward the United States, though at times leaders in Tokyo felt it necessary to undertake hedging strategies, including a turn toward Russia. At the same time, Russia also seemed briefly to be turning toward the United States as part of the anti-terror coalition, but leaders in Moscow knew advantages could be gained by maintaining good relations with both China and Japan. And even in the wake of deteriorating relations between Moscow and Washington in 2006–07, Russian leaders did not rush into the arms of China in search of a bulwark against the United States.

Korea re-emerges

In the analysis of quadrilateral relations one must not overlook another important player that factors into the Japanese-Russian relationship. By the 1990s the Korean Peninsula, once the locus of Japanese-Russian confrontation at the end of the nineteenth century, had again become a key factor in the bilateral relationship. In the late 1980s and early 1990s both North and South Korea had roles (however small) in the domestic political agendas of the Soviet Union and Russia, and also in the domestic and international policies of Japan. This peripheral role has also impacted the bilateral relationship between Moscow and Tokyo. Leaders in Moscow in the 1980s and 1990s saw in South Korea a potential partner that could diversify Russia's sources of economic assistance, particularly as Japan was seen as being intransigent over the territorial issue. Certain Japanese officials, concerned about Seoul's diplomatic initiatives toward Beijing and Moscow (and by the instability created thereafter in North Korea), hoped to create new dialogue with the leadership in Pyongyang (see Chapter 2). Both the Japanese and Russian diplomatic initiatives toward the Koreas were temporarily shut down by the North Korean nuclear crisis, and the South Korean economic crisis in the late 1990s. Early four-power talks (US-China, the two Koreas) on Korean Peninsula security issues excluded Japan and Russia, and when the Asian financial crisis hit South Korea, aid to Russia dried up.[80]

In the late 1990s the issue of Korea once again came to exert influence on Japanese-Russian relations. It was not as if leaders in Moscow and Tokyo viewed South Korea or North Korea as potential strategic partners. Instead, the emerging tense situation on the Korean Peninsula made leaders in Moscow and Tokyo aware of just how marginalized their nations had become politically in Northeast Asia.

By the late 1990s in Japan there was great concern for homeland safety once it was clear that North Korea had missiles capable of reaching Japan. North Korea test fired its first Nodong-1 missile into the Sea of Japan, halfway between the Korean Peninsula and the Japanese coast, in 1994. In the summer of 1998 the North Koreans tested a Taepo-dong missile by firing it in a trajectory over the Japanese archipelago. North Korean boats made repeated incursions into Japanese waters during the 1990s, prompting an armed Japanese response on one occasion (the first since the Second World War). In addition, fears of a massive influx of Korean refugees into Japan in the event of a war were well founded. In Russia, there were similar fears of North Korean refugees pouring into Russia's Far East, either as a result of war or the sudden collapse of the North Korean regime. Furthermore,

Russian leaders were concerned about the effects of an unpredictable war breaking out on the Korean Peninsula that could potentially involve large numbers of US and/or Chinese troops operating right up to their border, less than one hundred miles from Vladivostok.

As US, Chinese, North Korean, and South Korean negotiators sat down to discuss the sensitive issues of security on the Korean Peninsula in the mid-1990s, it was immediately apparent that Japan and Russia were not welcome at the table, in spite of the vital security link to their homelands. In the case of Japan, Japanese money was welcome, but not a diplomatic presence. Russia simply no longer factored into the equation in Northeast Asia, politically or economically, in the mid-1990s. The idea to include Japan and Russia in the four-party talks had been broached by the United States, but South Korea made it clear that it was not interested. Leaders in Seoul did not want to give North Korea the opportunity to play either a Japanese or a Russian 'card' at the table.[81] Neither Moscow nor Tokyo was happy with this situation. Furthermore, after the August 1998 Taepo-dong missile incident, Japanese officials were angry because the United States refused to delay the final signing of the KEDO (Korean Peninsula Energy Development Organization) agreement to supply a nuclear reactor to North Korea in compensation for Pyongyang abandoning its domestic nuclear program. Japan had agreed to provide $1 billion, but wanted to hold up all assistance and punish North Korea for having carried out the missile test. The United States decided to go ahead with the signing and to carry out the agreement. Japanese officials let it be known they were dissatisfied with the arrangement.[82]

Japan's concerns about being excluded from Korean affairs are more than just about political marginalization. Japanese leaders know they can count on the cooperation and protection of the United States. US forces in Japan in a regional conflict, however, could turn out to be a double-edged sword. During the Cold War the US security blanket was a comforting device that allowed Japan to brush off fears of a strategic Soviet attack, and to concentrate on the business of rebuilding the country economically. In the event of war in Northeast Asia, however, whether on the Korean Peninsula or in the Taiwan Strait, US forces based in Japan could become a lightning rod. North Korean missiles, if fired in anger on Japan, would very likely be aimed at US bases, many of which are located in the densely populated Tokyo-Yokohama area. Reports of North Korean nationals arrested while infiltrating nuclear facilities in the Russian Far East have reinforced Japanese fears.[83]

Of perhaps greatest concern to Japanese leaders are the potential scenarios of Korean reunification. A war in Korea (or even a peaceful North Korean collapse) would have to be followed with a massive reconstruction effort. No doubt Japanese economic assistance would be requested. In 1997–98 in the wake of the South Korean financial crisis, the Japanese Ministry of Finance pledged $10 billion as part of an IMF package to help restructure the South Korean economy. Given the likely costs of reunification in the future, the Japanese government may be in no position to offer assistance. Of course, after reunification the new Korean government may or may not want Japanese aid. If they do request it, and Japan is seen as giving anything less than complete assistance, the already notorious image of Japan in Korea would suffer even more. On the other had, Korean leaders might refuse Japanese aid, damaging bilateral relations even more. When unification does come about, Korea eventually will emerge economically healthy, militarily strong, and politically united with a population close in numbers to Japan's population. Some long-range planners in Japan feel that a newly unified Korea will potentially be hostile to Japan.[84] The South Korean government is, in fact, every bit as worried as the Chinese government about Japanese intentions in the region in the future, especially given the hesitance in Japan to address wartime issues from the 1930s and 1940s.[85]

Japanese-Korean acrimony is more than just about Japan's colonization of Korea in the early twentieth century. The traditional enmity dates back centuries. In the fourteenth century constant attacks by roving Japanese pirate bands helped bring down the Koryō Dynasty in Korea. On two occasions in the late sixteenth century the Japanese Shogun Hideyoshi launched unsuccessful invasions of the Korean Peninsula. In the late nineteenth century, taking a lesson from US Commodore Matthew Perry (whose 'Black Ships' helped to open up Japan), Japanese warships forced open Korean ports to foreign trade. The 1910–45 colonization period was accompanied by a systematic Japanese attempt to wipe out Korean culture. Japan's refusal to come to a clean reckoning with the historical issue with Korea has left a festering wound on the bilateral relationship. Culturally, most Japanese have always recognized the greatness of China. But most Japanese view Koreans as poor cousins. The mutual dislike is deep-seated and visceral, whereas Sino-Japanese enmity really only extends back to the 1894–95 Sino-Japanese War. In this context it is not difficult to understand that some Japanese would come to the conclusion that a unified Korea will be no friend of Japan. Though relations had appeared to be improving in the late 1990s, by the early 2000s they were again acrimonious.[86]

Many Japanese leaders are concerned that the United States will one day disengage militarily from not only the Korean Peninsula but from East Asia as a whole.[87] When the day does eventually come Japanese planners recognize that strong relations with Russia will be a key to regional stability vis-à-vis a potentially hostile Sino-Korean bloc. Japanese leaders would like to see an active Russian role in the region to counter-balance China's growing role and, eventually, a hostile, unified Korea.

In the 1990s Russia thought it was being slighted in Korea, and felt that it was being pushed to the margin in Northeast Asian affairs. The failure to be included in the negotiations following the 1994 nuclear crisis highlighted Russia's strategic decline in East Asia following the collapse of the Soviet Union. As part of the KEDO agreement Moscow had suggested that it supply the nuclear reactors to North Korea. South Korea and the United States rejected this idea out of hand, though the North Korean government had supported it. In 1996 a top Korea expert in Russia wrote, "In this geopolitical environment, which for the time being is quite unfavorable for Russia, efforts are being made to isolate [Russia] from settling the key problems of the region."[88]

Much of Moscow's influence on the Korean Peninsula had been lost when Mikhail Gorbachev recognized Seoul and 'dropped' Pyongyang. Russia maintained relations with North Korea through the 1990s, but economic relations drastically fell off. In 1993 Russia announced that it was no longer bound to the defense assistance clause of the 1961 Soviet-North Korean Friendship Treaty. Yet during the 1994 nuclear crisis Russian officials argued forcefully against sanctions on North Korea. Whether or not US officials heard these arguments, sanctions were not imposed, and the crisis was settled peacefully. "In this sense," wrote Georgii Kunadze, "the resolution of the North Korean nuclear problem cannot be ruled out as a success for Russian policies, even though the final stages of the settlement were conducted without Russia."[89]

In the second Yeltsin term after 1996, Russia slowly extended a hand to North Korea, although it was clear Russia had little to offer. The 1961 Treaty of Friendship had expired in 1996, so in 1999 the two sides signed a new treaty, minus the mutual defense clause. Yeltsin was criticized by conservatives in Moscow that Russia's policy of neglect of Pyongyang had not only denied Russia leverage with South Korea, but it had also poisoned relations with one of the few nations in the world still desiring Russian machinery and spare parts for its Soviet-age factories. Meanwhile, Moscow's relations with Seoul were troubled by 1998. Not only had the economic crisis in South Korea cut off much trade and investment going into the Russian Far East (two-way trade fell from $3.2 billion in 1997 to $2.4 billion in 1998), but a spy scandal also poisoned political relations and a number of diplomats were expelled from both countries. This gave impetus to those in Moscow calling for stepped up relations with Pyongyang.[90]

Vladimir Putin was quick to heed calls to improve relations with North Korea. In July 2000, several months after his inauguration as president, he visited Pyongyang. Putin hoped to use this trip not only to reactivate economic and political ties with Pyongyang but also as a stepping-stone to a Russian re-emergence in Northeast Asian affairs.[91] Putin had reportedly discussed with DPRK leader Kim Jong-il a North Korean offer to desist from the development of ballistic missiles, in return for economic assistance in the form of free satellite launches. Putin hoped he had the key to disarming one of the American rationales for building a missile defense system. But no sooner had Putin arrived home (after the Okinawa G-8 summit) than Kim Jong-il announced his discussion with Putin should be taken as a "joke."[92] Though this last statement put into a whole new context the Putin-Kim meeting, it was clear that in spite of his image as a 'Westernizer' Vladimir Putin was actively seeking a new place for Russia in the East Asian region. Putin's visit turned out to be premature, as Russia has little to offer North Korea, but he demonstrated his ardent desire for Russia to become a player again in Northeast Asia.

In February 2001 Putin made another trip to the Korean Peninsula. This time Putin's destination was Seoul. In meetings with South Korean President Kim Dae-jung he stressed Russia's interest in economic cooperation (via a trans-Korean railroad linking up with the Siberian mainline). In August 2002 Putin met Kim Jong-il in Vladivostok and again stressed Russia's interest in cooperating with both North and South Korea in building a rail link to Siberia, thus linking the Asian and European markets by an overland route. North Korean support for the railroad was important so that the Russian Far East would not be bypassed by a

Sino-Korean rail route that would go through Mongolia and only enter Russia further to the west. Although a trans-Korea rail link would cost at least $3 billion, Russia has clearly evinced its desire to have a stake in the $600 billion annual trade turnover between Asia and Europe. Such a rail link could bring up to $1 billion in annual transit fees to Russia, roughly the same amount Russia receives from arms sales to China.[93] Japan also has shown great interest in the trans-Korea rail link to the trans-Siberian railroad, because leaders in Tokyo recognize that a Korean rail link to Russia (via Nakhodka and Vladivostok) has implications for the Siberian oil pipeline route. As one Japanese analyst has pointed out, were a trans-Korea railroad to be linked to a Chinese route via Mongolia, this would help realize the Taishet-Skovorodino-Daqing oil pipeline and would be a blow to those hoping to realize the ESPO pipeline.[94] By 2006 once the ESPO pipeline construction began, however, these fears appeared unfounded.

Putin has also reiterated Russia's desire to be involved in any Korean peace settlement process, explaining that Russia's close relations with North Korea would be of great benefit to the process. As one analyst writes, Russia may play the role of an 'outside balancer' in Korea, choosing to sway between a China-Korea combination on the one hand, and the US-Japan alliance on the other. Additionally, Russia could be the balancer of choice for a unified Korea that has chosen to remain equidistant from the other powers.[95]

Moscow has encouraged South Korean participation in the construction of a far eastern energy complex. Vladimir Putin visited Seoul in late 2006 and brought along a coterie of officials from the Russian energy industry. During his visit with Putin, Industry and Energy Minister Khristenko asked South Korean business leaders to join the ESPO pipeline project by advancing a loan to Transneft.[96] Russian Foreign Minister Lavrov visited Seoul in the summer of 2007 and also discussed energy cooperation. Korean business leaders are again beginning to see the potential for economic cooperation with Russia after the setback of economic relations in the 1990s.

Both Japanese and Russian leaders saw in Korea the manifestation of the respective political marginalization of their two countries.[97] Realizing their similar positions, leaders in Moscow and Tokyo have supported one another's bids to become further involved in the affairs of Korea. In 1998 Japanese Prime Minister Keizo Obuchi asked Washington to support six-power talks on the Korean Peninsula, stressing Japan's and Russia's desire to participate.[98] In a 2000 visit to Tokyo, Russian Foreign Minister Igor Ivanov stressed that Japan and Russia should have expanded roles in Korean security summitry.[99] This was also the case in the 2002–03 Korean nuclear crisis.[100] Russian officials have also urged their Japanese colleagues to increase bilateral cooperation in Korea, at one point urging Japanese financial assistance for North Korea.[101] After the 2006 DPRK test exploded a nuclear device, both the Japanese and Russian governments roundly denounced the test. Both governments continue to routinely support one another's position in the talks.

Russia's influence on the Korean Peninsula began to reawaken in 2007, and not just in the energy sphere. In order to restart Six-Party Talks and to freeze the DPRK's nuclear program, Pyongyang demanded that funds ($25 million) it held in a bank in Macao be unfrozen and transferred back to its own coffers. The United States government had placed a hold on these funds due to Pyongyang's intransigence in the nuclear talks. For four months a bureaucratic and political knot held up the transfer. Many banks balked at the idea of transferring the funds for fear of being associated with and tainted by this "dirty money" (it was frozen as a result of accusations of money laundering and illicit activities). At the eleventh hour the Russian government offered its services. In a roundabout way, the money was transferred from Banco Delta in Macao to the Federal Reserve Bank of New York, thence to the Russian Central Bank. In Russia, the Far Eastern Commercial Bank (Dalkombank) in Vladivostok agreed to transfer the $25 million back to Pyongyang.[102] This gesture was the most productive act that Russia carried out as a participant in the Six-Party Talks up to 2007.

Most of the analysis of Japanese-Russian relations has ignored the Korean factor in the bilateral relationship. The Korean factor, however, is important in understanding not only the wounded psyche of the Japanese and Russian people as to the political marginalization of their countries, but also in understanding how Japan and Russia hope to position themselves strategically in the future in Northeast Asia. As the Russian diplomatic journalist Andrei Piontkovsky writes, "It seems ... that the relations to be formed in the coming century between these countries [China, Japan, Russia, the United States] and a new player – a unified Korea – will have to be a defining factor in the long-term global geopolitical structure."[103] Leaders in Japan and Russia seem to concur.

Eurasia's Silk Road

An entirely new aspect of Japanese-Russian relations is the political emergence of the new states of Central Asia along Eurasia's southern periphery. The birth of eight new states in Central Asia and the trans-Caucasus region has created new opportunities for leaders in Japan, who have viewed their nation's energy import dependence in the twenty-first century with increasing worry. For Russia, Central Asia's independence created a host of new problems that have built on traditional insecurities. Recently, however, Russia has taken a new view of regional developments that has translated into a new partnership in the region – but not with Japan.

Although Japan's arrival into the 'Great Game' was somewhat late, in the space of a few years Japan has become one of the largest bilateral aid donors to the energy producing nations of the region.[104] Along with most every country engaged in Central Asia, Japan is interested in tapping into the energy riches of the Caspian basin. Japan relies heavily on the Persian Gulf for its crude oil imports – more than 80 percent, which is a higher percentage than Japan imported prior to the first oil shock in 1973. At present, energy imports from the Persian Gulf region alone account for 10.5 percent of all Japanese imports (including energy and all other sectors).[105] Japanese policymakers know that Japan needs to diversify its sources in its elusive, never-ending quest for energy security, and they look upon the Central Asian region with great hope as a source of diversification.

Japanese companies began investing in energy projects in Azerbaijan,

Kazakhstan, Turkmenistan, and Uzbekistan in earnest around 1996 when Ryutaro Hashimoto became prime minister.[106] A year later he exhorted Japanese companies to become involved in the region in his speech to the *Keizai Dōyuukai*. In February 1998, on the occasion of the visit of Azerbaijani President Heydar Aliyev to Tokyo, Hashimoto announced that Japan would extend close to $240 million in aid to Azerbaijan, not coincidentally one of the major energy producers in the region. Shortly after Aliyev's Tokyo visit MITI reportedly formulated an 'action plan' for developing a comprehensive strategy for Central Asia.[107] In September 1998 it was announced that the trading companies Marubeni and Mitsubishi had signed four contracts (three of them energy related) in Kazakhstan totaling $2 billion. In the same month the Japanese company Indonesia Petroleum Company agreed to pay $500 million in cash to the Kazakh government for stakes in the 2,000-square-mile oilfield Kashagan. In what would be by far the largest investment, Japan's Mitsubishi Company is interested in a vast project involving a gas pipeline from Turkmenistan, across China to the Pacific coast, and through underwater pipes to Japan. "Japan's gas needs will double by 2010, so we need the energy," stated one Japanese official at the time.[108] These trends demonstrate Japan's genuine desire to diversify energy sources, not only out of fear of political instability in the Middle East, but also from the realization that rising energy demands in Asia will have a large impact on Japan's ability to acquire cheap and reliable sources of energy from traditional areas.

Officials from METI (MITI), the Ministry of Foreign Affairs, and the Ministry of Finance have publicly stated that Japan's interest in Central Asia is based on the elusive quest for energy security.[109] One official from the Ministry of Foreign Affairs stated that, "Whereas economic interests are important, I believe that Japan's policy toward the Caspian Sea Basin stems mainly from long-term strategic considerations."[110] One MITI official privately concurred with this assessment and insisted that Tokyo was looking to play an active role in the emerging, new Central Asian 'Great Game.' This would include skillful diplomatic maneuvering among China, Russia, and the United States.

Although Japan's interest in Central Asia is part energy and part politics, the link with Japan's Russia policy is still somewhat tenuous, in spite of what some analysts say.[111] To be sure, Japanese leaders desire good relations with Russia in Central Asia, but thus far there has been neither cooperation nor confrontation between Moscow and Tokyo in the region. In fact, at least one Japanese analyst has lambasted the Japanese government for its lack of a strategy in the region, accusing it of rushing headlong into Central Asia with no long-range planning.[112] Japan and Russia do share a common view of engaging Iran, in contrast to the United States. Japan and Russia also are unlikely to be happy with a massive Chinese presence in the region. Although this provides Moscow and Tokyo common ground for cooperation in Central Asia, thus far there has been very little.

Russia's interest in Central Asia is a natural continuation from the days when it occupied and governed the region. A large Russian population continues to live in the new nations of Central Asia (mainly Kazakhstan). Russia, meanwhile, has lately been primarily preoccupied with geopolitics in Central Asian. Russia's giant energy companies are also engaged regionally, particularly in seeing that pipelines from Kazakhstan and Turkmenistan bring gas destined for Europe through Russia and not through the US backed Caspian pipeline. The Russian government, however, has maintained a keen interest in political and strategic developments in the region. At first, it appeared Russian leaders were most concerned about Chinese intentions in the region.[113] The American presence seemed more peripheral, though perhaps no less worrying, to the Russians. The United States had rushed quickly into the region in the early 1990s. But soon disillusionment in Washington with the democratic progress of the Central Asian 'Republics,' combined with the oil glut of the late 1990s, forced leaders in Washington to make a reassessment. From this point on the United States was engaged primarily with Azerbaijan, the starting point for the Baku-Ceyhan Caspian pipeline. It appeared that Moscow was given a free hand in the Central Asian states, to the east of the Caspian Sea. Moscow and Beijing created the Shanghai Cooperation Organization in 2000, and convinced the other non-Caucasus states of Central Asia to join. At this point Moscow's major concern was political instability and the threat of terrorism spreading through Afghanistan into Russia proper, via Central Asia.[114]

After September 11, 2001, the equation dramatically changed. By the beginning of 2002 the United States had thousands of soldiers deployed in three former Soviet republics and had become the regional policeman. Although China was left out of the regional agenda and the American position was dominant, Russia's leaders were seemingly less concerned than might be expected. The United States was in the region combating a sworn enemy of the Kremlin, the Taliban. Many Russian observers suggested that the United States was simply doing Russia's dirty work: "why not let them do it for us?"[115] Putin in the beginning never wavered in his support for the United States in the war against terrorism, and up until 2003 he never publicly repudiated the US military presence in Central Asia. Following the launching of the war in Iraq and subsequent US policy in the Mideast and Central Asia, Putin has been less supportive of the United States. And by 2004–05 tensions between Moscow and Washington were beginning to take their toll on the US position in Central Asia. Leaders in Moscow were concerned about what they viewed as the US penchant to "export" revolution across the Eurasian heartland. Changeovers in government in Georgia, Kyrgyzstan, and Ukraine during this two-year period brought US-Russian competition for influence in Eurasia and Central Asia to new heights. Russian leaders, concerned about Chinese influence in the SCO and about NATO intentions in the region, looked to bolster the Collective Security Treaty Organization (CSTO) as the premier political and military bloc in Central Asia.

Japan and Russia, for different reasons, each have a big stake in Central Asia. So far, however, the cooperation between the two is virtually non-existent. Although they may share the same views of the future course for the region, the two nations have demonstrated neither the willingness nor the capability to merge these visions and work together. As a result the region has had little impact on the bilateral relationship between Moscow and Tokyo.

International factors have always played a critical role in the periods of Japanese-Russian *rapprochement* during the twentieth century. In the late 1990s it appeared that China's rise and the strategic ambiguity of the United States were pushing Japan and Russia closer together. The dominant role of China and the United States on the Korean Peninsula further caused Japanese and Russian leaders to reconsider bilateral relations. Strategic factors have created a favorable environment in the bilateral relationship. But these factors, as important as they are to the relationship, have often been overshadowed in the past by the other two major factors, domestic and ideational. This cycle of history seemed to repeat itself in the 11 years between 1996 and 2007.

5 The domestic political context

1996–2007: Internal political dynamics in Moscow and Tokyo

As the international system began its transformation with the end of the Cold War in the 1990s, Japan and Russia were two societies also in the process of internal transformation. Changes were in large part brought about by the demise of the Soviet Union and the ending of the bipolar confrontation that had so shaped the global political order, as well as the domestic systems in place in Japan and Russia for the previous five decades. Dramatic domestic change in either country would have been difficult to undertake in the atmosphere of the Cold War. The Cold War acted as a brake on political development in each nation, as the necessities of maintaining a strong posture in the face of global confrontation (in Japan's case under the wing of the United States) circumvented debate about transforming domestic processes and institutions that were clearly outdated.

A global transformation spurred by the revolution in information technology that was under way in the 1990s further accelerated domestic change in both countries. Japan and Russia were forced to adapt to a new world economic and political order dominated by the United States. Neither nation could be expected to competently meet these challenges without a thorough reordering of its political and economic systems. The political systems in Japan and Russia were changing at the end of the previous period of *rapprochement* in 1991–92, and yet the transformations in each nation are still far from complete in the early years of the twenty-first century. Political change in Japan was on a much smaller scale than that in Russia, in both relative and absolute terms.

Japan

Japan was on the path of domestic change in the early 1990s, although it was on a smaller scale and more subtle than in Russia. In the summer of 1993 Japanese voters put to an end the 38-year uninterrupted hold on political power of the Liberal Democratic Party. The LDP-dominated '1955 system' had been put in place by Japanese conservatives at the height of the Cold War. It had served the Japanese people well, as Japan experienced almost four decades of continuous economic growth, relative political stability, social cohesion, and a standard of living unimaginable to most people there in the dark, chaotic days of the late 1940s. Although the LDP's defeat in the Diet parliamentary elections in the summer of 1993 was a shock to many, the re-emergence of LDP power after 1996 (more on this below) is a testament to the confidence in and loyalty toward the LDP among large segments of the Japanese electorate. The necessity for change was apparent to almost all, but many groups in Japanese society were reluctant to 'fix' a system that had given them so much.[1]

The new political dynamics in Japan were much more than just about replacing the LDP at the top of the political hierarchy. The two other sides of the 'iron triangle' that had dominated Japanese politics since the 1950s – the bureaucracy and big business – were also in desperate need of reform by the 1990s. The sudden halt in economic growth that came on the heels of the collapse of the Japanese stock market and real estate bubbles in the early 1990s forced most business leaders to come around to the inevitability of structural change. This transformation also necessitated administrative reform. This meant that Japan's almighty bureaucracy had to be reformed. This has proven difficult to accomplish, as Japanese bureaucrats have clung zealously to their privileges and to their political power. Japanese political leaders are not always helpful either. In fact, up to 2007 it can be said that administrative reform made little headway in Japan. As Japan struggles to define a new role for itself in the international system, it is still struggling with domestic political and economic reform.

Meanwhile, Japan's political system is still in the process of righting itself after the epochal 1993 Diet election in which the LDP lost power to a coalition government of various reform parties. The leaders of many of these 'reform' parties were in fact defectors from the LDP, and so their reform credentials were often suspect. Two successive coalition cabinets governing from the summer of 1993 lasted in power no more than a year before their two prime ministers were forced to resign (one on corruption charges). A strange coalition was cobbled together in June 1994 between the LDP and its long-time rival the Socialist Party. The Socialist Diet member Tomoichi Murayama was named to head the cabinet. This agreement lasted until the LDP was able to place its leader Ryutaro Hashimoto at the head of a cabinet in January 1996. In 1998 the LDP was able to cast off its erstwhile Socialist allies by building a coalition with the conservative-religious party, *Kōmeitō*. This coalition still existed when Koizumi left his post as party leader in 2006. Japan's political realignment, however, is still ongoing. Since 1993 Japan has had nine prime ministers and over a dozen foreign ministers, making it extremely difficult to implement a coherent and consistent foreign policy. Beyond the political system, Japan is still undergoing painful economic readjustments and societal pressures. Pressing issues such as a major demographic crisis are starting to rip at the very fabric and identity of the nation, making the beginning of the twenty-first century a most painful time for Japan.

Needless to say, Japan's policy toward Russia, just in the last decade, has seen significant shifts. Traditionally, strong players in Japan's Russia policy have continued to play important roles. The highest profile player continues to be the Ministry of Foreign Affairs' Russia School, although its standing was greatly damaged by the Suzuki scandal. The Office of the Prime Minister (the Cabinet Office, or *Kantei*) also maintains a high profile in Japanese-Russian relations, a tradition dating back to the tenure of Ichirō Hatoyama in the mid-1950s. Politicians in the Diet, in particular LDP members, continue to play an influential role in bilateral relations with Moscow, especially those politicians with a link to an electoral base that has an interest in relations with Russia, such as in northern or western Japan. Business groups, in league with the Ministry of Economy, Trade and Industry (formerly MITI, now METI), also have maintained a deep interest in Russian affairs. These include the ever-powerful *shōsha* (Japanese multinational companies), the *Keidanren* (a powerful grouping of the major *shōsha*), smaller businesses, as well as fishing industry associations. The role of the Japanese media is also crucial. Japanese dailies and weekly journals continue to closely monitor the relationship with Russia, and they are not shy about offering their opinions, which are often heard at the highest levels of the government.

It is of interest to note the emergence of new players in Japan that are exerting increasing influence on bilateral relations with Russia. Among these are the Ministry of Finance, the Japanese Defense Ministry, think tanks, academics, and various individuals. Certain individuals, whether they are part of the political establishment or not, have taken a personal interest in Russian affairs and their influence has come to belie the traditional view of Japan as a consensus-oriented machine grinding out policy through a methodical and monotonous process. Also significant is the new rift within the Ministry of Foreign Affairs, which further puts the lie to the popular notion of the consensus-oriented Japanese bureaucratic machine. Not only has a divide on Russia policy opened up within the different divisions of the Ministry, but also within the Russia School itself, long the final power broker in Japan's Russia/Soviet policy.

The Ministry of Foreign Affairs

The influence and the role of the Ministry of Foreign Affairs, or MOFA, in Russia policy have remained strong since the end of the Cold War. For the last two decades of the Cold War MOFA's grip on Soviet policy had become vice-like. However, the internationalization of Japan's economy in the 1980s saw the encroachment of other ministries and political actors onto MOFA's traditional territory.[2] Japan's policy toward the United States during the Cold War was shaped by a myriad of actors and institutions, and the influence of MOFA was weakened significantly over the years. By the 1970s China policy became influenced more and more by the economic ministries and by big business, as did policy toward Southeast Asia, South Korea, and Taiwan. Soviet policy, however, throughout the 1970s and 1980s remained the undisputed domain of MOFA.[3]

During the early 1990s, MOFA policy toward Russia seemed frozen in place at times. In spite of radical changes in the international system and at home, as well as in Russia, Japan's Foreign Ministry continued along in its dogged pursuit of 'justice' and the return of the disputed 'Northern Territories' with little sign of new thinking or creative diplomacy. The steadily shrinking power of the MOFA in Japan's political hierarchy caused MOFA bureaucrats to cling even tighter to control over Russia policy in the 1990s.

The first signs of change in Japan's Russia policy in 1996–97 bore the marks of just about every major political player in Japan. When Prime Minister Hashimoto delivered his policy speech on Japan's new 'Eurasia diplomacy' in July 1997 the focus was on strategy and energy, not the return of the 'Northern Territories.' As mentioned, bureaucrats from MITI had a large hand in drafting Hashimoto's speech, as did the Prime Minister's Cabinet Office (the *Kantei*). MOFA was handed a copy of the speech and asked to make additions only after MITI and the Prime Minister's Office had first access to the draft.[4] Yet, the influence of two MOFA officials over Hashimoto's new policy toward Russia soon became apparent. Deputy Vice Minister (later Ambassador to Moscow) Minoru Tamba had a strategic vision for Japan that was shared by Hashimoto.[5] Hashimoto and Tamba were worried about China's rise and Japan's political marginalization in Northeast Asia.[6] For Tamba building bridges with Russia made sense for Japan. Another influential MOFA official was Kazuhiko Tōgō, Director-General of the influential Treaty Affairs Bureau (later Director-General of the European and Oceanic Affairs Bureau). Tōgō was instrumental in the early 1990s in separating the territorial problem from the advancement of relations with Russia in other areas. That Hashimoto avoided overemphasizing the territorial issue in his early meetings with Yeltsin and in his 1997 speech reveals the influence of Tamba and Tōgō.[7]

Curiously, Tamba and Tōgō had been on different sides of the debate in the Russia School during the early 1990s when policies were being formulated with respect to the 'new Russia.' At this time the split within the Russia School was just becoming apparent. Tamba favored the *iriguchi-ron*, or the 'entry-point' strategy insisting on the return of the islands as the key to improved relations. Tōgō favored the *deguchi-ron*, or the 'exit-point' strategy that insisted on better relations as a strategy leading to the return of the islands. By 1996–97 Tōgō's policy seemed to have the upper hand, and by then Tamba had also become a proponent of economic assistance programs and political exchange as a prelude to the handover of territory. That the hard-liner Tamba had come around says much about the major strategic reassessment under way at that time in Japan. Clearly, Russia had assumed a place of great importance in Japan's emerging post-Cold War diplomacy. Other key players at this juncture in the Foreign Ministry included Mutsuyoshi Nishimura (Director General of the European and Oceanic Affairs Bureau), Kenji Shinoda (head of the Russia division within the European and Oceanic Affairs Bureau), Kyoji Komachi (a MOFA official working at the Japan International Cooperation Agency), and Masaru Satō (a staff member of the Intelligence and Analysis Bureau).[8] Satō was a lower-level official whose analytical skills were highly prized at all levels of the government. Foreign Minister Keizo Obuchi, though not a career diplomat, also played a key role in the formulation of Russia policy at this stage (1997–98). As Hashimoto's successor to the premiership, Obuchi would continue to stress the importance of strengthening relations with Russia.

While Hashimoto served as Prime Minister, Tamba's influence over Russia policy remained strong. When Hashimoto was forced to step down in the summer of 1998, Tamba remained at his post and continued to shape Russia policy, though by this time Russia's economic crisis impeded any dramatic improvements.

Ironically, when Tamba was appointed ambassador to Moscow one year later in the summer of 1999, his influence over Russia policy had waned. By this time there was a turnover in personnel in the Foreign Ministry. Many in MOFA who had helped launch Japan's new 'Eurasia diplomacy' were moved elsewhere.[9] Tamba had moved on to Moscow out of the corridors of power in Tokyo. Mutsuyoshi Nishimura was replaced by Tōgō, and Jirō Kodera succeeded Kenji Shinoda as the director of the Russia division.

By late 2000, when the Foreign Ministry was frantically trying to push relations with Russia forward to reach a settlement, Tōgō was the undisputed leader of the 'two-track' policy (negotiating the return of two islands in line with the 1956 Joint Declaration, and the continuation of talks on the status of the other two). The push toward this new approach was clearly spelled out in the series of articles authored by 'Mr. X,' and published in *Sankei Shimbun* in December 2000. Tōgō was in fact the author of these articles, which were based on leaked MOFA documents. The articles were met by a wave of criticism from all circles in the political establishment, but especially from the right. But Tōgō had four powerful allies within the government: MOFA official Masaru Satō, LDP lawmakers Hiromu Nonaka and Muneo Suzuki, and Prime Minister Yoshiro Mori.

Satō's reputation within the Foreign Ministry and the government had reached new heights by 2000. Satō had initially made his mark in MOFA's Russia School when during the 1991 coup attempt in Moscow he was able to gain access to the leadership of both the coup and anti-coup factions. He was able to report back to his superiors in the Moscow embassy with valuable real-time intelligence about the progress of the quickly unfolding events. Satō initiated his long-standing working relationship with Muneo Suzuki in 1991 when the latter made a tour of the Soviet Baltic republics as a young lawmaker. Satō served as Suzuki's guide and interpreter, and the two experienced tense moments as Soviet authorities moved to crack down on the independence movement in Vilnius, Lithuania. Their relationship was bonded by this experience. By the end of the 1990s, Satō's knowledge of Russian affairs was so revered within the government that he was referred to in certain Japanese circles as 'Rasputin.' Satō also supposedly served as an information conduit for the CIA and the Mossad, and had carefully cultivated a network of contacts within the Kremlin that included trusted Putin aide and Chairman of the Security Council Sergei Ivanov (later Minister of Defense and now Deputy Prime Minister).[10] Although Tōgō and Satō agreed on the need to revise MOFA's long stagnant Russia policy, Tōgō felt uncomfortable with Satō's close relationship with LDP lawmaker Suzuki. By this time Suzuki (then Deputy Chief Cabinet Secretary) exercised immense influence on the *Kantei* and on MOFA's Russia School, and it was due in great part to Satō's support. During a December 2000 visit to Moscow to firm up dates for a visit of Prime Minister Mori to Irkutsk in Siberia, Suzuki had a closed door meeting with Sergei Ivanov, where Satō served as the interpreter. Tōgō had flown to Moscow expressly to meet with Ivanov, but was told by Suzuki that he "did not need to attend" the meeting. Tōgō only learned afterwards through his own personal contacts in the Russian government what had been discussed at the Suzuki-Ivanov meeting. The extent of the rift within MOFA's Russia School was demonstrated by the fact that people supposedly on the same side often hid information from one another. Foreign Minister Yohei Kōnō was also reportedly upset and criticized the 'one-man' diplomacy of Suzuki.[11]

Others within the Foreign Ministry were upset not only by the domineering attitude of Suzuki and 'Rasputin,' but also by the dogged pursuit of the 'two-track' approach that had not yielded any noticeable results. The Russian government had made it clear that even if there were an agreement based on the 1956 Joint Declaration, Russia would not be obliged to immediately hand over two islands. Two former Foreign Ministry officials, Sumio Edamura and Nagao Hyōdō, launched scathing attacks on the 'two-track' approach and on Suzuki's one-man diplomatic show.[12] One MOFA official said that the resentment against Suzuki in the Foreign Ministry was building up pressure like "magma."[13] Jirō Kodera, head of the Russia Division within the Bureau of European and Oceanic Affairs (now the Eurasian Bureau), was reportedly so upset by the 'abandonment' of the principle of returning all four islands at one time (*yontō ikkatsu henkan*) that he had become insubordinate to his boss Tōgō, and after the Irkutsk summit Kodera was sent off to a posting in London unrelated to Russia. Kodera's discontent was shared by two high-ranking members of the Foreign Ministry: Yukio Takeuchi, Deputy Vice Minister in the Policy Planning Bureau and Chikahito Harada, a Russia expert, and Deputy Director of the North American Affairs Bureau. In 1999 they had both argued in favor of criticizing Russia for its actions in Chechnya, thus challenging Suzuki and Satō, who had been arguing to treat the matter as a Russian internal affair. The Suzuki-Satō line won out, but resentment had lingered and Takeuchi and Harada continued to oppose Suzuki at every juncture.[14]

When the Suzuki scandal (see Chapter 3) hit the Diet in early 2002, confidential papers and secret documents from the Foreign Ministry were leaked and produced by Diet members as damning evidence against Suzuki and Satō. Those forces in the Foreign Ministry opposed to the Suzuki-Tōgō-Satō line were no doubt responsible for these leaks. In the spring of 2001 before the Diet deliberations against Suzuki opened, Foreign Minister Makiko Tanaka, with Prime Minister Koizumi's blessing, had recalled Kodera from London. Tōgō had been moved to the Netherlands as Ambassador to that country. Clearly, the Koizumi cabinet had come out in opposition to the 'two-track' diplomacy with Russia.[15] Foreign Minister Tanaka did much to undermine the 'two-track' approach and the diplomatic effort that had been put forth leading up to the Irkutsk summit. By the beginning of 2002 the Foreign Ministry's Russia School was purged of the last vestiges of the group of diplomats that had painstakingly promoted an improvement of relations with Russia since 1996–97. Tōgō was forced to retire from the Foreign Ministry, and Satō was arrested on suspicion of the misuse of MOFA funds in league with Suzuki. Other officials were demoted or also forced to retire. Finally, Suzuki was arrested in June 2002. The methods of this group were considered too radical for the tastes of the Foreign Ministry's old guard and for the people of Japan.[16] By 2002 all talk of 'two-track' diplomacy vanished, and the Foreign Ministry returned to its traditional *yontō ikkatsu henkan* (four islands at once) line.[17] Henceforth, any creative thinking about Russia within the Japanese Foreign Ministry was quickly quashed for fear of any backlash created by the Suzuki scandal. This sordid affair truly poisoned Japanese-Russian relations for the next few years. It was in this context that Putin's offer for a compromise in late 2004 was barely even received in Japan.

Other ministries and agencies

In part due to the rifts opening within the Russia School, by the late 1990s the Foreign Ministry's grip over Russia policy had weakened. Different government ministries and agencies began to play an important role in facilitating the *rapprochement* with Russia, including MITI/METI, the Ministry of Agriculture, Forestry and Fisheries, the Economic Planning Agency, and the Ministry of Finance. Other actors also assumed an important role in the new Russia policy, most notably the Japan Defense Agency (to be upgraded as the Japanese Defense Ministry in 2006) and the National Police Agency. In the past both had been quite hostile toward the Soviet Union and Russia.

Officials at MITI (today METI, or the Ministry of Economy, Trade, and Industry) had been attracted to the idea of Siberian energy development for decades. The brief *rapprochement* of 1972–73 under Prime Minister Kakuei Tanaka was driven in large part by the desire to develop alternative energy resources in Western Siberia and the Russian Far East. MITI's role in Soviet policy during this juncture was crucial (Tanaka, like Hashimoto, had previously served as MITI Minister – see Chapter 2). Hashimoto was named Minister of MITI in mid-1994. In November 1994 MITI officials drew up a 'Support Plan for Russian Trade and Industry,' which came to be known as the 'Hashimoto Plan.' Under this program Japanese technical expertise was sent to Russia and Russian engineers came to Japan for training.[18] When Hashimoto became Prime Minister in 1996 the role of MITI increased even more. As mentioned, MITI had pressed for market reform in Russia and paid special attention to Russian industrial policy in an advisory capacity. Several MITI officials (most notably section chief of the Policy Division Kenji Isayama) helped Hashimoto formulate his plan for a new diplomatic agenda in Russia and Eurasia. The chief goals of MITI officials were to promote energy development in Russia, to diversify Japanese sources, and to help Japanese *shōsha* become involved in pipeline projects in Russia and Central Asia.[19] The Agency of Natural Resources and Energy (ANRE) within MITI/METI has also been active in the Sakhalin energy development projects. In fact, some Japanese commentators cynically deduced that the *rapprochement* with Russia was designed by Hashimoto and MITI specifically to help bail out the insolvent Japan National Oil Corporation (JNOC) by creating a new revenue source through Sakhalin oil and gas development.[20] The fortunes of MITI in developing Russia policy closely followed those of Hashimoto and when he resigned as Prime Minister in the summer of 1998 following the LDP defeat in the Upper House Diet elections, MITI's influence on Russia policy waned. MITI/METI, however, will continue to play a large role in Russia policy, particularly as large-scale energy projects in Russia continue to operate, and as Japanese companies begin investing more in Russia.

The Ministry of Finance (MOF) assumed an important role in Russia in the

1990s as the Japanese government began extending economic assistance to the Russian government. Any credit or assistance extended to Russia ($6.3 billion up to the year 2000) must be vetted by traditionally stingy MOF bureaucrats. Former top MOF officials often serve as Governor of the Bank of Japan (Japan's Central Bank), and MOF officials dominate Japanese representation in many international financial organizations that have a large say in multilateral aid to Russia, such as the IMF and the World Bank.[21] In addition, MOF has a large role (along with MOFA, METI, and the Economic Planning Agency) in the policy formulation of the Japan Export-Import Bank, which has overseen most of the bilateral aid programs to Russia.[22] The Ministry of Finance's role in Russia policy can be expected to grow as long as Russia continues to require Japanese economic assistance. The urgency in Russia for economic assistance died down with the run up in oil revenues beginning in the first years of the 2000s. By 2007, as mentioned, Russia was in little need of assistance in the development of the Russian Far East.

Japanese Overseas Development Assistance program (a body administered by MOFA and MOF) cannot be officially extended to Russia because it does not qualify as a developing country. Yet, small amounts of ODA have been extended to the 'Northern Territories' through a Foreign Ministry committee known as the *Shien Iinkai* (the Committee for Assistance), though this program was suspended in the wake of the Suzuki scandal. The amount of aid administered to the disputed islands is small and covers necessities such as power generators, hospitals, and schools (as well as the infamous 'Muneo House'). Some in Japan even support the extension of ODA to the Russian Far East, via regional Japanese governments (so that regulations could be bypassed and to avoid the stigma in Moscow that it is relying on largesse from Tokyo).[23] Nevertheless, regional leaders in the Russian Far East are wary of Japanese aid that would be tied to not only Japanese goods but also to political concessions. In the end, they feel that accepting aid could strengthen Japan's claims over the islands.[24] But, again, the Russian government saw less need after 2004–05 to attract large-scale foreign assistance, even in the Russian Far East.

The Japanese Defense Ministry, or JDM (Japan Defense Agency or JDA prior to 2006) is also interested in improving relations with Russia, not least due to concerns about China. Previously the JDM had opposed other ministries that favored improving relations with Russia.[25] This opposition today has disappeared. Furthermore, the JDM has assumed a large profile in the bilateral relationship. Since the mid-1990s JDA/JDM officials, both civilian and military, have actively promoted bilateral programs with their Russian counterparts. The parade of official exchanges from 1996 to 2001 is enumerated in Chapter 3. Some of the highlights include the first visit to Russia by the head of the JDA, Hideo Usui, in April 1996; the first friendly visit by a Japanese warship to Russia waters in July 1996; and the first joint military exercises on the high seas in the summer of 1998. The list of 'firsts' in defense contacts is an extensive one. Civilian officials at the highest level of the defense establishment actively pushed for new contacts between the two defense organizations, and uniformed officers of all three Japanese services eagerly took up the call. Japanese officers also actively participated in the meetings of the Trilateral Forum on North Pacific Security (see Chapter 4). The Japanese delegations to the Trilateral Forum normally included more officials from the JDA than from any other Japanese government agency or ministry. Some of the more enthusiastic proponents of a close Japanese-Russian defense relationship included Ryuichirō Yamazaki, the Director-General for International Affairs of the JDA, Noboru Yamaguchi, a Colonel in the Defense Planning Division of the Ground Self-Defense Forces, and scholar Masashi Nishihara, a Director at the National Institute for Defense Studies, a JDM think tank in Tokyo.

In line with the JDM, the National Police Agency has also been a strong advocate of stepped up relations with Russia. The Police Agency's interest in cooperation with Russian counterparts is mainly out of necessity. The sarin gas terrorist attack in the Tokyo subway in 1995 by the religious cult group Aum Shinrikyo was a wake up call for both nations, because the group maintained a large following in Japan and in Russia.[26] Japanese and Russian organized crime are said to dominate large portions of the fishing industry that revolve around the undocumented transaction of fish and shellfish from Russian boats in Hokkaido ports.[27] Drug and illegal immigrant smuggling have also been a grave problem for law enforcement authorities in both nations.[28] The dangers of organized crime to both nations were highlighted when in the spring of 2002 a high official in the Russian Border Guard Service was the target of a firebomb attack in Sakhalin. He was flown to Japan for treatment but later died. Russian authorities linked the attack to the 'fish mafia' operating throughout the Russian Far East.[29]

The Japanese Ministry of Agriculture, Forestry, and Fisheries (MAFF) has long played a highly visible role in relations with Russia, dating back to the 1950s. Japan imports large amounts of marine and forestry products from the Russian Far East. Since 1992 roughly 40 percent of the total of exports from the region has gone to Japan. More than 3,500 Russian fishing vessels visit ports in Hokkaido every year, and the turnover (much of it undocumented) in marine products is said to exceed $1 billion per year. The Russian government claims to lose more than $2.5 billion annually in the illegal sale of marine products from the waters of the Russian Far East.[30] In spite of the importance of trade with the Russian Far East, MAFF officials have been consistently critical of the Russian government. They denounce the exclusion of Japanese fishing boats from the waters surrounding the disputed territory, and the frequent impounding of Japanese fishing trawlers. These officials praised the 1998 agreement (wherein MAFF was deeply involved) allowing Japanese boats a quota of fish and shellfish in these waters, but a later agreement allowing South Korean boats to fish these same waters was severely criticized. Therefore the role of the Ministry of Agriculture, Forestry, and Fisheries has been two-sided: forestry and marine products are an important aspect of Japan's trade with Russia and as such they are promoted, but MAFF officials are highly critical of the less than helpful attitude of Russian officials vis-à-vis fishing.[31]

The Prime Minister's Office

The sudden turn toward Russia in 1996–97 was orchestrated most of all by the Prime Minister himself, Ryutaro Hashimoto. Past Japanese prime ministers had also single-handedly taken charge of Russia/Soviet policy, most notably Ichirō Hatoyama in 1955–56 and Kakuei Tanaka in 1972–73, and so there was a precedent. Hashimoto's successors Keizo Obuchi and Yoshiro Mori were also deeply involved in shaping the bilateral relationship, and were equally determined to improve relations with Russia in hope of bringing about a full normalization of relations. The two subsequent prime ministers, Jun'ichiro Koizumi and Shinzō Abe, were much more careful about Russia due primarily to the backlash of the Suzuki scandal. These two focused on the territorial issue, even as the East Siberian energy bonanza was coming to its fruition.

As mentioned before, each successive post-war Japanese prime minister has attempted to make a mark for himself in the diplomatic arena. The standard was set by Prime Minister Shigeru Yoshida, who negotiated a peace treaty with the United States, and more importantly helped negotiate an end to the US occupation in 1952 while formulating the 'Yoshida Doctrine,' which is still the operating principle for much of Japanese foreign policy (advocating strong security ties with the United States). Prime Minister Nobusuke Kishi renegotiated the US-Japan Security Treaty in 1960. In the late 1960s and early 1970s Prime Minister Eisake Satō oversaw the reversion of Okinawa to Japanese rule. As outlined in Chapter 2, Kakuei Tanaka pushed for the normalization of relations with China and the Soviet Union. He was successful in advancing relations with Beijing, if not Moscow. In 1978 Prime Minister Takeo Fukuda oversaw the signing of the Sino-Japanese Treaty of Peace and Friendship. Under Prime Minister Yasuhiro Nakasone Japan assumed its duties as one of the G-5 (later G-7) countries in 1982. In 1992, Prime Minister Kiichi Miyazawa dispatched Japanese peacekeepers to the UN mission in Cambodia, the first time since the end of the Second World War that Japanese troops had been sent overseas on an official mission. Hashimoto was perhaps eager to secure his own place in Japanese history, and hence his attempt to normalize relations with Russia and secure the return of the 'Northern Territories' can be partially attributed to this. The motives for Keizo Obuchi and Yoshiro Mori were probably no less grandiose. Koizumi and Abe were more concerned about strengthening ties with Washington than with seeking accommodation with Russia, even in the face of worsening ties with China.

Hashimoto's main interests in Russia were based on strategy and energy. As head of MITI in 1994–96 he had promoted the 'Hashimoto Plan.' Senior officials in MITI and MOFA say that Hashimoto developed his Eurasia initiative without consulting them initially. Hashimoto worked painstakingly himself on the draft of the 'Eurasia Diplomacy' speech in the summer of 1997 before turning it over to MITI and MOFA officials for rework.[32] The Krasnoyarsk and Kawana summit meetings with Yeltsin came off well because of the personal chemistry between Hashimoto and Yeltsin, and because Hashimoto was determined to see results in his Russia policy. Politically, he had much invested in the Russia relationship and wanted to follow the process through to full normalization.

Hashimoto had assembled in his cabinet a team around him that was deeply involved in this delicate diplomatic process, including Foreign Minister Obuchi, the Director-General of the Japan Defense Agency Hideo Usui, and Director of the Hokkaido and Okinawa Development Agency Muneo Suzuki. Hashimoto was also not hesitant to use non-governmental expertise, including the services of Ichirō Suetsugu, the head of an influential Tokyo think tank, and a man with deep connections throughout Japan and Russia. Most of these actors would continue to play an influential role in the ongoing process, even after Hashimoto stepped down from office in the summer of 1998.

In his Russia policy Hashimoto was responding to big-picture issues, and claims to have been influenced by the urgings of US President Bill Clinton.[33] Some observers feel that Hashimoto had become increasingly involved in the process of improving relations with Russia as the domestic economic and political situation in Japan deteriorated in the last months of his premiership, almost as a sort of escape.[34] Yoshihiro Sakamoto, a former high-ranking MITI official, claimed that Hashimoto's approach to Russia was part of a strategy aimed at gathering US support for his shaky coalition government. By agreeing with Clinton's calls for stepped-up assistance to Russia, Hashimoto hoped to win trade concessions from Washington that would bolster his standing in Tokyo.[35] In the end, Hashimoto's great accomplishment in bringing about a *rapprochement* with Russia was unable to save him domestically and he resigned in July 1998, the very day of the arrival in Tokyo of Russian Prime Minister Sergei Kirienko.

Keizo Obuchi was no less eager than his predecessor to normalize relations with Russia, and the team assembled around him included a majority of the players Hashimoto had relied on. By late 1998 Muneo Suzuki's influence over much of Japan's Russia policy was keenly felt. Suzuki's chief ally in the cabinet was Hiromu Nonaka. Chief Cabinet Secretary Nonaka had become an influential cabinet member, and was known as an advocate of stepped-up relations with North Korea (a large community of North Koreans lived in his home district in the Kansai region near Kyoto). Although Nonaka was a novice in Russian affairs he was an inveterate back-room negotiator and a power broker in the LDP. He was one of the group of LDP politicians that tapped Mori to be prime minister when Obuchi collapsed into a coma in the spring of 2000, hence his moniker by some as the new 'Shadow Shogun.' At times Suzuki, who would become Deputy Chief Cabinet Secretary (equivalent to the Deputy Chief of Staff in the White House), acted beyond his authority but was rarely reprimanded by Obuchi or Nonaka. He invited himself to meetings with Russian leaders, and monopolized the time he spent at these meetings, often to the exclusion of top Foreign Ministry officials. Foreign Minister Yohei Kōnō complained about this to Obuchi and to Mori on several occasions, but to no avail. Suzuki maintained his unique position as special envoy to Russia with the power to personally negotiate with Vladimir Putin and Sergei Ivanov.[36]

When Mori became Prime Minister in the spring of 2000, Deputy Chief Cabinet Secretary Suzuki's influence increased even more. Suzuki wanted to bypass the normal channels between the Foreign Ministries in Moscow and Tokyo, and build a 'fat pipe' (*futoi paipu*) between the executive branches centered on himself and Sergei Ivanov. In July 2000 after the Mori-Putin meeting in Okinawa, Hiromu Nonaka, newly promoted to Chief Cabinet Secretary, shocked the Japanese public by stating that the territorial dispute should not stand in the way of a peace treaty between

Japan and Russia, thus laying the groundwork for the 'two-track' formula. In fact, according to Ichirō Suetsugu, Suzuki convinced Nonaka to include the statement in a long speech, whose substance mostly had little to do with Russia policy. It was a trial balloon for the 'two-track' formula. In Suetsugu's words Suzuki "duped" Nonaka.[37] Although the probe was initially unsuccessful, by the time of the Irkutsk summit in March 2001, the 'two-track' formula had become official Japanese government policy. Suzuki, who represents the district in northern Hokkaido opposite the disputed islands (and where many former islanders now live), claims to have made the return of the 'Northern Territories' his life work. Others have a more cynical view of Suzuki's motives. University of Tokyo historian Haruki Wada points out that Suzuki was somehow able to convince other Japanese lawmakers and bureaucrats to extend economic assistance to the disputed islands, even though Russia does not qualify for Japanese Overseas Development Assistance.[38] Bureaucrats in the Foreign Ministry feared Suzuki's power and his ability to have them transferred in the event they were to oppose him. Previously, the idea of assistance for the islands was taboo in Japan. It was thought that aid would not only amount to tacit recognition of Russian administration of the islands, but that an improved economic situation on the islands would also make the Russian inhabitants less amenable to a handover of the islands to Japan. Many observers have deduced that Suzuki hoped to financially profit from development contracts in the Russian Far East and the 'Northern Territories.'[39] Similar accusations came out in the spring 2002 Diet deliberations, and Suzuki was forced to resign from the LDP and was arrested and indicted several months later on charges of bribery and bid rigging.

This scandal around Russia policy affected Japanese and bureaucrats to the extent that few people connected with Prime Ministers Jun'ichirō Koizumi or Shinzō Abe were prepared to expend political capital promoting relations with Russia. As such, between 2002 and 2007, the driving force behind the *rapprochement*, the Prime Minister's Office, was no longer willing to take up the leadership baton in Russia policy. Koizumi made several half-hearted attempts to improve relations with the Kremlin, but when Putin made a firm offer for negotiations in late 2004, Koizumi declined to become involved.

The Diet and regional governments

Although the Diet was made up of a hodge-podge of political parties and movements from 1996 to 2007, the LDP continued to dominate Russia policy, primarily through the Hashimoto-Obuchi faction that was instrumental in the push toward Russia at the highest levels. Other political parties and individual politicians took an interest in Russian affairs only as it was deemed fit to their political agenda, which most often took the form of attack against the LDP's 'soft' policies toward Russia.

The role of *zoku* (literally translated as 'tribes', but a more apt rendering is 'policy caucuses') in the policymaking process within the LDP has been documented *ad infinitum*.[40] *Zoku* are composed of groups of senior LDP lawmakers with expertise in certain core areas (agriculture, construction, defense, etc.) that have been able to exert tremendous influence over policy in their area of expertise. In league with bureaucrats and business groups whose careers also overlap these certain areas, these groups often dictate which policies should be formulated and implemented in the Diet. In the past when the LDP had an overwhelming majority in the Diet, policy was often made by *zoku* in league with bureaucrats at LDP meetings before it ever reached the floor of the Diet for deliberation. Although the *zoku* lost much of their influence after the fall of the LDP from power, they still have clout. The energy *zoku* had heavy representation from the Hashimoto-Obuchi faction and was a powerful force in the Diet. From 1996 to 1998 the energy *zoku* included Yukihiko Ikeda, former Minister of Foreign Affairs, Seiroku Kajiyama, former Chief Cabinet Secretary, and Shinji Satō, former Minister of MITI.[41] The fisheries *zoku* demonstrated its power when it nearly sabotaged Japanese-South Korean relations when an agreement could not be reached over fishery zones in the Sea of Japan.[42] The fisheries *zoku* was also extremely unhappy with South Korea's 2001 fishing agreement with Russia.

The LDP continued to dominate Russia policy, but other parties in the Diet were not shy about letting their views be heard. Diet members often took the opportunity to lambaste Prime Minister Mori over his willingness to 'surrender' to Russia two islands. Critics ranged from the Democratic Party to the Communist Party. Politicians from the Democratic Party and the Communist Party led the attack on Suzuki in the Diet in the spring of 2002. It was Communist Party member Shōken Sasaki who produced the damning Foreign Ministry documents that pointed to Suzuki's misuse of MOFA funds and his undue influence in Russia policy.

Local governments, though eager for economic cooperation with Russia, have had very little influence over Russia policy, unlike their regional Russian counterparts. Governors of prefectures in the northern and western areas of Japan (such as Hokkaido, Kanazawa, and Niigata) have sponsored regional conferences, encouraged local businesses to invest in the Russian Far East, and have taken their agendas for the development of *Ura Nippon* (the 'back of Japan') to national leaders in Tokyo. For the most part their voices have gone unheard and their plans for regional economic integration were undermined by the near impossibility of conducting small- and medium-scale business in the Russian Far East. Smaller, regionally based Japanese companies have often lost their entire investments due to corrupt partners in Russia, or to hostile or unresponsive local governments. Some observers even claim that the push for regional economic integration has only created further bad blood between Japan and Russia.[43]

Big business and trading companies

Much as Ryutaro Hashimoto's personal diplomacy toward Russia mirrored that of Kakuei Tanaka, the role of big business (the *zaikai*) was every bit as instrumental in the initial push of 1996–97 as it had been in 1972–73. Certain Japanese trading companies (*shōsha*) had long been proponents of Russian Far Eastern and Siberian energy development programs. Japanese interest in Russia's natural resources dates to the 1920s and Japanese oil companies operated in Sakhalin until 1945. In fact, the first significant oil field on Sakhalin was a Japanese concern in the town of Okha in the 1920s.[44] The rising demand for energy in East Asia in the 1990s forced many Japanese politicians, bureaucrats, and businessmen to take a long hard look at the potential for oil and gas reserves in Siberia and the Russian Far East. Japanese companies in the fishery and lumber industries also viewed the Russian Far East as a potential prize area of cooperation with Russia in the 1990s. Some large Japanese manufacturing companies still see opportunities in Moscow and other areas of European Russia. These companies have worked hand in hand with MITI/METI and LDP leaders from the various *zoku* to argue for improved political relations with Russia in order to promote economic cooperation. Meanwhile, smaller Japanese companies have met with little success in the Russian market, particularly the Russian Far East, and so unlike the larger *shōsha* they cannot necessarily be said to be proponents of stepped-up relations. Despite the interest evinced by large Japanese companies in the late 1990s, by the year 2000 Japan ranked tenth on the list of foreign investors in Russia with just $372 million in direct investments.[45]

Things did start to change after the year 2000 as the energy projects got off of the ground in Eastern Siberia. Japan's largest investments in Russia have undoubtedly been the Sakhalin oil and gas projects. A consortium of Japanese companies (that includes the Japan National Oil Corporation, a METI firm) known as SODECO owns a 30 percent stake in the Sakhalin-1 project. The *shōsha* Mitsui and Mitsubishi owned a 25 percent and 20 percent share respectively in the Sakhalin-2 project, although in 2007 they sold off half of their interest (see Chapter 3).[46] Still, in the coming years the level of investment can be expected to reach billions of dollars.[47] In addition, the Japanese Sakhalin Pipeline Study Consortium (consisting of Itochu, JAPEX, Marubeni) and Exxon/Mobil are undertaking a feasibility study of developing a pipeline linking Sakhalin and Hokkaido. All of these firms are heavyweight players in Japan with deep political connections, and they all have an interest in a smooth bilateral relationship with Russia.

Japan's energy interests are not confined to the oil reserves in the Russian Far East. A Japanese consortium comprising the *shōsha* Mitsui, Sumitomo, and Itochu under the guarantees of MITI/METI and the Japan Bank for International Cooperation (the Japan Export-Import Bank) received a contract worth $400 million to provide 300,000 tons of steel pipe to the 'Blue Stream' gas pipeline project in the Black Sea in 1999.[48] Japanese *shōsha* including Itochu, Nissho-Iwai, and Sumitomo covet Russian metal deposits, particularly aluminum ore and Russian-refined aluminum because there is no longer an aluminum-refining industry in Japan. Kōji Hitachi, formerly of the firm Nissho-Iwai, says that though there is still great interest in investing Russia, it is often pushed from the top, by officials in METI and the government. Hitachi feels that Japanese companies are less interested in long-term investments than in the extraction of resources. He adds that while Russia is pouring foreign finance into the big industries, it is ignoring the economic fundamentals such as a solid infrastructure and a solid social system (education, health care, etc.). These fundamentals affect all aspects of Russia: the economy, the political situation, and social stability. On the other hand, Takeshi

Kondo of the Itochu Corporation believes that long-term investments can work, particularly if they are energy related. Itochu is interested in developing gas pipelines from Yakutia, linking China, Japan, and Korea to Russian resources. Kondo also believes that Central Asia could be a key energy supplier one day for Japan, and feels that Japan has "immense interests" in Central Asia, both strategic and commercial.[49]

All of the large Japanese trading companies are represented in the *Keidanren* (the Federation of Economic Organizations) and the *Keizai Dōyuukai* (the Association of Corporate Executives). These two organizations maintain strong political ties through fundraising and lobbying activities in Japan, and through their commercial contacts overseas they can exert strong influence on Japanese foreign policy. It was no coincidence that Hashimoto announced his strategy for a new 'Eurasia diplomacy' at a speech given at the *Keizai Dōyuukai*. The *shōsha* that are represented on these organizations maintain staff doing research and analysis for each country in which the firms do business. Often the information that they gather is considered better than anything the Ministry of Foreign Affairs produces, and so the *shōsha* can influence government policy in this regard.[50]

A big problem even for the major Japanese companies has been the inability of the Russian government to come up with an effective law on production sharing agreements (PSAs). Although President Putin promised to give this issue his utmost attention, he has assigned the portfolio to German Gref, Minister of Trade and Economic Development, and it appears to have made little progress. Thus far only three of the 28 projects brought up for review in the Duma had been cleared for production (two of them the projects on Sakhalin Island).[51] Meanwhile these three PSAs came under review in 2006 within the Russian government, as the Kremlin began to indicate that it would review "unfavorable" projects "imposed" on a weak Russia in the mid-1990s.[52]

Marine resources continue to occupy a central position in Japanese-Russian trade relations. The waters around the southern Kuril Islands are one of the world's richest fishing grounds. As mentioned, much of the trade in marine products is undocumented, and yet even discounting the unreported trade, the official level is still substantial. Fishery groups in Japan have a strong lobby in the government and for years they were united behind their demand for fishing rights around the disputed islands. This was finally granted in 1998, but now the consensus that once defined the industry seems to be unraveling. Some fishermen in Hokkaido have expressed concern that now that Japanese companies have the green light to fish the waters over-fishing may become a problem, depleting the rich sea beds and squeezing out the smaller fishermen. Although companies that have been granted a license to fish the waters are supposedly held to a quota, this does not mean the quota will be adhered to, given the corruption in the Russian regional governments and the criminal elements in the fishing industries of both countries.[53] This conflict within the Japan fishing industry greatly affects their lobbying power in Tokyo and their ability to influence decisions on Russia policy.

Resource extraction extends to the lumber industry as well. Two Japanese lumber concerns in the Russian Far East, STS Technowood and PTS Hardwood, have been touted as success stories. These joint-investment companies have been able to establish modern factories utilizing local management and have been able to maintain a profit.[54]

Manufacturers have slowly been making their mark on the Russian market, though often indirectly. Japanese electronic goods have a strong reputation in Russia, and it seems that many homes in Russia have at least one Japanese-made television, VCR, camera, or tape player. The reason why trade figures do not show this Japanese presence is that many of the goods are sold through third countries, where Japanese firms either have a manufacturing presence, or where Japanese firms export these goods. For example, between 1994 and 1997 as many as ten million Japanese television sets may have been sold to Russia (for a value of $1–2 billion). Goods such as these, and even automobiles, are sold through Dubai, Finland, Malaysia, and even Hokkaido, where Russian fishermen or tourists often buy goods and take them back to the Russian Far East.[55]

These manufacturing firms have little influence on the formulation of Russia policy. If anything the unsuccessful track record of Japanese companies operating in Russia, especially the small and mid-sized regionally based companies, may come to harm the relationship by poisoning the perception of Russia as a reliable partner for Japan.[56] Conversely, it should be pointed out that the export of large amounts of Japanese capital, investment, and goods has often created tensions between Japan and various nations and regions (most notably the United States and Southeast Asia). Therefore an overwhelming amount of Japanese investment in Russia could actually cause a backlash, especially in the Russian Far East.[57]

Think tanks, media, and grass-roots organizations

Most think tanks and 'second-track' organizations in Japan are in fact heavily funded by the Japanese government. MITI/METI, MOF, and MOFA all have their own institutions dedicated to research, and though they are deemed non-governmental, in fact much of the research that is produced is often in line with government policy. Good private foreign policy think tanks in Japan number around half a dozen or so, and their influence on the policymaking process is minuscule.

In the case of Russia, however, certain powerful individuals and organizations with a deep interest in Russia or the 'Northern Territories' have come to play an active role in the Russia policy debate. Perhaps no one individual outside of the government could be said to have had as much influence over government policy as Ichirō Suetsugu, who exercised extraordinary influence over Russia policy for close to three decades. Suetsugu was Chairman of the Council on National Security Problems (CNSP) from 1968 until his death in 2001. CNSP is a private organization composed of prominent scholars and specialists founded in 1968 to help lobby for the return of Okinawa (in fact CNSP's original name was the Council on the Okinawa Base Problem). Once the United States handed over administrative control of Okinawa in 1972, Suetsugu's organization dedicated itself almost exclusively to the reversion of the 'Northern Territories.' Suetsugu was trained in military intelligence during the Second World War. At war's end he contemplated taking his own life, but instead vowed to dedicate his remaining years to resolving the two major outstanding issues left from the war – the reversion of Okinawa and the return of the 'Northern Territories.' Suetsugu personally reached out to help Japanese POWs returning from the Soviet Union readjust to life in Japan after the war. Later Suetsugu built CNSP through business and political contacts. His main power base was centered in the Diet, and especially in the executive branch. He cultivated relationships with powerful LDP leaders for more than four decades. Many of them would become prime minister, including Kakuei Tanaka, Yasuhiro Nakasone, Ryutaro Hashimoto, and Keizo Obuchi. Others such as Shin Kanemaru and Hiromu Nonaka would exercise power over the LDP in the backrooms. Suetsugu also maintained good relations with bureaucrats in the Ministry of Foreign Affairs (though MOFA officials and LDP leaders often disagreed on Soviet and Russia policy). Suetsugu could always be counted on to maintain secrecy, hence his ability to not only build a powerful support network in the government, but also to exercise so much influence without ever drawing the attention of the Japanese public. Up until his death in 2001 his name today was well known only in the corridors of power.[58]

As he was building his organization and his contacts in Japan, Suetsugu was also able to build an extensive network of contacts in the Soviet Union. Starting in the 1970s Suetsugu established a connection with several outstanding academicians in the Soviet Academy of Sciences. Later these individuals would also come to wield political influence in the Soviet Union under Gorbachev, and in the new Russia under Yeltsin and Putin. Among these were Aleksandr Yakovlev, a member of the Politburo of the Central Committee under Gorbachev and one of the chief architects of *perestroika*, as well as Evgenii Primakov, future Foreign and Prime Minister. Both Yakovlev and Primakov served as Director of the Institute of World Economy and International Relations (known by its Russian acronym as IMEMO). Another academic at IMEMO with strong connections to Suetsugu was Georgii Kunadze, future Deputy Foreign Minister. CNSP and IMEMO began holding symposiums and conferences in the early 1970s and the relationship has continued up to the beginning of the twenty-first century. Both CNSP and IMEMO were also deeply involved with the Trilateral Forum on North Pacific Security, in conjunction with US partners. Through these high-level contacts Suetsugu was able to maintain a direct personal pipeline between the Japanese and Russian governments through the 1990s.[59] Suetsugu was also instrumental in organizing the Japan-Russia Friendship Forum 21, a forum gathering top Japanese and Russian business and political leaders to discuss policy recommendations for their respective governments. LDP Diet member Yoshio Sakurauchi, Chairman of the Japanese Parliamentary Deputies Association for Friendship with Russia, was named President of this organization.

Although Suetsugu was a strong proponent of improving relations with the Soviet Union and Russia, in the late 1990s he maintained his firm commitment to holding out for the return of all four islands together at the same time (*yontō ikkatsu henkan*). Suetsugu was a strong opponent of Suzuki and of the 'two-track' formula, and he fought to persuade members of the Mori cabinet, but his influence was reduced after the death of Obuchi. Among the Japanese academics and experts on Russian affairs who maintained a close relationship with Suetsugu were Shigeki Hakamada, Hiroshi Kimura, and Masamori Sase. To the extent that they maintained a close relationship with Suetsugu, their opinions on Japanese-Russian relations (normally hard-line) were often heard at the highest levels of the Japanese government. Conversely, academics who did not maintain a close relationship with Suetsugu and who were seen as taking a softer line toward Russia, such as Haruki Wada and Nobuo Shimotomai, were considered within the expert community in Japan to have little influence on Tokyo's Russia policy.[60] With Suetsugu's death in 2001, the chairmanship of CNSP passed on to Sase, with Hakamada and Kimura maintaining advisory roles.

Another influential think tank in Japan is the Japan Forum on International Relations (JFIR), whose Chairman, Takashi Imai, served as president of the *Keidanren*, and whose President, Ken'ichi Itō, is a respected and influential businessman. Although the JFIR is not necessarily heavily involved in the Russia debate, in the spring of 1999 a group of experts with the JFIR authored a report entitled, "Japan's Initiatives towards US, China and Russia." The report called on the government to build a relationship with Russia that would "serve the strategic interests of both countries." The report also called for Japanese industry, in league with the government, to help develop the Russian Far East and Siberia. "There is no other way in the long term for these regions, forsaken by European Russia, to survive economically than to engage in economic exchanges with the Asia-Pacific region, especially Japan," the report states. Although recognizing that Japan does not need economic cooperation with Russia, the report points out that improved relations could result in the handover of the 'Northern Territories' with the result that

> Japan will not only have succeeded in finally resolving the security concern of its northern borders that has been at issue since the end of the Edo Period, but will have also expanded the range of foreign policy options open to it geopolitically and strengthened its bargaining position in the international political and economic arenas.

The report also strongly urged the Japanese government to lobby for the creation of a six-power forum to discuss security issues on the Korean Peninsula, including Japan and Russia.[61]

The Japan Association for Trade with Russia and Central-Eastern Europe (known by its Japanese acronym ROTOBO) is a non-profit organization engaged in the promotion of trade and business cooperation between Japan and the former countries of the Soviet Union, including Russia. It publishes a monthly journal on trade and investment patterns in Russia and in its regions. It is a recognized source of economic expertise on Russia and as such has some influence, though the task of attracting Japanese investment to Russia is a daunting one.

The Japanese media have played an extraordinarily active role in national debates surrounding foreign policy. The major conservative dailies include the *Yomiuri Shimbun* and the *Sankei Shimbun*. The *Asahi Shimbun* and the *Mainichi Shimbun* are the two largest, more liberal dailies. The regional *Hokkaido Shimbun* has also assumed a large role in the Russia debate. It is the only Japanese regional paper to maintain bureaus in Moscow and the Russian Far East. All of the major dailies have remained unified in their demands for the reversion of all four islands together. The newspapers, meanwhile, try to outdo one another and uncover major scoops concerning Russia policy. Often this has upset government officials in both Japan and Russia, especially when the revelations are the result of official leaks from one side that makes the other side look bad. Japanese dailies have also been guilty of building up false expectations that often did not measure up to the realities of the complex situation. At times the Japanese public was led to believe that normalization was just around the corner, while at the negotiating table diplomats were still miles apart. Weekly and monthly journals have also joined in the lively debate over Russia since the mid-1990s, some of these such as *Aera, Chuuō Kōron*, and *Sekai* are owned by the major dailies and often feature articles by top government officials. Others such as *Foresight Magazine* and *Seiron* have maintained a somewhat independent line, even if the journalism is sometimes lacking.

Grass-roots organizations with the largest voice in Russia policy include the association of former residents of the 'Northern Territories' and the association of former Siberian POWs. The influence of these organizations reached a peak in the early 1980s when Japanese-Soviet relations reached their nadir. In 1981 the Japanese government established February 7 as 'Northern Territories' Day. February 7 is the anniversary of the 1855 Shimoda Treaty that originally gave Japan control of the southernmost Kuril Islands, known as the 'Northern Territories.' Large banners appeared in front of LDP headquarters and several of the ministries in Tokyo calling for the return of the disputed islands. Today such banners still fly in many locations around Japan. Nevertheless, of the 17,000 former residents less than 9,000 remain today, and this generation will decrease in large numbers over the next decade. There is less interest in the issue among the younger generation in Japan, even the children and grandchildren of former residents and veterans.[62]

Another organization that tries its best to influence Russia policy is the right-wing organization known as the *Uyoku* (literally 'right-wing' in Japanese). This group of ultra-nationalists has consistently criticized the Soviet Union (and now Russia), and likes to drive its sound trucks, whose speakers blare nationalistic messages, in front of the Russian embassy in Tokyo. The *Uyoku* comes out in force during high-level visits of Russian officials, and manifested a particularly large and noisy presence during Putin's visit to Tokyo in September 2000. Although most observers discount the impact of the *Uyoku*, Japanese politicians, especially LDP members, are careful not to cross paths with this organization. Muneo Suzuki expressed concern about *Uyoku* backlash during his campaign to have two islands returned first. In fact, in 2001–02 *Uyoku* trucks often blasted out criticism of "Suzuki the traitor (*kokuzoku*)."[63]

Russia

As in Japan, change in Russia is ongoing. But unlike in Japan, the change in Russia has been dramatic and fundamental. The political hierarchy, though perhaps better defined than in 1991–92, is still being settled. Between 1995 and 2000 Russia went through two presidential elections and two elections for the State Duma. These were among the first successful democratic elections carried out in Russian history. Although the definition of 'success' is a relative one, the fact that the elections were carried out (although there was the threat that both of those presidential elections would be postponed), and the results generally accepted by the Russian public and world community is testament to the progress in Russia. Vladimir Putin was re-elected in 2004 by an overwhelming majority of votes.

For almost the entire decade of the 1990s Boris Yeltsin and a small team of his closest advisors, known as the 'family', dominated the political agenda in Russia, though opposition to the family was vocal and fairly widespread. The traumatic and violent events of the fall of 1993, in which Yeltsin ordered federal troops to attack and arrest government opposition members occupying the Russian White House (the seat of the Federal government), led to the creation of an 'imperial' presidency in Russia. After the attack on the White House Yeltsin ordered the disbanding of the Russian Congress of People's Deputies (that had opposed him) and a referendum on a new constitution (which was passed by a vote of 58.4 percent). This new constitution gave the Russian president great power, in both domestic and foreign affairs. A new bicameral legislature was established, made up of the representative State Duma and the Federation Council (composed of regional leaders). The inaugural State Duma elections in December 1993 shocked the Yeltsin team and dismayed Western observers. The ultra-nationalist Liberal Democratic Party of Russia (LDPR) led by Vladimir Zhirinovsky captured more votes than any other party, winning 22.9 percent of the total vote. The Democratic Choice of Russia, led by Yeltsin's reformist (though former by 1993) Prime Minister, Yegor Gaidar, captured just 15.5 percent of the vote. The Communist Party gained 12.4 percent of the vote to come in a close third. The results were a reflection of the mounting frustration over the political and social situation in Russia.[64]

Progress in economic and social reforms was slow in Russia throughout the next two years. Yeltsin's political standing at home still rested on a shaky foundation, and at the end of 1994 Russian forces were engaged in serious fighting in the southern republic of Chechnya. During 1995 Yeltsin suffered two heart attacks within a few months, and as the 1996 presidential elections loomed, there was a question as to whether Russia could right itself before the political situation deteriorated into a crisis inviting all types of frightening scenarios. The December 1995 Duma elections were even more of a shock for many observers than had been the 1993 elections. This time Zhirinovsky's LDPR gained only 11.1 percent of the vote, but the Communist Party garnered 22.3 percent of the national vote. Prime Minister Viktor Chernomyrdin's Our Home Russia movement received a paltry 10 percent of the vote. What led to worry among liberals in Russia and observers in the West was the fact that the presidential elections were only six months away and the Communists appeared to be gathering momentum. Meanwhile the war in Chechnya dragged on with no end in sight. Yeltsin did manage to win the 1996 presidential election over Communist Party leader Gennady Zyuganov, but it was a close run affair, and the support of the semi-nationalist politician, ex-soldier Aleksandr Lebed was crucial for Yeltsin. Lebed was asked to join the administration and helped negotiate an end to the first Chechen War by the end of 1996.

The economic crisis of August 1998 put to an end rosy prognostications for Russia and its economy that had been prevalent in 1997. Yeltsin's health and competence continued to be questioned throughout this period. In November 1998 Yeltsin was unable to meet with Japanese Prime Minister Obuchi during the latter's visit to Moscow. The economy was still in shambles after the meltdown in August 1998. Meanwhile, between the spring of 1998 and the summer of 1999 Russia had five different prime ministers. The last one appointed by Yeltsin in the summer of 1999 was a relatively unknown former *razvedchik* (intelligence agent) from St. Petersburg, Vladimir Putin. Duma elections were scheduled for the end of 1999, and the presidential election in the summer of 2000. An opposition group led by ex-Prime Minister Yevgenii Primakov and Moscow Mayor Yuri Luzhkov appeared more viable than any of the anti-Yeltsin movements of the past. In the fall of 1999 Russia found itself engaged again in the Chechen quagmire, and many questioned whether Yeltsin would use this as a pretext to cancel the Duma and/or the presidential elections. But Duma elections were duly held in December 1999, and the result was a victory for Yeltsin and Putin. An effective, if dirty, PR campaign against Primakov and Luzhkov held their vote to 13.3 percent of the total, while the Unity movement closely identified with Putin won 23.3 percent of the vote, coming in just behind the communists with 24.3 percent of the vote.

Now that the conservative opposition had been crushed in the Duma elections, Yeltsin's resignation on December 31, 1999 paved the way for Putin's election as president in March 2000. Putin was elected with 50 percent of the vote on the first ballot and was inaugurated in May 2000. Since this time Putin has effectively maintained a strong support base in Russia, with approval ratings always hovering around the 70–80 percent level. As mentioned, he was re-elected in 2004, and since that time the economy dramatically rebounded, thanks in part to the historically high price of oil (the price surpassed $50 per barrel in 2004, and stood at over $70 in 2007).

In spite of these dramatic political vicissitudes Russia has not experienced the merry-go-round of changing leadership that Japan has since 1993. Since this time Russia has had only two presidents and three foreign ministers. Although Russia has had six different prime ministers, the influence of these men on Japan policy has been minimal, with the exception of Yevgenii Primakov. The impact of Russian domestic factors on the formulation of policy with Japan is more difficult to summarize because much of the policymaking process in the new Russia is often carried out in the same Byzantine manner as it was in the days of the Soviet Union. Decision-making is still centered in the Kremlin, and it is still difficult to understand how the Russian President arrives at his policy decisions, especially when it comes to Japan.

The Kremlin

It can be said that Japan policy begins and ends in the Kremlin. Most foreign policy initiatives begin in the Kremlin, and there they either die or they continue on to become official policy. Japan policy was undertaken as a Yeltsin initiative, but the Russian President had very little support from below. It was not only that the territorial issue could potentially be a dangerous one for any Russian politician to become involved in, but also because no one in the Russian government seemed to have a genuine interest in Japan, apart from a few individuals. Treaty or no treaty, Yeltsin took it upon himself to improve relations with Japan, and his successor Vladimir Putin seemed to have taken Japan policy on by himself as well.[65] Normalization with Japan probably meant territorial concessions, and no Russian politician was prepared to hand over territory, no matter how beneficial the normalization of relations with Japan may have been.

Although the initial push for a *rapprochement* in 1996–97 came from Tokyo, Yeltsin met the new Japanese policy with open arms. The statement issued at the Krasnoyarsk summit that Japan and Russia would do their utmost to sign peace treaty by the year 2000 originated with Yeltsin.[66] Whereas Soviet leaders had in the past been wary about reaching out too closely to their Japanese counterparts, Yeltsin seems to have been enamored of Hashimoto and truly cherished their friendship. Only German Chancellor Helmut Kohl and US President Bill Clinton could be said to have been closer to Yeltsin among foreign leaders. Hashimoto continued to keep up with Yeltsin until Yeltsin's death on 23 April 2007, and visited him in Moscow on several occasions.

Yeltsin's warm response to Japanese overtures was due primarily to economic factors.[67] Japan was seen as one of the few countries able to extend to Russia significant economic assistance. Japan had thus far given little in the eyes of the Kremlin. Now that Germany was unable to accommodate both internal reunification and the rebuilding of Russia, it was thought by many in Russia and the West that it was Japan's turn to ante up. In 1997 Japanese capital accounted for only 1.1 percent of the total foreign investment in Russia. Yeltsin wanted this figure to go up – dramatically.[68] No less than the economic and political survival of the Russian Far East was at stake. In line with this thinking, Foreign Minister (later Prime Minister) Yevgenii Primakov looked to Japan for strategic reasons. Primakov, known as an Eurasianist (as opposed to a Westernizer), looked to balance the global dominance of the United States by improving relations with China and other nations in East Asia, including Japan and the two Koreas. In fact, some Russian commentators feel that Russia completely ignored Asia until Primakov became foreign minister in 1996.[69] In 1997–98 Russian policy was increasingly moving in a direction that was strategically closer to China. By 1998 China and Russia had cemented a 'strategic partnership' that was partly aimed at the United States.[70] Primakov, however, believed that improved relations with Japan were also a key to Russia's resurgence. All the more important, the vulnerability of the Russian Far East to China's demographic pressures led Primakov and others to believe that Japan could be an effective counterweight to China in the Far East. In a 1997 visit to Japan Primakov noted that "both countries share a concern about China."[71]

Others within the presidential administration, though not necessarily enthusiastic about Japan, were wary of China's rise, including Kremlin spokesman Sergei Yastrzhembsky, and Yeltsin foreign policy advisor Sergei Prikhodko.[72] After the

1996 Russian presidential elections Chairman of the Security Council Aleksandr Lebed also expressed interest in improving relations with Japan. After he left the presidential administration he continued to publicly voice his concern about the state of Russo-Japanese relations, and when he visited Japan in the fall of 1997 he called for normalization to spur economic cooperation and to shore up Russia's strategic position. Lebed was especially worried about China's rising power and the vacuum that existed in the Russian Far East.[73] The Security Council, though it rarely addressed issues concerning Japan, remained an important advisory body, and under Vladimir Putin (and Sergei Ivanov) its influence would increase even more.

Other Kremlin advisors to Yeltsin were more favorably disposed toward Japan for purely economic reasons. This group of reformist 'financiers' included close Yeltsin advisor First Deputy Prime Minister/Minister of Finance Anatolii Chubais, and Chubais' protégé First Deputy Prime Minister/Minister of Fuel and Energy Boris Nemtsov.[74] Nemtsov was appointed in April 1997 to lead the program for the economic development of the Russian Far East and the Trans-Baikal. His portfolio included economic policy toward China and Japan. Nemtsov (and Chubais) appeared to be much less interested in cooperation with China than with Japan.[75] Chubais and Nemtsov conducted several high-profile visits to Japan between 1997 and 2000, as did Sergei Kirienko, former Prime Minister, and a political ally to Chubais and Nemtsov. In each case these visits were conducted with the goal of arranging financing for large energy projects in the Russian Far East. It is not clear to what extent this latter group of advisors influenced Yeltsin's thinking on Japan, but it should be pointed out that Nemtsov was one of the few people who accompanied Yeltsin to both of his informal summits with Hashimoto at Krasnoyarsk and Kawana, so it can safely be assumed that he carried at least some influence with Yeltsin's Japan policy. This group was seen in the West as 'reformers' and 'liberals.' In Russia, however, their interests were seen as suspect and their ties to oligarchs in Russia's energy industry led some to question how reform-minded they really were (especially the recognized head of this group, Chubais).[76] It seems that many of these so-called 'reformers' saw Japan as a potential partner with whom Russia could not only develop energy resources, but also prop up the crumbling economic and social systems in the beleaguered Russian Far East. Later, as head of the Russian energy monopoly, Unified Energy Systems (UES), Chubais continued to lobby for improved relations with Japan under Putin, and he visited Japan in the fall of 2000 to promote the Sakhalin-Hokkaido 'Energy Bridge.'[77]

None of these advisors, with the exception of Lebed, was ever bold enough to suggest to Yeltsin that he should compromise territory for the sake of better relations with Japan. Kozyrev and Kunadze had tried doing this in 1991–92 and they only succeeded in stirring up intense opposition in the Congress of People's Deputies, in the Russian Far East, and in Russian society in general. By 1996–97 Yeltsin's advisors were not about to compromise their political position at home for the sake of patching up relations with Japan, no matter how they may have thought about the issue privately.

Although the impetus for the 1996–97 *rapprochement* with Japan may have originated elsewhere (whether in Tokyo or among Yeltsin's advisors), the final decisions rested solely with Yeltsin. This is not to suggest that Russia's Japan policy was coherent. In fact, Yeltsin went through many stages in his attitudes on the 'Japan question.' As an opposition politician in the late 1980s and early 1990s Yeltsin was thought to favor territorial concessions to Japan. Once he became president he wavered, and opposed any type of concession. By 1997, however, Yeltsin seemed to have shifted back again, and though never favoring territorial concessions he promised to explore all avenues leading to the signing of a peace treaty. But by the end of 1998 Yeltsin had swung back again to intense opposition against any type of concessions to Japan.

President Yeltsin preferred operating at his own pace, and was often inaccessible to even his closet advisors. Advisors from institutions and ministries outside the Kremlin had to wait in a long line in order to get Yeltsin's attention. Some of Russia's top Japan experts were completely ignored when it came to the formulation of Japan policy, especially those experts from the Academy of Sciences and other academic and research institutes, as well as experts from the Ministry of Foreign Affairs. As Hiroshi Kimura pointed out, "there were no 'formalized institutional mechanisms' through which information and policy-related suggestions by specialists on international affairs and on Japan were channeled into the Yeltsin leadership."[78]

In the end, even had Yeltsin agreed to return the disputed islands to Japan, it would have been difficult to implement any such agreement. The Russian constitution of 1993 states that the President cannot change Russia's borders without the consent of the Federal Assembly, as both Boris Nemtsov and Sergei Yastrzhembsky were quick to point out after the Krasnoyarsk summit.[79] During Yeltsin's second term it seemed impossible to imagine that Yeltsin could have pushed through a territorial agreement, short of disbanding the Duma and the Federation Council. According to Sergei Blagovolin of IMEMO: "The sky would have fallen before this happened." He added that the Russian public is still not ready, even under Putin.[80] Public opinion in Russia was aligned strongly against any type of territorial concession to the Japanese, especially given the fact that the period of *rapprochement* with Japan (1996–2007) overlapped with the two wars in Chechnya.[81] Although Yeltsin was master in the Kremlin, his position domestically was much less secure. He obviously was not keen on expending too much energy toward normalizing relations with Japan. Nevertheless, given Yeltsin's mercurial temperament and his history of violently suppressing opposition, a territorial agreement satisfactory to the Japanese side could never be ruled out.

Vladimir Putin, on the other hand, throughout his two terms of office was in firm control and had the support of most every segment of Russian society. Putin clearly evinced interest in the Japan issue early on in his administration. One of the first foreign leaders he hosted in St. Petersburg, even before he was inaugurated, was Japanese Prime Minister Mori in April 2000. Japanese observers were impressed with Putin's quick grasp of the diplomatic, historical, and legal issues surrounding the territorial issue.[82] He clearly seemed to be in control of Japan policy. On two occasions he boldly supported a settlement based on the 1956 joint declaration (in 2000 and 2004), and although he never promised to return even two islands, it seemed clear to many Japanese that this was a man who would be unlikely go back on his word, once an agreement was reached. Putin's ability to push through a deal on territory with the Chinese in 2004, in spite of the intense local opposition, demonstrated his ability to deal with difficult issues directly and with authority. In his policy toward Japan Putin seemed to be influenced (like Gorbachev and Yeltsin before him) by the chance for economic rewards. Instead of seeking investment for the Russian Far East, Putin looked for investment in the far eastern energy complex from a wide variety of sources, including Japan. There also seemed to be from the beginning a strategic element to Putin's Japan policy. Recognizing the challenges posed by China and the United States, Putin felt that Japan offered new options for Russia.[83] As such Putin made it a point to closely study the Japan issue, and along with advisor Sergei Ivanov's encouragement, he took the issue on in a positive manner.[84] But by the end of 2005 Putin seemed uninterested in further pursuing negotiations with the Japanese.

In spite of Putin's strong domestic support, he was anxious not to be seen as 'soft' on Japan, particularly as the bloody stalemate in Chechnya continued. Although many in Japan have long hoped that a strong Russian leader could deliver to them the coveted 'Northern Territories,'[85] such a leader, in fact, would undermine his own domestic power base by conceding territory. Some saw Putin as a hostage of his own popularity ratings, especially early in his presidency.[86] If he were perceived to be weak on foreign policy (especially territorial issues), he stood to lose public support. Putin's rise from obscurity was not only meteoric, but it was in large part due to the vigor with which he prosecuted the war in Chechnya and the reaction it touched off among the Russian public, which was under a siege of terrorist bombings in the summer and fall of 1999. He recognized this and he was no doubt reluctant to dismount from the horse that he had ridden so far.

Unlike Boris Yeltsin, Putin is not shy about seeking advice from experts, but much of his thinking is clearly influenced by his KGB background. Putin surrounded himself with the so-called *siloviki* (men from the 'power' ministries – defense, emergencies, intelligence, interior – and from the Security Council). Many of these men hail from Putin's hometown of St. Petersburg. One of Putin's closest advisors on foreign and security affairs is Deputy Prime Minister (formerly chairman of the Security Council) Sergei Ivanov. It was with Ivanov that Japanese lawmaker Muneo Suzuki established such a close connection. Suzuki and his trusted confidante Masaru Satō quickly understood where the center of gravity lay in Putin's Kremlin. During his trip to Tokyo in September 2000 Putin stayed in the Japanese government guesthouse at the Akasaka Palace, surrounded by many of the more prominent *siloviki*. Normally, the visiting president invites his foreign minister and other cabinet-level officials to stay in the guest residence. This time Foreign Minister Igor Ivanov and others stayed in a hotel. It seems that Putin felt most comfortable with his Petersburg security group.[87] Other influential advisors from among the *siloviki* in the first years of Putin's presidency included Minister of the Interior Boris Gryzlov, head of the Federal Security Service (the FSB, analogous to the FBI) Nikolai Patrushev, and Emergencies Minister Sergei Shoigu. Foreign policy, however, seemed to very much be the domain of Sergei

Ivanov. Putin also tended to rely on a relatively little-known aide in the Kremlin, Sergei Prikhodko, for foreign policy matters that included Japan and Asia. Putin does occasionally consult outside experts (this issue to be addressed below), but the foreign policy buck always stops in the Kremlin.

Putin is a product of the KGB and the old power apparatus of the Soviet Union. This does not mean, however, that he is incapable of thinking beyond the parameters of an ex-KGB agent. In fact, Putin has also sought advice from a number of people that are classified as economic reformers, including Minister of Trade and Economic Development German Gref, Minister of Finance Aleksei Kudrin, and formerly from the special presidential advisor on economic affairs Andrei Illarionov. These economic reformers and the *siloviki* reportedly 'battled over the soul' of Vladimir Putin in the early years (although it seems that by 2005–06 the *siloviki* had won out).[88] Putin's policies incorporate elements of each side's strategy. The *siloviki* have pushed for a severe policy in Chechnya. Putin's push to the West, and his eagerness to reform the Russian economy, could be partially attributed to the influence of Gref, Kudrin, and Illarionov (who themselves did not always see eye to eye). Putin's decision to back the United States in Central Asia and in the war against terrorism appeared to have been a personal decision. His *siloviki* were reportedly not too happy with his acquiescence in an American military presence in Central Asia, but it is unlikely that the economic 'reformers' were involved in this debate on foreign policy. The subsequent turn in policy against the 'high-handed' policies of the Bush administration in 2004–05 also seemed to come directly from Putin, although his team of *siloviki* undoubtedly urged a change in US policy. Similarly, Putin's decision to recognize the validity of the 1956 Japanese-Soviet Joint Declaration at the Tokyo summit in September 2000 and again in December 2004 was also probably a personal decision. Shigeki Hakamada feels that Putin's decision to make a conciliatory gesture to Japan in 2000–01 was partially attributable to the desire for Japanese economic assistance, and partially due to the harsh rhetoric of the Bush administration toward Russia during the presidential campaign and in its first months in power.[89]

Putin's initial desire to improve relations with the West and to move Russia closer to Europe was viewed positively in Japan. Russia's great 'Westernizer', former Foreign Minister Andrei Kozyrev, normally included Japan with the West.[90] For the same reasons that Putin hoped to integrate Russia with the West (economics), he saw Japan as a valuable partner. And even after Putin became tougher against the United States and NATO after 2004, he was prepared to offer some sort of compromise with the Japanese government.

From the beginning Putin evinced a strong interest in the Russian Far East and visited the region regularly (as often as once a year), something no Russian or Soviet leader had ever done with regularity. This interest in the region was enough (along with his love for judo), evidently, to pique his curiosity about Japan. His landmark speech in Blagoveshchensk in 2000 alluded to the Asianization of the Russian Far East, and it was clear that he was not speaking in positive terms. It was also clear to anyone able to read between the lines that Putin was not speaking necessarily about Japan or Korea, but about China. Putin dreamt of a large program of economic and social development in the region for the purposes of national security. He met with local officials, listened to scholars from the Siberian branch of the Academy of Sciences, and backed local authorities who pushed social programs. His plan for a far eastern energy complex fit perfectly into this vision.[91] As one scholar put it, "the socio-economic advancement of these territories is perhaps even more important than energy exports. The VSTO [ESPO] oil pipeline project and the expansion of the gas transportation infrastructure [in the Russian Far East] ... are ... instruments of economic development."[92] In this context, Japan's potential positive role in this development was in the back of Putin's mind.

The Kremlin will probably continue to dominate Japan policy as long as Russia's presidents continue to draw policy expertise primarily from their personal entourage. Russia's democracy must have time to mature and flourish before all of the major institutions are brought into the policymaking process.

Russian ministries

Russia's ministries weathered the tempestuous days of the Yeltsin administration and have seen clearer weather under Vladimir Putin. In the early days of the Yeltsin presidency in 1991–92 the Foreign Ministry under the leadership of Andrei Kozyrev and Georgii Kunadze had been granted wide leeway to formulate and implement Japan policy. The subsequent backlash in 1992–93, however, doomed the chances of normalization with Japan, and also damaged the standing of the Russian Foreign Ministry (known by the Russian acronym, MID) in the political hierarchy. The Ministry of Defense maintained a fairly strong influence on foreign policy in Russia (*viz.*, Chechnya), which is remarkable considering the withered state of the Russian military and the lack of state funding for defense in the 1990s. This says more, however, about the siege mentality prevalent among all Russians than about the political skills of Russia's top soldiers. The influence of economic ministries seemed to come and go like the health of Boris Yeltsin, although under Putin they grew in stature. Russia's new 'oligarchs', representing financial-industrial groups, sometimes acted in unison with the economic industries, and when they did so this presented a unified, formidable force. But often the oligarchs did not care to involve themselves with the ministries, when they often could go straight to the president. Unlike in Japan where business and bureaucracy often join forces, Russia's economic ministries have little clout.

The Russian Ministry of Foreign Affairs has consistently maintained a fine cadre of Japan experts. The Second Asia Department (under which Japan and Korea fall) saw its influence peak when Georgii Kunadze was Deputy Foreign Minister in the early 1990s. By the mid-1990s, however, the department had little or no say in the implementation of Japan policy. During the Cold War the Japan team at the Soviet Foreign Ministry, led by Mikhail Kapitsa, had taken a consistently hard line toward Japan. The Russian public and government officials were perhaps caught off guard by the dramatic reversal of the Foreign Ministry's Japan policy, led by Kunadze. According to Aleksei Bogaturov, Kunadze was too ambitious and alienated the wrong people, especially in the presidential administration.[93] By the mid-1990s, Kunadze had been sent off to Seoul and Japan policy, at least in the Foreign Ministry, seemed dead.

When Yeltsin traveled to Krasnoyarsk in November 1997 to meet with Hashimoto, the Foreign Ministry was not asked to do any preparatory work. Yeltsin shocked most Russians and the Foreign Ministry with his announced goal of signing a peace treaty by the year 2000.[94] The Foreign Ministry, and the Second Asia Department in particular, had fallen to such a low position in the Yeltsin political hierarchy that during the Krasnoyarsk summit its sole representative was an interpreter. A spy scandal that resulted in the arrest of a Korea expert in the Second Asia Department in 1997 and subsequent diplomatic expulsions in 1998 further damaged the reputation of the Foreign Ministry's Asia team. Relations with China had improved dramatically, but this was due more to the presidential administration and the Ministry of Defense.[95]

After the Kozyrev-Kunadze years, top officials and Japan experts in the Foreign Ministry had learned to be cautious and any dynamic, creative type of diplomacy was unlikely to emanate from there. The Foreign Ministry's Japan experts were still very good, but they were gun shy. Aleksandr Panov (Ambassador to Tokyo from the end of 1996 to 2002) epitomized the muffled diplomacy coming from the Foreign Ministry. Fluent in Japanese and very knowledgeable about East Asian affairs, Panov has been careful to follow the ever-changing line coming from the Kremlin. His capable successor to the post as head of the Second Asia Department, Mikhail Galuzin, is equally cautious. The deputy foreign ministers, first Grigorii Karasin, and then Aleksandr Losyukov, have handled much of the bilateral negotiating with Japan, and they too have demonstrated caution. Neither Karasin nor Losyukov were trained as Japan experts (although Losyukov was named ambassador to Tokyo in 2003 and stayed for four years), and they were hesitant to push an ambitious agenda with Japan. After Losyukov, Mikhail Bely was named ambassador to Tokyo in early 2007, after spending most of his career in Southeast Asia. Foreign Ministry officials were hesitant to make bold approaches toward Japan, given the events of the early 1990s. Losyukov, as plenipotentiary diplomat to the Six-Party talks on Korean Peninsula security issues, often commented on Japan affairs, but it was normally a re-hash of policy already formulated in the Kremlin.

The Defense Ministry demonstrated that it has teeth when it comes to Asia policy. This is possibly in response to its lack of clout in traditional spheres of influence, such as relations with NATO and the United States. In the 1992 battle between 'Arbatskaya and Smolenskaya' (Foreign Ministry vs. Defense Ministry – see Chapter 2) the Defense Ministry emerged victorious and helped pressure Boris Yeltsin to postpone his visit to Tokyo.[96] Officials at the Defense Ministry and in the military eyed China not only as a strategic partner but also as a prize market for Russian weapons systems. China had surpassed India as the primary importer of Russian weapons, and the Russian defense establishment increasingly relied on this income.[97]

Since that time the Defense Ministry has continued to push for close economic relations with China, but it recognizes that China presents not just economic opportunities but certain risks. In a 1996 speech Defense Minister Igor Rodionov referred to China as a "potential threat." Russian air force and naval officers feel that their Chinese counterparts have weapons systems that are in better condition than their own.[98] More ominous for many of these officers is that China now has an overwhelming conventional superiority in the Russian Far East, and within a decade China may have superiority in nuclear weapons as well.

Attitudes in the defense establishment toward Japan softened somewhat by the late 1990s, when contacts between the Japanese and Russian defense establishments reached new heights. Most of Russia's top defense officials, unlike in Japan, are military officers. Clearly, many officers in the Russian military were interested in stepping up defense contacts with their Japanese counterparts. Actually, Soviet officers first suggested stepped-up defense contacts in the late 1980s and early 1990s, but this was more of a ploy to draw Japan away from the United States, a tactic attempted on several occasions throughout the Cold War. But during the late 1990s Russian officers were interested in increasing defense contacts, partially to hedge against China's growing power, and to seek out new markets for Russian weapons systems. Japanese pilots trained on Russian warplanes at a base just south of Moscow in 1997–98. Russian officials hoped the Japanese could be convinced to purchase air and naval systems, such as airplanes, destroyers, and missiles.[99] When JDA officials expressed concern to visiting Russian Defense Minister Pavel Grachev about Russian sales of advanced fighter planes to China, Grachev responded with an offer to sell the planes to Japan instead. The JDA did buy some Su-27s but it was only a small number and they were subsequently dismantled to study the avionics system to aid in improving Japan's indigenous fighter program.[100]

Nevertheless, the Russian defense establishment is not about to turn its back on a Chinese market that has demonstrated an enormous appetite for Russian weapons systems. In line with their Chinese colleagues, some Russian officials began expressing concern about Japanese participation in US plans for a ballistic missile defense system in Asia. Meanwhile, on several occasions Russian military aircraft flew through Japanese air space beginning in 2001 and through 2007, after a six-year hiatus begun in 1995. Concurrently in 2001, the Russian Far Eastern fleet began a tour of the Western Pacific. This raised concern in Tokyo, though the probes were no doubt aimed more at testing US defenses in the Northern Pacific and Japan.[101] Some old habits seem to die hard, especially in the Russian military, which still sees a threat in the United States. As one Japanese officer reported in the late 1990s, "Both of us [Japan and Russia] still have Cold War wariness."[102]

Russia's economic ministries (the Ministry of Trade and Economic Development and the Ministry of Finance) have an obvious interest in improving relations with Japan. Both German Gref (former Minister of Trade and Economic Development) and Aleksei Kudrin (Minister of Finance) are known to favor integration with the West, although they have shown no special interest in Japan. They do, however, view Japan as one of the G-7 nations with whom Russia must cooperate. In an interview with the *Nikkei Shimbun* in 2000, Gref stated that he believes that Russia and Japan possess enormous potential for economic cooperation by combining Japan's financial and technological might with Russia's natural and intellectual resources.[103] Russia's Finance Minister in the late 1990s under Yeltsin, Mikhail Zadornov, visited Japan on several occasions, and later was named chairman of the Russian Parliamentary Deputies' Association for Friendship with Japan. Zadornov was generally viewed as a moderate on the territorial issue.[104] First Deputy Prime Minister (former Deputy Finance Minister) Viktor Khristenko led the Russian delegation to meetings of the Japanese-Russian intergovernmental commission on trade and economic problems in the late 1990s. Khristenko later became Minster of Industry and Energy, where he lobbied for far eastern energy projects with Japanese participation.

Top officials in the Ministry of Industry and Energy also have expressed the need to improve the relationship with Japan. This includes former Industry and Energy Ministers Boris Nemtsov and Sergei Kirienko, as well as their mentor Anatolii Chubais.[105] Later Minister Khristenko would make the same appeals. These men approach Japan from a purely practical standpoint, and harbor no special fondness for Japan (apart with the possible exception of Nemtsov, who has shown a familiarity with Japanese culture). To the extent they maintained influence within the government (limited under Yeltsin, but bigger under Putin) they could be expected to push for an improvement in bilateral relations to help further their economic agendas in Russia. None has ever advocated, however, a territorial compromise.

Within the Russian economic ministries, bureaucratic infighting over the route of the ESPO pipeline continued to involve a number of important ministries in the Russian government. By early 2006 both the Russian Defense Ministry and the presidential envoy to the Far Eastern Federal District Kamil Iskhakov (who had replaced Konstantin Pulikovsky in January 2005) came out against the ESPO pipeline terminal on Perevoznaya Bay, ostensibly for environmental reasons.[106] Both were backed by the Federal Environmental, Engineering and Nuclear Safety Agency. In response Transneft proposed instead a terminus on Kozmino Bay, 50 kilometers from Perevoznaya.[107] Russian Railways (RZD), a state-owned monopoly, and its powerful director Viktor Yakunin, are opposed to any pipeline that would decrease the amount of oil it ships to China.[108] In July 2007, Deputy Prime Minister Sergei Naryshkin was named by the Kremlin as chief coordinator for the ESPO pipeline project. Interestingly, the announcement came on the heels of a trip to Tokyo by Naryshkin, where he met with Prime Minster Abe, former Prime Minister Yoshiro Mori, and Foreign Minister Taro Aso.[109] Like Putin, Naryshkin hails from St. Petersburg, and he was rumored to have been a KGB agent operating in Belgium. In 2007 Naryshkin was rumored to be a potential dark horse candidate to succeed Putin as president.

Officials at the State Fisheries Committee view Japan not as a potential partner, but as a nemesis.[110] The Committee has been able to exert a negative influence on Japan policy, though it has been by no means been a decisive player. Two former chairmen of the Committee, Yurii Sinel'nik and Evgenii Nazdratenko, have been vocal critics of Japan. What worries them most is the possibility of the handover of the disputed islands, whose waters they claim provide more than $1 billion annually in revenue to Russia.[111] Nazdratenko not only rails against the idea of a territorial handover, but he also is against the licensing of foreign firms to fish in

Russian Far Eastern waters. Nazdratenko successfully pushed for the exclusion of foreign vessels fishing for pallock in the Sea of Okhotsk, and he would prefer all foreign fishing concerns to vanish from Russian waters.

> The experience of recent years has shown that foreign investors enter the Russian fishing industry with one goal: to get access to Russia's marine biological resources .... In doing this foreigners are not in the least interested in the preservation of our resources, or in the life of our villages, and less so in Russia's food security.[112]

One reason for the Committee's approval of the 1998 fishing agreement with Japan was that it wanted to be able to better document fishing hauls in the Far East. Over-harvesting of certain stocks in Far Eastern waters (especially crab) has been a concern among Russian officials. For example, in 1997 Russian statistics showed exports of marine products to Japan amounting to $92 million. The same year Japanese statistics showed imports of marine products from Russia amounting to $1.1 billion.[113] Some Russian analysts say the problems lie with the Russian fishing industry, and that Japan has not profited unduly from fishing agreements with Russia. Officials in the Committee are also thought to be among the most corrupt of Russian officials.[114] According to Georgii Kunadze, however, the fishing industry, backed by the State Fisheries Committee, wields strong influence in the Duma and has a very good lobby in all sectors of the government.

The State Duma and regional governments

Russia's legislature and regional governments also maintained a negative influence on Russia's Japan policy throughout the 1990s and into the twenty-first century. As noted in Chapter 2, hearings at the Congress of People's Deputies helped force Boris Yeltsin to cancel a planned trip to Tokyo in the fall of 1992. Under Yeltsin the Duma was one of the Kremlin's greatest critics. Duma deputies from the left, right, and center were not afraid to criticize Boris Yeltsin. They have shown less of a tendency to criticize Vladimir Putin since he became President. Policy formulation rarely ever begins in the Duma, and the passage of legislation through the State Duma and the upper house, the Federation Council, is often merely a rubber-stamp process to enact laws that were already decided on long before their formal introduction. But the Duma has at times been successful in blocking legislation and policy initiatives. For example, Boris Yeltsin was never able to fire the obstreperous governor of Primorskii Krai Evgenii Nazdratenko because the Federation Council (composed at the time of governors and republic presidents) would not allow it.

Japan policy is one area where Duma deputies have made their presence felt. The Duma has let it be known that it would never pass any agreement ceding territory to Japan. Neither Yeltsin nor Putin has ever tested this stand. Putin, however, was able to push through a territorial settlement with the Chinese in 2004. Every few years the Duma holds hearings on the status of the southern Kuril Islands, and invites in experts to give advice. The hearings are usually more of an opportunity to advertise why Russia will never 'give up the Kurils.' Far less energy is expended on bilateral cooperation and the future of Russia's Japan policy. Often lawmakers still bring up the military and strategic value of the Kuril Islands, in spite of all expert evidence to the contrary.[115]

The Communist Party of the Russian Federation has been a consistent critic of Japan's persistence in bringing up the territorial issue in all bilateral negotiations. Not surprisingly, among the bigger critics of Japan are members of the LDPR, including its leader Vladimir Zhirinovsky. Zhirinovsky went on record in the early 1990s as saying that were he elected President, he would have Russian armed forces attack Japan with nuclear weapons and seize Hokkaido, if Japanese leaders were to keep demanding the return of the disputed islands.[116] Dmitri Rogozin, also a member of the LDPR, was the head of the Duma's International Affairs Committee in the late 1990s, and was a critic of those calling for a reversion of territory. As far as a general improvement in bilateral relations, LDPR lawmakers seem uninterested. Recently, however, Zhirinovsky has shown some interest in Japan and appears to have reversed his position at least partially. During a trip to Tokyo late in the spring of 2002, he suggested that Japan could have the islands back in exchange for "big investments in the Russian economy." He mentioned the figure $100 billion. What makes this statement interesting is that under the Putin administration Zhirinovsky tamed his wild pronouncements, often heeding the government line, and is actually seen by some analysts as a figure who might launch trial balloons for the Kremlin.[117]

The *Yedinstvo*, or Unity, movement was seen as the proto-nationalist party of power in the Duma because it had the support of Vladimir Putin during the 1999 Duma elections, and because one of Unity's leaders, Sergei Shoigu, was a political ally of Putin. Unity has since merged with another political movement led by Moscow mayor Yurii Luzhkov and is now known as the Unified Russia party. Although Luzhkov was a member of the Japan-Russia Friendship Forum 21, he showed little interest in Japan apart from investment opportunities for Moscow. Fellow members of the Unified Russia party have also demonstrated little passion for Japan policy.

Some right-leaning, economically liberal politicians in the Duma have shown an inclination to cooperate with Japan to the extent that this can bring about economic benefits to Russia. As mentioned, those politicians associated with Russia's energy industry showed a keen interest in shoring up relations with Japan. These included people such as Boris Nemtsov, Yegor Gaidar, and Sergei Kirienko. Their acknowledged leader was Anatolii Chubais, now the head of the energy monopoly Unified Energy Systems (UES). Irina Hakamada was also one of the young leaders of the liberal right, though as a half-Japanese woman she was very careful not to be seen as too pro-Japanese.

One of the few elected officials in Russia that has called for a return of the disputed islands to Japan was Grigorii Yavlinsky, head of the liberal Duma faction Yabloko. During the 1996 presidential elections Yavlinsky went on record as saying that he supports the reversion of this territory. After that he became less vocal on the issue, but he is said to still support this position.[118] Two other Yabloko deputies did support an improvement in relations with Japan during the early Putin years. Former Deputy Duma Speaker and former Ambassador to the United States Vladimir Lukin recognized the strategic necessity of good relations with Japan, not just to balance against China, but also to keep the United States from dominating the entire region. But, Lukin pointed out that solving the territorial issue would not necessarily improve relations with Japan. Japan, he stated, needs to show more interest in Russia, beyond the disputed islands.[119] Yabloko member Aleksei Arbatov, Chairman of the Duma Defense Committee, is one of the only other elected Russian officials to have called for reversion of the disputed islands to Japan. Arbatov felt that China's rise as a potential threat to Russia was one reason to improve relations with Japan. He felt that the 'islands for credit' approach would miss the mark: China will one day be a threat to Russia, and given the US penchant for strategic ambiguity in East Asia, a *rapprochement* with Japan is necessary for strategic reasons. "Resolution of the problem with Japan over the Kurile [*sic*] Islands would provide Russia with a much more advantageous political position and greater freedom of maneuver in the Western Pacific."[120] At the same time Arbatov pointed out that Japan's expanded military program could pose long-term risks for Russia.[121]

Although Duma members can be obstructionist in Japan policy, none of the members who have demonstrated a constructive attitude toward Japan has been able to further their agenda in the Duma. Vladimir Putin has demonstrated great power in the Duma, and were he to decide to push the bilateral relationship, then certain Duma members might follow in line. Given the ever-fluid political situation in Russia, however, this may not hold true for long.

Opposition to a dramatic warming of relations with Japan can easily be found in the Russian Far East, and given the right conditions it can be whipped into a frenzy. The governors of the various far eastern regions have consistently (with one major exception: Viktor Ishayev of Khabarovsk) come out against territorial reversion to Japan and they have been able to mount effective national campaigns through the Federation Council, on which they once sat, and whose representative from their district they can now appoint. In the late 1990s two governors led the opposition to Moscow's Japan policy. One was the late Igor Farkhutdinov, Governor of Sakhalinskaya Oblast until his death in 2003. The other was Evgenii Nazdratenko, former Governor of Primorskii Krai, whose capital is Vladivostok.

Among Russian leaders in the Far East, Farkhutdinov was by far the most consistent critic of Japan and its calls to have the 'Northern Territories' returned. Farkhutdinov was the successor to Governor Valentin Fedorov, who had succeeded so well in turning the attention of the Russian nation onto the Kuril Islands in 1992. Farkhutdinov was every bit as vocal, if not as flamboyant, as Fedorov in opposing territorial concessions to Japan. In response to Japan's persistent demands to return the 'Northern Territories' Farkhutdinov was on record as saying,

> Let them raise [the issue] if they want to. I believe that there should be no ultimatums in politics. And the Kurils should not be a bargaining counter in

> the course of the signing of a peace treaty between our countries .... These are Russian lands and they need the state's attention.[122]

Farkhutdinov, however, recognized the need for Japanese investment in the region, not just in Sakhalin's energy projects, but in other areas as well, including in the disputed islands. He was careful not to be overly antagonistic with the Japanese, and he even visited Japan on occasion with the presidential administration at summit meetings, or on his own. Farkhutdinov, however, feared that Japan would tie economic credits to the return of territory. Authorities in Moscow also recognize this and have made it a custom to visit Sakhalin on the way to or the way back from meetings in Japan to reassure local authorities of their steadfastness. The successor as governor to Sakhalin, Ivan Malakhov, followed a similar line to Farkhutdinov.

Local officials in Yuzhno-Sakhalinsk, the capital of Sakhalinskaya Oblast, were also of one mind with Farkhutdinov. They wish to attract Japanese investment to the region, but they do not want to see any strings attached. They have been on the whole pessimistic about the chances for a territorial settlement. Nikolai Vishnevsky, representative of the Kuril Administrative District in Yuzhno-Sakhalinsk, stated that though Russia is the rightful owner of the Kuril Islands, the region cannot get along without Japanese help. Sergei Kastornov, the local representative of the Russian Foreign Ministry in Yuzhno-Sakhalinsk, said that the issue cannot be resolved "without one side seeming to come out the loser." Vitalii Yelizariev, head of the Department of Foreign Economic Relations of the Sakhalin government, claimed that the two governments of Japan and Russia have been equally guilty for the economic situation in the region, as one side ignores the issues (Russia), while the other (Japan) harps on only one issue. There is a clear lack of trust, he adds, on both sides.[123]

Public opinion in the disputed territories appears to be divided. Three of the disputed islands (Habomai, Shikotan, and Kunashir/i) are administratively grouped into the South Kuril region, while Etorofu/Iturup is placed within the Kuril administrative region (both regions are administered within Sakhalinskaya Oblast). This administrative grouping also reflects economic and political differences between the islands. The citizens of Etorofu/Iturup generally look northward to Sakhalin and to the Maritime Province and even beyond for economic, political, and even cultural guidance. According to Nikolai Vishnevsky the US presence on Etorofu/Iturup is "highly visible." US imports and products are regularly seen in markets there, many of them originating in Alaska. In the cannery factories on that island "Made in the USA" labels can be seen on the machinery. Etorofu/Iturup exports 60 percent of its marine products to the United States, and 30 percent to Japan. The local cannery factory run by the firm Gidostroi employs a large number of islanders and the island practically has no unemployment.[124] According to recent reports, a very rare metal, rhenium, has been found in the volcanoes on the northern half of the island. Rhenium is used in the manufacture of certain electronic components, spacecraft and missiles, and also in high-grade octane fuel. If this development is successful, the citizens of Etorofu/Iturup could see positive economic benefits.[125] The economic situation there has already improved to the point that the local newspaper has commented that locals who once fled the island are now returning. The population has even grown over the past few years, something that is rare in all of Russia. The people on the island are reportedly very 'cool' toward Japan and toward the idea of being incorporated politically into Japan.[126]

On Kunashir/i, on the other hand, the citizens view Japan favorably. Vladimir Zema, the former Administrative Head of the South Kuril region, claims that Japan's popularity on that island is growing by the day, and that many people favor political union with Japan. The economic situation there was grim in the 1990s. Power plants did not have enough fuel, and power outages are a daily reality. This can be dangerous in the winter. On Shikotan the situation was dramatically worse. Japan recently had to deliver 2,000 tons of diesel fuel to the beleaguered islanders. In 1994 the population there was 6,543; by 1999 it stood at 3,746. On all four of the disputed islands the population has dropped by 18,700 since 1989.[127] The situation after 2003–04 improved but it is comparably worse than elsewhere in the Russian Far East.[128]

Polls conducted in the 1990s on the four islands generally confirmed this split among the islanders. Residents on Etorofu/Iturup overwhelming opposed reversion; a majority of residents on Shikotan generally supported reversion.[129] But two long-time American residents of the region, Michael Allen (formerly of the American Business Council in Sakhalin) and Russell Working (formerly of the Vladivostok News), warned that one must treat any opinion polls conducted on the islands with great care. They both point out that the results of polls conducted by Japanese tend to favor their viewpoint, while Russian polls generally cite figures opposing reversion.[130] The fact of the matter is, however, the residents of the islands exert no influence in Moscow, and are unable to make themselves noticed there, unlike leaders in Yuzhno-Sakhalinsk.

Evgenii Nazdratenko was not only a thorn in the side for the Japanese, but also for Boris Yeltsin, and to a lesser extent, Vladimir Putin. Nazdratenko (formerly also the Chairman of the State Fisheries Committee) served as the iron-fisted governor of Primorskii Krai from 1993 until he was asked by Vladimir Putin to head the State Fisheries Committee in 2000 (serving until 2003, when he then took up a post on the Security Council). Nazdratenko openly defied Boris Yeltsin, who tried to fire him on more than one occasion. On each occasion Yeltsin knew the Federation Council would not support him, as it could set a precedent wherein each governor would be at the mercy of Yeltsin's whims. The roots of the 'war' between Yeltsin and Nazdratenko lay in the deteriorating economic situation in Primorskii Krai in the 1990s. Yeltsin blamed Nazdratenko for incompetence and mismanagement; Nazdratenko blamed the center for not providing sufficient funding. Both parties were probably right. The Soviet Far East had become overly dependent on handouts from Moscow. It had been made into a bastion of defense during the Cold War, and was therefore shut off from the dynamic economies of East Asia from the 1950s to the 1990s. The new Russia could not afford to subsidize Far Eastern residents and Yeltsin essentially cut off most programs, exhorting Russia's republics and provinces to take as much sovereignty as they deemed necessary.

All of the Far Eastern provinces suffered during the 1990s, and Primorskii

Krai was no exception, in spite of its location next to China, across from Japan, and adjacent to the Korean Peninsula. Even in the twenty-first century the region suffers from a crumbling infrastructure, power outages, and water shut-offs, not to mention natural disasters, and the abandonment by the central government. Adding to the depressed situation are frequent worker strikes, mutinous soldiers, and increasing lawlessness.[131] Incredibly, in 1996 Moscow classified Primorskii Krai as a donor region (as opposed to debtor), meaning that it was one of the few provinces expected to pay more in taxes to the center than it would receive in federal subsidies.[132] Nazdratenko, meanwhile, ruled in a dictatorial fashion, consolidating his power base by dismissing local leaders who did not bow to his every demand, including the former mayor of Vladivostok, Viktor Cherepkov. Authorities in Moscow could not turn their backs on Nazdratenko while the situation in Primorskii Krai deteriorated to the extent that local citizens were begging for Moscow's intercession. Nazdratenko and Presidential Advisor Anatolii Chubais engaged one another in a series of high-profile battles that mirrored Chubais' war with Yeltsin strongmen Korzhakov and Soskovets in 1996. In fact the latter two were political allies of Nazdratenko, and tried to prevent his dismissal. Chubais saw Nazdratenko as an opponent of market reform and an enemy of democracy in the new Russia. Nazdratenko blamed Chubais and his experiments with market reform for the economic and energy crises gripping his region. Although Chubais won his war in Moscow (Korzhakov and Soskovets were dismissed), he was unable to unseat Nazdratenko in the Far East.[133] Nazdratenko also engaged in battle with other local governors in the Far East, all of whom were fighting to attract not just subsidies from Moscow but also foreign investment. Perhaps sensing an economic opportunity, Nazdratenko tried to wrestle away administrative control of the southern Kuril Islands from Sakhalin to Primorskii Krai.[134]

Nazdratenko further irritated Moscow by interceding in Russia's foreign affairs. He severely criticized the series of Sino-Russian border agreements in the 1990s that in some cases gave territory along the Ussuri River (Primorskii Krai's border with China) to China.[135] Nazdratenko was not as vocally critical of Japan as he was of China, and despite initial reluctance he visited Japan on several occasions trying to drum up investment for Primorskii Krai, even claiming that, "Japan is a very important partner to us."[136] Nazdratenko, however, has consistently opposed territorial reversion, and as pointed out, as Chairman of the State Fisheries Committee he also criticized Japanese fishing operations in the waters around the disputed territories.

The governor of the third major Far Eastern administrative body Khabarovskii Krai, Viktor Ishayev, is much more accommodating with Japan. Ishayev was the longest tenured of the far eastern governors (still ruling in 2007), attesting not only to his support in the region, but also to his political acumen in playing the frequently shifting politics of the Kremlin. He frequently traveled to Japan to meet with local leaders and to lobby for investment in Khabarovsk. He even made the politically risky move of declaring his support for the return of the disputed islands to Japan.[137] Ishayev is far more concerned about China, as his province is adjacent to more than 70 million Chinese residents in Heilongjiang Province. As such Ishayev favors Japan and Korean investment to keep a balance against China's potential demographic and economic encroachment.

Konstantin Pulikovsky was appointed Presidential Plenipotentiary Representative in Russia's Far Eastern Federal District by Vladimir Putin in 2000. He was one of seven such presidential representatives in Russia's provinces. He was a former general in the Soviet and Russian armies, and a member of the *siloviki*. Pulikovsky's duties in the Russian Far East were somewhat ambiguous, and he had little pull within the presidential administration (otherwise he would be in Moscow). Pulikovsky remained active trying to attract Japanese investment to all parts of the vast region, and he traveled to Japan individually and with the Kremlin administration. He championed the idea of constructing a nuclear power plant on one of the disputed islands and exporting some of the generated electricity to Japan.[138] Like Ishayev, Pulikovsky was seen as a moderate in relations with Japan. He was replaced by Kamil Iskhakov in late 2005. Iskhakov seemed to possess little interest in Asia apart from attracting investment to the far eastern energy complex.

Like the ministries, the Duma and regional administrative bodies, while unable to positively influence Japanese-Russian relations, can often made their voices heard enough to obstruct bilateral relations, especially on the issue of territory.

Russian industry, think tanks, and grass-root movements

Non-governmental organizations in Russia have not developed the same level of influence in foreign relations as have similar organizations in Japan. Ironically, under the Soviet Union some quasi-governmental institutions (especially the Academy of Sciences) did play a role in Japan policy (see Chapter 2). Today this cannot be said to be the case. Of all the non-governmental institutions and actors, Russian business is the only one with any kind of influence on policy outcomes, but most often this is reflected in domestic policy, and only peripherally in foreign policy.

Russian industry, and by extension Russian oligarchs, have little interest in Japan, other than how they might profit through cooperation with Japanese partners or with the Japanese government. This mirrors the stance of Russia's economic and energy ministries. Private Russian firms have little interest in political affairs with Japan, except perhaps energy and fishing firms. These latter two maintain profitable relations with Japan, and they do not wish to see bilateral political issues spill into the economic arena. But they do not involve themselves at all in the political process, however attentively they may follow it.

There are several high-dollar, high-profile joint projects between Japan and Russian firms; most of these involve resource extraction. Japanese *shōsha* have partnered with Russian energy giants such as Gazprom in projects from the Caspian Sea to the Sea of Okhotsk. The Sakhalin projects dwarf all other bilateral trade or investment projects. The investment figures for each of the Sakhalin-1 and 2 projects will surpass $20 billion. The Russian partners Rosneft and Sakhalinmorneftegaz (SMNG) have grown in stature and, along with Gazprom and Transneft, will represent state control of the far eastern energy complex. Russia's energy monopoly UES, led by Anatolii Chubais, has long harbored ambitions of working side by side with the Japanese. The Russian metals giant Norilsk, the largest producer of nickel in the world, has signed long-term contracts to supply the Japanese market with palladium.[139] Russian metals firms also are the largest supplier of aluminum to Japan. None of these firms wishes to see political relations spoil economic relations. They are unlikely, however, to become involved in Japan policy because they know this is risky for themselves, and because they realize that Japanese firms will invest in Russia if economic conditions are good, whether the islands are returned or not.

The Russian fishing industry has strong connections in Moscow, through the State Fisheries Committee. Unfortunately, the Committee has exerted a negative influence on Japan policy. Furthermore, organized crime elements (*mafiya*) in the Russian Far East have come to dominate the two-way trade in marine products with Japan, even establishing links to organized crime in Japan (*yakuza*). Smuggling crab into Japan has become a huge moneymaker for the crime groups, whose blood feuds often explode into violence in Japan and Russia. As pointed out, a top Russian border guard official was murdered in Yuzhno-Sakhalinsk in the spring of 2002. Russian authorities linked the attack to the 'fish mafia,' because the official was reportedly trying to crack down on illegal smuggling operations.[140] Another Russian citizen, the captain of a Russian crab-fishing boat, was murdered in the Japanese port of Wakkanai in the summer of 2001. Reportedly, he had ties with the Russian *mafiya* and was involved in smuggling operations.[141] These groups have an interest in maintaining the status quo. Whatever influence Russian organized crime might have, they would undoubtedly use it to keep Japan from assuming control of the disputed islands. Meanwhile, their hold over the local fishing industry is likely to continue as long as the economic and social situation in the Russian Far East remains as depressed as it is today.

The Russian military-industrial complex (known by the Russian acronym VPK) became a big proponent of stepped-up relations with the nations of East Asia, not just China, but also Japan, South Korea, and the nations of Southeast Asia. The state's major defense contractor Rosvooruzheniye was created in 1993 out of several firms and it began an export offensive centered primarily on East Asia. The VPK and Rosvooruzheniye's best clients include China, India, Indonesia, Malaysia, and South Korea.[142] Leaders from the VPK made it clear on more than one occasion that they desired cooperation with Japan.[143] When it became clear that Japan was unlikely to place big orders and that the United States had a lock on this market, the enthusiasm waned.

Russian institutions specializing in international relations saw their heyday under Gorbachev, when academics and specialists on Japanese affairs were routinely consulted about policy decisions. Under Yeltsin these institutions, particularly the Academy of Sciences, suffered when funding was drastically curtailed. Forced to make their own ends meet, many resorted to renting space within their buildings to private firms, Russian and foreign. In IMEMO, where this author had an office from 1999 to 2001, almost half of the 18 floors were occupied by foreign firms, including the Finnish cellular firm Nokia.

Under Putin the situation improved somewhat, but this did not necessarily translate into good news for Japan. The Council on Foreign and Defense Relations headed by Sergei Karaganov was supposed to have some influence in the Kremlin. Nevertheless, it became clear that Karaganov, a protégé of Evgenii Primakov, did not share the same interest in Japan as his mentor. In early 2000 Karaganov published an article on the importance of economic relations in Russia's new foreign policy in the weekly *Moskovskie Novosti*. In his lengthy piece mention of Japan was conspicuously absent.[144] Gleb Pavlovsky who runs the Efficient Policy Fund is said to have influence on Putin but has never expressed interest in Japan. Aleksandr Dugin, head of the organization Eurasia Movement, argues for a strategic partnership with Japan, aimed against China. Putin is reportedly interested in many of Dugin's theories, although whether Japan is part of the interest is unclear.[145]

Under Putin Japan experts from the Academy of Sciences have been consulted on an ad hoc basis and are occasionally asked to submit policy recommendations. IMEMO experts were consulted by the presidential administration at times, although the institute is seen as being too soft on Japan. Before Putin's trip to Japan in September 2000 experts from the institute's Japan team were asked to submit a policy paper. The paper characterized relations with Japan as an "open wound" and called for a peace treaty with Japan, even if it meant a handover of territory. The paper outlined what Russia could gain from the normalization of relations with Japan: a larger role for Russia in East Asian affairs, including in Korea; a balance against China's rise; and substantial Japanese investment for the Russian Far East. The last could help not only rebuild the region's infrastructure, but also help to clean up the impending ecological catastrophe. In summation, IMEMO's experts suggested that Russia return the disputed islands to Japan for approximately $25–30 billion.[146]

The Institute of the Far East (known by the acronym IDV in Russian) of the Academy of Sciences was also consulted by the Kremlin occasionally, and its director, Mikhail Titarenko, was said to have close ties to individuals in the presidential administration. The Japan group at IDV was seen as more hard-line and felt that reaffirmation of the 1956 Joint Declaration was as far as Russia should go. Whether they had any influence or not, this is the direction that the Kremlin chose to go.[147] The accession of the Putin administration and the Kremlin's recognition of the necessity for the funding of social science research gave new hope to the private think tanks and the Academy of Sciences that perhaps they would once again become respected players in policy circles.

Grass-roots organizations with any influence over foreign policy are scarce and it is unlikely they will have any influence over the policymaking process in Russia other than in domestic issues. Those organizations with the potential to influence relations with Japan (such as environmental groups, fishing organizations, or manufacturers' associations) are either non-existent or are controlled (as is the case with fishing organizations) by organized crime elements. Other groups with interest in Japan (such as energy or metals associations, or veterans' organizations) can be grouped with industry or the VPK. Residents of the disputed islands could potentially develop an effective lobbying organization, but they seem rather divided themselves on the issue.

It would seem that the international changes and the exigencies of the tense strategic situation in East Asia in the late 1990s had brought Japan and Russia together to the brink of *rapprochement*. If international factors generally served to bring Japan and Russia closer together in the late 1990s, then why could not interested domestic actors and institutions in both countries help the process along during the early years of the twenty-first century?

It seems that certain domestic factors in Japan and Russia partially helped to bring about the *rapprochement* of 1996–2007, but domestic factors in each nation mostly played an obstructionist role. Also, it must be noted that Russia experienced a true domestic transformation, economically, politically, and socially. Change in Japan, on the other hand, was more modest. Many would argue that there was no true change. Domestic actors in Japan, still accustomed to dealing with Moscow as they had dealt with the Soviet leadership, were unable to recognize the dramatic opportunities that presented themselves with the remarkable transformation of Russian politics and society, or to think creatively enough to formulate new policies in this regard. The political change in Japan was marginal. In Russia, however, the democratization of the political system meant that Russian leaders were now somewhat beholden to public opinion, which was clearly aligned against the further reversion of territory. Strong leaders such as Boris Yeltsin and Vladimir Putin were hesitant to stake their political careers on an issue of questionable importance for Russia.

Japanese society was experiencing more change than the domestic political establishment in Tokyo. These societal changes, however, created more angst than hope. Change on the Japanese political scene was slow to come and was met with administrative and organizational resistance. Japanese politicians were slow to react to change. In Russia, the enormous task of bringing a society along together with the fundamental political transformation that had taken place left little room on the political agenda for foreign relations, particularly those issues that are peripheral to national security. A territorial dispute with Japan over four small islands was obviously low on the political agenda of Moscow.

6 The ideational context

1996–2007: The role of perception and historical memory

Perception and subjective historical memory are two key elements that influence public opinion and policymaking in nations.[1] Images of other nations and cultures exist at all levels of a society, from the elite policymakers down to the ordinary citizens. And these images do not always necessarily correspond to one another. Oftentimes, however, the perceptions toward another country are created at the top among the small circle of the political elite. Hence, perceptions are often viewed among the society at large through the lens of these elite groups. Throughout the twentieth century this has often been the case with Japan and Russia. Perceptions formed (or skewed) by groups of political elite are even more preponderant between Japan and Russia given the enormous geographic distance between the main population centers of the two nations, and the scant history of large-scale direct cultural interaction.[2] It is important to consider that as democracy caught hold, first in Japan, and then later in Russia, policymakers became answerable to public opinion. But in the case of Japanese-Russian bilateral relations, public opinion very much mirrored the government line, as perceptions and images had been planted by the governments into the public mind in previous decades, and there they still remain for the most part.

In this study, 'perceptions' are meant to portray the ideas of the general public, which of course includes the political elite. In Chapter 5 there are several references to the perceptions of the political elite in each nation (for example, how business "elites" in both countries viewed one another). This chapter portrays the wider images and perceptions that existed in the publics of both nations. It must be kept in mind that the Japanese-Russian relationship is influenced by the fact that interaction has been minimal for more than a century, and so the tenor of domestic politics in each nation still colors mutual perceptions. This began to change somewhat in the 1990s, but the negative stereotypes that existed throughout the twentieth century have continued to persist into the twenty-first century. In addition, although there are exceptions to the rule, the elite and the publics of both nations view the bilateral relationship in a strongly unified fashion. Opinion polls taken in both nations during the late 1990s and over the last few years clearly show that the publics of both nations were against a peace treaty if it meant sacrificing territorial integrity.[3]

This section will examine the four major issues that have colored mutual perceptions over the past few years: the role of historical memory, the end of the Cold

War, national identity in this time of international transformation, and the forces of globalization and democratization. These issues are all directly related to the failure to effectively address outstanding issues between Japan and Russia. As such, each of these four issues has negatively affected the chances for normalization. As in earlier periods of *rapprochement*, whatever the international or domestic exigencies may have been, ideational factors were a major hindrance to normalization efforts, even when top leaders may have felt that normalization was a necessary step.

The role of history and historical memory

History and historical memory, and how they affect perceptions in both Japan and Russia, have been important to the bilateral relationship for the better part of the past century. Perhaps no one issue has had more to do with national identity in both Japan and Russia over the past six decades than the Second World War. For both Japan and Russia the Second World War has been a defining event, shaping the political and social development of each nation over the past 60 years. These two nations, with the exception of China (and possibly Germany and Poland), can be said to have suffered more from the war than the other combatants. Japanese and Russian views of the war are contrary to one another, particularly in how they view Japanese-Soviet interaction between 1938 and 1945. Aleksandr Panov, Russia's former ambassador in Tokyo, wrote, "When examining [Russo-Japanese relations], inevitably one comes across 'negative historical memory,' a phenomenon that seems impervious to change and that frustrates all attempts to resolve long-standing antagonisms."[4] While discussions around the world center on the definitive ending of the Cold War, Japan and Russia are still mired in issues of the Second World War.

The influence of the Second World War on Japan remains enormous. Its political system still reflects the war experiences, particularly the pacifist constitution. Japan's relations with virtually all of its Asian neighbors are still affected by what transpired more than 60 years ago. Japanese politicians are constantly fending off criticism from Asian colleagues, not only about Japan's behavior in the region during the war, but also about statements issued by politicians even today. As mentioned, Ryutaro Hashimoto is on record stating that it is "matter of definition whether Japan committed aggression against Asian nations during the Second World War."[5] Hashimoto, Koizumi, and other Japanese prime ministers have regularly visited Yasukuni Shrine, the site where Japan's war dead, including convicted war criminals, are memorialized. Asian nations decry Japan's behavior – past and present – and demand if not restitution, then at least an apology. Both China and South Korea have been issued apologies (of varying degrees), and the governments of both nations agreed to desist from demands for further reparations (South Korea after 1965, and China after 1978). Yet somehow the two nations seem unsatisfied with Japan's atonement. Japan and Japanese politicians spend much time defending their nation's record and the specter of the 'Great East Asian War' hangs over the country.

But the people of Japan have also developed their own 'victim mentality' (*higaisha ishiki*) in relation to the Second World War, and this has greatly affected relations with Russia.[6] The Soviet Union invaded Japan in the closing days of the war (August 8, 1945) and seized territories that it had been promised (by the United States and Great Britain) at Yalta. The capture and imprisonment of hundreds of thousands of Japanese civilians and soldiers in camps in Siberia and Central Asia is an image that still haunts many Japanese today.[7] Many of those imprisoned soldiers and civilians never returned, and the Japanese feel that a forthright explanation and apology has never been issued from the Russians. According to one historian, Tsuyoshi Hasegawa, the Japanese suffer from "historical amnesia," and to many Japanese "the Soviet-Japanese War served as a psychological means by which the Japanese acquired a sense of victimization, which served as a major excuse to avoid atonement for the Pacific War."[8] Japanese conservatives earlier used this anger to build a strong domestic political platform based on anti-communism. Meanwhile, the black and white nature of the Cold War years allowed this negative image of the Soviet Union to take root in Japan and spread through all levels of society. As Hasegawa points out, there has been no serious reckoning in Japan concerning its action in the rest of East Asia. And there also has been no resolution concerning the guilt of the Japanese government in not ending the war sooner. While lamenting the suffering of Japanese POWs at the hands of the Soviets, Japanese refuse to examine how they treated POWs (civilians and soldiers) from China, Korea, Southeast Asia, and the Allied forces, and comfort women from all over the region. This is not to mention how poorly the Okinawans fared at the hands of the Japanese Imperial Army.

In his seminal history of Japan in the immediate post-war years, John Dower wrote that in the post-war period Japanese regarded themselves "victims rather than victimizers." Dower goes on to point out that since the war Japanese have looked to find scapegoats, rather than examining themselves (as many Germans have done).[9] Some have even questioned whether all Japanese desire a resolution of the territorial dispute. It can be speculated that if the Japanese people were to demand redefinition of the boundaries of the Kuril Islands as part of a 'righting' of injustices from the war, then they would also have to come to terms with the injustices committed by Japanese troops across all of East Asia. Japanese leaders and Japanese society may not be ready for this. The Japanese government insists on putting the war behind it in discussions with Asian neighbors, but with Russia it insists on achieving justice and vindication from the war. The ghost of the Second World War will continue to haunt Japan as long as its people and its leaders refuse to come to a full accounting of what transpired during those painful years. Relations with Russia very much hinge on whether Japan can come to a complete reckoning with this historical specter.

Russians are as caught up in the history of the Second World War as the Japanese. Russians also harbor somewhat of a 'victim mentality' from the war. Most Russian families lost at least one person in the 'Great Patriotic War.' But in Russia the war is celebrated as one of the greatest events in Russian history. As Tchaikovsky wrote an overture to celebrate victory over Napoleon, Shostakovich wrote a symphony immortalizing the brave citizens of Leningrad, who withstood a siege of Nazi forces for almost three years. Artwork, films, and literature all celebrate the war, and Red Army veterans still proudly walk the streets of Russia's cities with medals hanging from their lapels. May 8 Victory Day parades in Russia are as common as parades on the Fourth of July in the United States. For Russians May 8 is the true national holiday. Even Vladimir Putin has confessed that his hero growing up was the fictitious spy from the film *Stirlitz*, which portrayed a Soviet agent who infiltrated the highest ranks of the German Wehrmacht in the Second World War.

Russians justify what happened in Northeast Asia in August 1945 by saying that they came to the aid of an ally, and were helping to bring the war to a quick end. Although this explanation may sound self-serving (especially in light of Stalin's strategic designs in East Asia), almost every Russian feels comfortable with it. They feel that Russia won the war and that the territory it gained at the end was earned through sacrifice. In today's Russia, given the unsettled state of the economy and society, the humiliations associated with losing the Cold War, and the territorial loses in Europe and Central Asia, the victory in the Second World War is truly the last great twentieth century historical event that Russians can point to with pride. Hence Russia's reluctance to give up territory it gained from Germany (Kaliningrad) and Japan (the southern Kuril Islands). In fact, most of the territory returned from Russia to other former Soviet states in the 1990s was land that did not belong to the Russian Republic under the Soviet constitution. It was the territory of other republics (Ukraine, Belarus, etc.) and giving it away did not compromise the territorial integrity of the Russian Federation, at least on paper. Russia has clung fiercely to its own territory, as the war in Chechnya demonstrates. In fact, besides Kaliningrad and the southern Kuril Islands, Russia has territorial disputes with Finland, Estonia, and Latvia, and it has not budged an inch in any of these disputes.[10] Many Russians feel that by giving back any of the Kuril Islands they would be retreating from gains made by sacrifices during the Second World War. Many consider this too high of a price to pay.

Politicians in Japan and Russia have used the war legacy to great effect domestically. But the inability to move past wartime issues has proven to be a major stumbling block that directly influences the territorial dispute. The territorial dispute may indeed be a 'symptom' rather than a 'source' of the troubled relationship, but before any disease is cured, the symptoms must normally be suppressed.

The end of the Cold War

The demise of the Soviet Union (and the subsequent ending of the Cold War and the bipolar system that had been in place for almost five decades) was an epochal event matched during the twentieth century only by the shifts in the international system at the conclusion of the two world wars. The end of the Cold War presented Japan and Russia the opportunity to finally normalize relations after decades of acrimony. Leaders in both nations seemed to grasp this and steps were accordingly taken, but they were quickly dashed in 1991–92. It was not until the late 1990s that either side seemed to make genuine efforts at reconciliation. In spite of the realization at the highest levels of leadership in each country that normalization would do much to help their nations in a time of uncertainty, a peace treaty was never signed. Disputes dating back decades continued to plague the relationship, holding up any *rapprochement* before it could reach its full fruition. Why were the two nations unable to take advantage of the tremendous opportunity presented by the end of the Cold War to normalize relations? Although leaders in both countries understood the necessity of improved relations, they were unable to explain to all political circles and to their respective publics why normalization made sense. Many of the prejudices that had been planted in the consciousness of the publics during the Cold War by the two governments in fact impeded later efforts by officials to improve relations. The publics in Japan and Russia were unwilling to compromise the position that they were taught by their respective governments to back for the previous five decades. All through the Cold War, each side was told to dislike and distrust the other; now it is difficult for the people of each nation to unlearn these ideas.

While the rest of the world applauded first Mikhail Gorbachev's *perestroika* and then Boris Yeltsin's *demokratizatsiya*, Japanese leaders and the public merely seemed to yawn and shrug. One Japanese diplomat was reported to have said, "Of course, they [the Russians] have abandoned Communism. But for us, it is basically the same country. The people are the same. And I think we have a difficult neighbor again."[11] The excitement engendered around the world by Russia's transformation never really took full root in Japan. There was euphoria among the Japanese in the early 1990s, but this was centered more on the belief that the new Russia would likely hand over the disputed territory. In other words, the new Russian government was initially viewed positively in Japan simply because it was believed to be more apt to return territory, not because the transformation boded well for Japanese-Russian relations or for Japan's national and strategic interests.

Some in Japan were able to grasp the larger significance of Russia's transformation. Leaders such as Ryutaro Hashimoto and Keizo Obuchi genuinely desired normalization with Russia, but the support they received was scattered and ineffective. As pointed out, several groups such as a small coterie of bureaucrats in the Foreign Ministry, officers in the Japan Self-Defense Forces, major trading companies, and others also supported a fundamental transformation in relations with Russia. But even among these groups there were differences as to how far *rapprochement* should go. Although the Cold War had ended, many within Japan considered the principal issues in East Asia left over from the era of bipolar confrontation unresolved, particularly the territorial dispute over the 'Northern Territories.'[12] One Russian historian argues that some groups in Japan profited immensely from the Cold War, and are reluctant to give up their positions. Even though the era of US-Soviet confrontation has passed, Japan has gained no tremendous advantage, and in fact many Japanese feel that the nation is now faced with even greater threats.[13] This idea may not be far from the mark.

Although the military threat of the Soviet Union disappeared virtually overnight, Japan did little to bring this about (at least in the minds of most observers), and so there was no clear sense of victory or vindication in Japan, as there was in the United States. Japanese conservatives were left without an enemy and without the rationale for an alliance that had defined Japan's security strategy for five decades. And ultimately the territorial dispute remained unresolved, something perhaps even more troubling to the people of Japan than to the politicians. For decades conservative politician had used this dispute to maintain an unshakeable hold on domestic power. Additionally, unlike in the West many in Japan saw continuity, not change, between Tsarist Russia, Soviet communism, and Russian democracy.[14] Russia was still the same bad neighbor; only the nature of the security threat from the north had changed slightly. Russia's strength was no longer a major factor of insecurity in Japan. Instead, Russia's weakness created a host of different problems including the threat of an authoritarian backlash, lawlessness and 'loose nukes' in the Russian Far East, and environmental degradation that threatens Japan and its citizens even more directly than Soviet nuclear weapons did.

Russia's economic weakness and social problems, especially in the Russian Far East, might give many in Japan cause to reconsider their position. Japan could be a key in preventing the further deterioration in Russia's economic and social situation in the Far East, which left unchecked could cause even greater threats to Japan than the Cold War-era Soviet military threat. But among the Japanese public there has been little support for a revitalization of relations with Russia. Public attitudes toward Russia have been slow to change. The Japanese Prime Minister's Office sponsored a series of annual surveys conducted by the Shin Jōhō Center between 1993 and 1999 asking Japanese respondents how closely they felt toward Russia. Never more than 15 percent of respondents said they felt close toward Russia (this highest figure in 1999), while the percentage that did not feel close to Russia consistently topped 80 percent, reaching a high of 87 percent in 1996. In the eyes of the Japanese public, relations with Russia offer Japan "no visible [or] viable national interest either in the negative or positive sense."[15] Polls taken in 2005 seemed to be no different.[16] Japanese political leaders (such as Hashimoto and Obuchi) may have understood why relations should be improved, but in spite of their best efforts, they failed to convince their own citizens.

Both Japanese political leaders and the public have been hard-pressed to consider the last vestiges of the Second World War laid to rest until the 'Northern Territories' have been returned to their rightful owners – the Japanese. According to the *iriguchi-ron* argument, Japan should not improve relations until the disputed territories have been returned. But a Catch-22 situation presents itself: how can the territorial dispute be settled without an improvement in relations? Kazuhiko Tōgō and other proponents of the *deguchi-ron* argument understood that relations must be dramatically improved before a territorial dispute favoring the Japanese side could be reached. But Tōgō, Hashimoto, and other supporters were too busy trying to outmaneuver political opponents of their policies to take the agenda to the people, where with persuasive arguments they might have been able to establish widespread support.[17] Unlike in the West where the press was sometimes accused of being overly sympathetic toward Gorbachev and Yeltsin, the Japanese press – both conservative and liberal – was never fully convinced of Russia's transformation and so it never fully explored all the options open to Japan in Russia policy.[18] Most of the major dailies consistently stuck by the creed of no compromise until all islands were returned, especially after Prime Minister Mori, backed by Muneo Suzuki, unveiled the 'two-track' approach in 2001. Even writings that describe the crumbling situation in Russia rarely touch on the implications for Japan. They seem more focused on highlighting how poorly the people live on the disputed islands in order to justify Japan's claims.[19] Efforts by Hashimoto, Obuchi, and the top leadership to use the press to expound their convictions and views were half-hearted and often aimed at political opponents. Tōgō's views were expressed in the conservative *Sankei Shimbun*, but his name was kept a secret and the controversial articles were published under the *nom de plume* 'Mr. X.' Japanese leaders and the Japanese public seem unable to move beyond this obsession with the territorial dispute, which has been referred to as the "Northern Territories syndrome."[20]

On the other hand, the Russian government and the Russian public seem more favorably disposed toward putting the vestiges of the past behind them, but they are unwilling to pay too big of a price to normalize relations with Japan. For Japan, the Cold War meant confrontation with the Soviet Union. For the Soviets, the Cold War was mostly about America and Europe. Thus, for most Russians, the Cold War has become a memory, albeit a fresh memory. Although the territorial dispute with Japan remains, very few Russians will tell you they believe the Cold War continues because of this.

Russia's top political leaders have been somewhat ambivalent about Japan. Although Japan was seen as a great economic power in the later years of the Soviet Union, many Russians consider Japan's political prowess vastly underdeveloped in proportion to its economic standing. Russian political leaders understand the advantages of improved relations with Japan. Yet thus far they have been unwilling to compromise Russia's sovereignty and their political position at home to further the *rapprochement* with Japan, thus no territory has changed hands. The Cold War is indeed over for Russia, and Russia lost. Russia's top leadership knows this, but political leaders are not about to endure further territorial loses to improve a relationship from which Russia may derive only economic benefits (the extent of which are still unknown).

The Russian public is probably more open to improved relations with Japan, because they sense the possibility of tangible economic benefits. Opinion polls taken in 1992 and 1998 revealed positive attitudes among those surveyed toward Japan. Approximately 73 percent of Russians surveyed in 1992 had a favorable image of Japan.[21] By 1998 this figure had fallen, but a majority (56 percent) still expressed a liking for Japan.[22] More recent surveys have suggested that these levels still hold. Surveys taken in 2007 suggested that more than half (56 percent) of Russians interviewed thought relations with Japan were "good," while 83 percent felt that relations with Japan were important for Russia.[23] Traditionally, Russian (and Soviet) people have been presented with conflicting images of Japan. During the Cold War the Soviet propaganda machine created an image of Japan on the verge of remilitarization.[24] But often, at the same time, the Soviet press often insisted that the Japanese people were 'peaceful' and desired good relations with the Soviet Union.[25] It should also be pointed out that the memory of Japan's undeclared attack on Port Arthur in 1904 still negatively resonates in Russia today. Later, the economic miracle of Japan was widely touted in the Soviet Union, and Japan was even seen as a potential model for *perestroika*.[26] This uneven attempt to paint a negative image of Japan still colors perceptions in Russia today. Japan is viewed favorably by many, but it is still a country far away in the minds of the average Russians living in European Russia, while in the minds of those in the Russian Far East Japan is often viewed as a nation with territorial 'pretensions.'

Japan and Russia view the end of the Cold War through completely different lenses. For many Japanese, the end of the Cold War has heralded neither peace nor stability. The international system held in place by the Cold War benefited the Japanese, and if anything the international system today presents more dangers and threats in the minds of many Japanese. Russian citizens, though no doubt nostalgic for the halcyon days of the Soviet Union, are happy that the Cold War system of bipolar confrontation that demanded such a high price in terms of economics and personal freedoms is over. One of the most lasting legacies of the Cold War, the US-Japan Security Treaty, is a marked example of the discrepancy in views in Japan and Russia. Ironically, just as many Japanese are beginning to wonder about the efficacy and the long-term viability of the US-led security alliance, some in Russia are beginning to see the benefits of this alliance. As the Cold War meant different things to the Japanese and to the Soviets for five decades, so the end of the Cold War has meant different things to the Japanese and the Russians over the past decade. Consequently, undertaking a fundamental reassessment of the territorial dispute has not been easy for either nation. Each considers its claims 'just' and 'legal,' and any attempt to compromise has been met with opposition both at home and in the bilateral negotiating process.

The ending of the Cold War offered a good opportunity for Japan and Russia to put aside the last vestige of the Second World War and to move forward with the bilateral relationship into the twenty-first century. Both nations failed to overcome the negative perceptions that colored the lenses through which the two publics viewed one another, and in failing to do so they failed to normalize relations.

National identity

Throughout the late nineteenth century and the entire twentieth century Japan and Russia have been in search not only for a national identity or mission, but also simply for a 'normal' place in the international system. Since the economic and political modernization processes got under way in Russia (in the early eighteenth century) and in Japan (in the mid-nineteenth century) the two nations have been in a continual game of 'catch up' with the West, against which all standards of development were set.[27] These standards were not only economic, but also often cultural, political, and social. While seeking economic or political models to follow, both Japan and Russia have also constantly sought a unique national identity for themselves. Prior to the First World War, Russia often to looked to its pan-Slavic heritage and the institution of the Tsar. Up until the Second World War, Japan built on its system of emperor veneration and military glorification, which evoked the earliest days of the Japanese nation-state. The Soviet Union, meantime, had built an identity around the ideology of communism, which in the beginning was purely international but which had mutated into a uniquely Russian institution by the early 1930s. Although communism later spread, the brand of socialism in the Soviet Union maintained its uniqueness. Japan rebuilt itself from the ashes of defeat in the 1940s and by the 1980s it was practicing its own unique brand of capitalism.

With the collapse of communism in Russia and the decline of Japan's economic prowess in the 1990s, both nations find themselves on the cusp of economic, political, and societal transformation. And once again the two nations find themselves in an uneasy position between the East and the West. Today the definition of East and West takes on a different connotation than even ten years ago. During the nineteenth century the East generally referred to the decaying empires from the Ottoman and the Persian to the Chinese. The West referred to Europe and to a lesser extent the United States. Russia was clearly in between – it was a multi-ethnic empire, although it was modernizing economically and militarily, if not politically. Its decay was not yet apparent to most. Japan was seen as a copy of China – an overripe fruit ready to fall from the tree. But its subsequent amazing modernization clearly placed it in a whole different category. During the Cold War 'the East' generally referred to the communist camp led by the USSR, while 'the West' referred to the capitalist camp led by the United States, with Japan firmly ensconced in the latter. Today 'the East' generally refers to East Asia, and 'the West' refers to the United States and the European Union. Both Japan and Russia are in the process of seeking a new national identity and a 'natural' place in the world, but for now they still occupy an uneasy ground between 'the East' and 'the West.' Ironically, in many situations this may cause both nations to view one another with more sympathy, but with Japan and Russia this has clearly not been the case.

Japanese leaders have the task of refurbishing their economy and their political system. Leaders are reluctant to tamper with a system that brought them to power. At the same time, Japan – its leaders and its people – seems to be groping for some sort of an international role and a position for itself that can help define the national course for the century ahead.[28] Among the issues debated today in Japan include constitutional revision and the nation's right to exercise collective self-defense. These issues strike at the very core of the ideals set forth in the (US- authored) constitution written in the late 1940s. Most Japanese citizens have held dearly Japan's identity as a pacifist nation, particularly those who suffered through the 15 years of war and the seven-year US military occupation. Many Japanese hoped that Japan would become the 'Switzerland of Asia.' The United States wanted Japan to become the 'Great Britain of Asia.' Ultimately the truth lies somewhere in between. Now a new generation of politicians and citizens are confronted with a new world and new choices. Given new security dilemmas (e.g. China's rise and the *perceived* ambiguous US security commitment) Japanese are closely exploring options that would greatly change the *modus operandi* that has been in place for five decades. Japan has already undertaken a significant modernization of its military forces, and it is beginning to call for a stronger political role both regionally and globally. Most attention in the press has been given to Japan's economic and domestic political transformation, but the economic troubles and political stalemate constitute only two parts of the troubling equation that confronts Tokyo. Japan's security policy is at a similar crossroads.[29] Some in Japan have even begun debating Japan's non-nuclear status, wondering whether a continuation of this policy is practical or realistic for Japan in the twenty-first century.[30]

Underlying these important questions of policy and strategy lays the fundamental issue of identity. Who are the Japanese? What does Japan represent? Japan has never gone out on a limb to support global ideals or systems such as democracy or free-market economics. It is debatable whether Japan has been fully embraced in the West, in spite of its status of a G-8 member. It has been said that Japan's economic policies are unique and not in line with Western capitalism. Nor is Japan fully embraced in Asia. It is seen as a rich nation, with a clouded past that has never made amends either to its East Asian neighbors or even to its own citizens. There is great angst among Japanese as to the nation's future course and well-being. As long as these issues remain unresolved it will be difficult to come to terms completely with Russia, in spite of the fact that improved relations with Russia may be a key to answering some of the dilemmas Japan now faces.

Russia also finds itself in a middle ground. Russia is in the midst of domestic political and economic transformation, and is confronted with new security dilemmas, involving not only China but also the increasingly unstable southern flank along the trans-Caucasus region and Central Asia.[31] Democratization has been a painful process for most Russians. Today, it seems clear that Russia's national interests are changing, and although NATO is still seen as a potential threat, the threat from the west has receded to a level unseen for centuries. Russia sees new threats from the East and the South for the first time in generations. But the enduring interests of national sovereignty and territorial integrity, though greatly compromised throughout Russia over the past decade, still exert a negative influence on relations with Japan, a nation that clearly poses no security threat to Russia. The people of Russia are not ready to give up four of the Kuril Islands (whose area comprises not much more territory than the metropolitan area of Moscow) when over the past decade the borders making up the territory governed from Moscow have shrunk to an area not seen since the days of Catherine the Great in the eighteenth century. Russia has given up sovereignty over Belarus, the Ukraine, and the Crimea in the Black Sea. And yet it clings defiantly, almost stubbornly, to the southern Kuril Islands, which it has ruled over only since 1945. Even more perplexing is that the Russian government agreed to give back disputed territory to China, a nation who many in the Russian Far East view as Russia's number one potential strategic threat in the twenty-first century. Meanwhile Japan, seen as a potential economic lifeline for Russia, is being pushed away because of Russia's intransigence over the territorial issue.

Russia stands at a more important crossroads in its economic and political development than Japan. No one can question Japan's economic footing in the camp of the most advanced industrialized nations. Many Russians, however, clearly live at third world standards, as anyone who has traveled in the regions of Russia can testify. Whereas Russia's unique position between East and West once qualified the nation to serve as a bridge between the 'advanced' West and the 'backward' East, now Russia lags behind many of its Asian neighbors economically. As such it has developed an inferiority complex of being both between and behind the East and the West.[32] Vladimir Putin's push to move Russia closer to the West economically has reopened the age-old issue in Russia that has divided the ruling and intellectual elite for centuries. Should Russia move to integrate with the West, or define its own unique path? This debate in Russia really began in the late seventeenth century during the reign of Tsar Aleksei, Peter the Great's father. In the nineteenth century the debate was formulated in the context of 'Slavophiles' versus 'Westernizers.' The term 'Eurasianists' has also been synonymous with Russians that support a unique position for their country in the world order, one that is neither Eastern nor Western, but uniquely Russian. These terms are used loosely and interchangeably in Western historiography of Russia, but Russians more generally use the terms '*Zapadniki*' [Westerners] and '*Pochvenniki*,' which could be translated as 'nativists,' and which comes from the Russian word '*pochva*,' meaning earth or soil.[33] Putin is said to have taken an interest in the writings of Aleksandr Dugin, head of the organization 'Eurasia Movement.'[34] Eurasia Movement argues for a strategic partnership with Japan, India, and Germany, aimed against China and the West.[35] Although 'Westernizers' are said to be winning out in Russia, the debates about Russia's identity and its future path are not about to die off, and it is certain that the '*Pochvenniki*' will be around for a long time in Russia. Questions involving Japan are still up in the air, because in spite of what Dugin says, it is not clear where Japan fits in the eyes of Russia's 'Eurasianists.' Is it part of the East or the West? Or does Japan occupy a unique middle ground like Russia (and India), thus presenting opportunities for Moscow? Once this issue is resolved, perhaps Japan's role in Russia's future will be somewhat clearer.

How do the two nations view one another as they grapple with the issues of economic and political development, security and strategy, and national identity? One scholar, Mari Kuraishi Horne, analyzed Japanese-Russian relations in the context of the historicist-development perspective. Horne argued in 1989 that the pattern in bilateral relations from the middle of the nineteenth century was rooted in

> the uncertain competition between two nations which were so clearly focused upon the First World [i.e. the West], emulating it, catching up with it, and proving themselves against it .... The uneasy competition is currently in the process of breaking down as the historical parity of their starting points falls away, and as the pretended superiority of the Soviet Union ... is discarded.[36]

Horne argued that Japan and Russia have long been locked in a race toward modernization and parity with the West, and that they viewed one another as competitors in this race. She surmised that with the end of the Cold War and Japan's unquestioned economic superiority, competition between the two will die out and the territorial dispute, which was more 'symptom than source' will cease to divide the two. This has obviously not proven to be the case. Nevertheless, the point that Japan and Russia have been uneasy competitors is something that should be taken into consideration in understanding what transpired in the latter half of the 1990s. If the Russians/Soviets viewed the Japanese as peer competitors, then this could help explain the reluctance of leaders in Moscow, and for Russians in general, to accept the Japanese demands for territorial reversion. Even more difficult to swallow would be to accept Japanese demands, and then turn around and accept Japanese economic assistance. It is hard enough to do so with the United States and Europe, but to do so with an Asian neighbor could be viewed within Russia as the ultimate sign of the nation's complete and utter defeat, and its inferiority before the developed world. Russians might have too much pride to accept such a scenario (especially given Russia's recent resurgence due to booming energy prices), and perhaps this is what keeps them from normalizing relations with Japan.

Similarly, Japan's failure to regain its lost territories after the collapse of the Soviet Union must also be viewed not only with indignation around the country but also with a measure of disgrace.[37] Russia accepted territorial compromises with its European neighbors, with China, and even with its neighbors in Central Asia. Japan, however, failed to come up with a solution to the territorial dispute. Not only did the government and the Foreign Ministry look impotent, but Japanese dignity surely took a heavy blow with its inability to resolve this last remaining issue from Japan's humiliating experience in the Second World War. The new quest for national identity and worth in each country at the dawn of the twenty-first century seems to preclude a solution to the primary issue dividing Japan and Russia.

Democratization, globalization, and the twenty-first century

Clearly Japan and Russia stand at a crossroads politically, economically, and strategically. At the same time global technological forces are also changing society and culture in both nations. American dominance in political and security issues, in the marketplace, and in global cultural affairs is perhaps historically unprecedented. Although Japan and Russia are adjusting to new realities in the "shadow of a rapidly rising China," in fact, the shadow cast by the United States is much larger.[38] The US influence, however, comes in a much different form than that of China. Both China and the United States are seen by many people in Japan and Russia as 800-pound gorillas. But the American factor also appears in other ways. Democratization and the free-market economy have become the ideology driving the foreign policy of the United States (during the Cold War a certain realism always pervaded US policy). Technical advances in high-speed communications, combined with America's traditional cultural exports (film, music, and television), present platforms with which to spread this ideology. Japanese and Russians are big consumers, not only of America's traditional exports, but also of this very ideology that was seen as the antithesis of everything Japanese more than 50 years ago and everything Russian merely ten years ago. Many Japanese and Russians have become eager apostles of the American way.

Nevertheless, the growing American dominance concerns many in Japan and Russia. It is not the United States *per se* that represents a threat, but rather the anarchic forces associated with globalization and free economic interaction that threaten to create instability in these two countries, just as they are attempting to reform and rationalize their political and economic systems.[39] "The United States presented both the Russians and the Japanese with a more pernicious problem .... A young, rich, rootless, arrogant country, swaggering around the globe, the United States came to represent the worst nightmare of traditionalists everywhere, including Russians and Japanese."[40] This passage was written to convey the image of the United States in Japan and Russia at the beginning of the twentieth century, although it is as appropriate for today's world. At the beginning of the twentieth century Japan and Russia felt threatened by America's vibrant growth, especially as the United States began moving into East Asian markets and as it became politically active in the region. In some ways it galvanized the two and brought them together on certain issues (see Chapter 1). Today, however, it is proving much less easy for traditionalists in either country to overcome America's 'soft power.' In fact, Japanese-Russian cultural exchanges remain at a minimum, partially because of the US cultural dominance that goes hand in hand with 'globalization.'[41]

In Japan the adjustment to the new forces of globalization has been much less traumatic than in Russia, but nevertheless it has been a painful process for many. Many of Japan's largest export industries (e.g. electronics, microprocessors) have been leap-frogged in terms of technological advances by American, Asian, and European competitors. Japan has long since lost its comparative advantage in many of its traditionally strong heavy industries (e.g. shipbuilding and steel) to neighbors such as South Korea and China. Taking advantage of the communications revolution and cheaper prices, many Japanese firms have moved manufacturing centers and assembly plants to Southeast Asia. Over the past decade there has been talk in Japan of the 'hollowing out' (*kudōka*) of Japanese industry.[42] With unemployment rates hovering at historically high levels and the ongoing economic stagnation, there is great concern among many whether Japan can wake up in time to meet these new challenges head-on, or whether it will once again be reduced to playing a game of catch up with the United States and the West.[43] The debate within Japan on the effects of globalization and how Japan should respond is only beginning, but it is widespread and can expect to continue for a while.[44] What direction Japan takes will have large consequences for its relations with Russia, which is also agonizing over the decisions forced upon it by the new forces of globalization.

The effects of democratization and the new economy on Russia cannot be overstated. The collapse of the Soviet Union was in fact at least partially spurred by the growing economic dominance of the West due to technological innovation. How Russians now feel about democratization and economic liberalism is a mixed bag. Some have indeed profited; but the vast majority of Russians can be said to be economically worse off than they were 20 years ago. The term 'globalization' is not heard widely in Russia, but its effects, viewed synonymously in Russia with '*demokratizatsiya*,' are well known. Without going into a full-scale analysis of Russia's economic performance over the past decade, suffice it to say that the recent Russian economic boom has not brought economic prosperity to all in Russia.

Much of the blame, rightly or wrongly, is placed squarely at the door of the United States. It is easier to blame scapegoats. The United States (and Mikhail Gorbachev, who is accused of having become too close with the West) is widely considered the main culprit. The movement against 'globalization' takes on a variety of forms in Russia – leftist political opposition, grass-roots movements (such as *pochveniki*), religious fundamentalist groups, and other such organizations – and it has so far made little progress.[45] But Putin's plan to integrate Russia economically with the West may meet more widespread opposition if Russia's economy does not register sufficient growth, and as Russia is asked to meet certain requirements in order to qualify for membership in organizations like the WTO. As in the debates between 'Westernizers' and 'Eurasianists,' it is unclear exactly where Japan fits in the Russian perceptions. Japan is viewed as an advanced economic power, and as such Putin's emphasis on good relations with the West and with economic revitalization should generally bode well for Japanese-Russian relations. But if Russia's economy can grow and become competitive, will Japan and Russia once again lapse into the "uneasy competition" that Horne described?

Discussions among intellectuals in both Japan and Russia about the 'national identity' appeared with the onset of the post-Cold War era of globalization. But at the same time arguments purporting to demonstrate the 'uniqueness' of each nation seem to have become less widespread with the onset of the economic and political liberalization and transformation.[46] Both nations once again confront a major challenge from the West. Both nations once again feel sandwiched between East and West, but greatly desire to be part of the West. In the past this parallel path of political and economic development did not bring the two nations closer together. The globalization and the challenges of the twenty-first century may be a new and different case, but an examination of history will demonstrate that there is little reason to assume these forces will bring Japan and Russia closer together.

The major international events of the 1990s helped pave the way for the *rapprochement* between Moscow and Tokyo in 1996–2007. Clearly, several of the top leaders in both nations recognized the value of improving relations with one another. Domestic forces below the top level of the leadership, however favorably they may have viewed *rapprochement*, were clearly aligned in helping to prevent a complete normalization of relations. The respective publics of both nations and many top officials never seemed entirely convinced of the necessity of normalization, even after 1996. Indeed, during this period the two nations barely registered in the public consciousness of one another. To a large degree this has to do with the fact that the peoples of the two nations were simply too preoccupied with turbulent societal and international change to notice one another. In addition, neither the Japanese nor the Russian public was prepared to give up many of the legacies – political, historical, ideological, and social – that had so strongly defined the bilateral relationship for the past six decades.

The history of Japanese-Russian mutual perceptions has been from the very beginning defined by misunderstanding and mistrust. The earliest contacts in the eighteenth and nineteenth centuries, though sometimes amiable, more often resulted in localized armed conflict. Throughout the twentieth century even when the two nations reached a political accommodation, mistrust was never far below the surface. Ironically, mutual respect between Japan and Russia may have been greatest during the Russo-Japanese War of 1904–05, and in the years immediately after the war. This mutual respect gave way to suspicion once again not long after the shooting stopped. The series of treaties of alliance, friendship, and neutrality signed between 1907 and 1941 (see Chapter 1) were never able to completely mask the latent hostility that existed in ruling circles and among the public in both nations. The eras of *rapprochement* during the Cold War (see Chapter 2) were short, and never achieved any lasting result, beyond mutual recognition in 1956. The territorial dispute was never recognized by the Soviet Union, and the Soviet annexation of territory in 1945 was never recognized by Japan.[47] Additionally, the public perception of one another in both nations suffered from attitudes shaped both by the Second World War and the Cold War. During these eras of *rapprochement* international, and in some cases domestic, factors forced leaders in both nations to recognize the utility of harmonious bilateral relations. But rarely during the twentieth century did the two publics recognize and acknowledge the benefits of *rapprochement*. The good intentions of several top political leaders in both Japan and the Soviet Union/Russia invariably ran into the wall of public and official opposition to normalization. The cultures, histories, and psychologies of Japan and Russia are vastly different, save perhaps in one important area: pride and the unwillingness to lose face or back down. "Both nations link this bilateral relationship to national dignity," noted one American scholar.[48] "The Japanese and the Russians are highly culture-bound peoples cordially disliked by most of their neighbors, including one another. It would be hard to name any pairs of peoples less well suited by temperament and culture to get along with each other," said another American scholar in the 1970s.[49] Although this description was used to summarize relations in the 1970s, it captures the perceptions of many in Japan and Russia today, as well.

The end of the Cold War has not ameliorated issues that continue to linger from the Second World War, and from even further back in time. Hiroshi Kimura wrote:

> Certainly, once a tradition or an image becomes fixed, it tends to persist for a very long time. As long as much stronger evidence to the contrary is not forthcoming, the power of beliefs and habits that sustain long-standing images, principles, and actions remains strong. The extremely limited nature of information exchange between [Japan and Russia] only spurs on the hardening of entrenched images and perceptions.[50]

Perceptions of the war (or subjective historical memory) and its long aftermath continue to poison the atmosphere among the populace of both nations. Indeed, the two nations have yet to sign a peace treaty ending the Second World War. The changes wrought by globalization add uncertainty to the volatile mix. Before the issues of national identity and the nation's place in the era of globalization are resolved among both peoples, it will be difficult for Japan and Russia to come to a mutual realization that normalization could be of immense benefit to both nations. The mutual images of one another in both countries have not had time to be reformed in line with more contemporary realities. This can only come with time, and in the case of Japan and Russia it could be a long time.

Conclusion

Japan and Russia in 1996–2007

The characteristic patterns of Japanese-Russian relations in the twentieth century repeated themselves over the past decade. The period 1996–2007 was similar to that of earlier periods of Japanese-Russian *rapprochement*. Like the earlier eras, the initial hopes and expectations that accompanied the *rapprochement* came to a grinding halt in the face of the numerous domestic obstacles and political realities in each nation.

A look back at each period of *rapprochement* during the twentieth century will reveal similar motivations for a *rapprochement*, similar obstacles, and similar reasons for the ultimate failure to bring about a complete normalization in relations. Prior to the Second World War Japan and Russia (and the Soviet Union) on several occasions were brought together because of factors related to the international situation. Leaders in both countries, in spite of mutual dislike and suspicion, recognized the expediency of amicable relations and dealt with the distaste of moving closer to one another for short periods of time. But in each instance the distrust was latent enough that it was merely a matter of time before relations again unraveled back to the original state of acrimony. The first period of *rapprochement* (1907–1916) might have conceivably developed into a more long-lasting stable bilateral relationship, but political developments in both nations prevented this from ever happening. This is one of the interesting historical speculations on what 'might have been' had the two nations developed along similar lines from oligarchic-dominated autocracies to liberal, constitutional monarchies. The other two pre-war periods of *rapprochement* (1923–25; and 1941–45) seemed doomed even from the start. In the 1920s Japanese and Russian leaders were openly critical of one another even though they recognized overlapping strategic interests in Northeast Asia and extended mutual diplomatic recognition. The neutrality of 1941–45 could be best summed up as the 'temporary neutrality.'[1]

The Cold War periods of *rapprochement* were also brought about because of international factors that were prone to change at any moment, thus exposing the two nations' deep distrust of one another. As with earlier periods, the *rapprochements* in the mid-1950s and mid-1970s were pushed from the top by leaders that were usually incommunicative with their respective publics. The leaders could recognize the expediency and the utility of *rapprochement*, but they seemed unprepared to contemplate a strategy to build a long-lasting, mutually beneficial relationship. One might assume that although leaders understood the big picture, they were uninterested in the long-term picture. No foundation for a strong bilateral relationship existed, and the leaders failed to build one. As such these leaders were unable to convincingly persuade their own people of the necessity of normalization. Attempts at true normalization repeatedly foundered due to an uncommitted leadership and an unconvinced people. It is also important to note that international structural factors, while giving rise to the eras of *rapprochement* in the 1950s and 1970s, also often were key in preventing normalization and bringing to a halt the brief periods of amicability.

If the improvement in relations during the latter half of the1990s and the early years of the twenty-first century held so much promise, why did the two nations fail to normalize relations? Why does this inability to normalize relations exist? Why are the two nations still unable to sign a peace treaty ending the Second World War despite the domestic changes in each country, and the global changes that have transpired since the end of the Cold War?

A distinguishing feature of the earlier periods of *rapprochement*, as pointed out, has been the initial push from the top levels of leadership due to international factors. In the late 1990s these same factors and motivations existed (as described in Chapter 4). Each period of *rapprochement* was an era of international turmoil or transformation. The 1990s were no exception. Indeed, the 1990s marked the end of the Cold War, the defining factor of the international system for five decades. The demise of this bipolar system had a tremendous influence in both Japan and Russia, and on their bilateral relations. Leaders in both nations, forced to adjust to the new geopolitical realities of the 1990s, scrambled to make reassessments of their respective national strategies. They realized that amicable relations were not only desirable but necessary. But, as in each of the earlier periods, the leaders of both nations failed to make convincing arguments to their respective governments and publics.

Domestic change was also a prominent feature in both nations during each period. Although the change that existed in each country to vary degrees was partly responsible for leaders in both nations to contemplate *rapprochement*, on the whole domestic factors exerted a negative influence on bilateral relations. In the late 1990s and into the twenty-first century, Japan and Russia both saw heated domestic opposition to any compromise over territory. Leaders in both nations who were convinced of the necessity of better relations for strategic reasons almost without exception met with opposition, even from members of their own political party or affiliation. These opponents of compromise, and hence normalization, were able to rally public support much more effectively than those leaders who favored *rapprochement*, even though the latter group may have been much more far-sighted. The domestic political scene in each country during the 1990s and into the twenty-first century clearly reflected this phenomenon once again.

In the period 1996–2007, as in each previous era of *rapprochement*, ideational factors were also a decidedly negative element. Negative stereotypes, misunderstandings, and misperceptions have generally been the rule in the history of Japanese-Russian relations, especially since the end of the Second World War.

Although some scholars point to the pre-Second World War era as a time of amicability between Japan and Russia,[2] the periods of *rapprochement* even then were short-lived, and the mutual distrust was thinly disguised. The acrimony after the Second World War grew even deeper, not only because of that war (the fourth in four decades between Japan and the Soviet Union/Russia) but also because of the legacy of the Cold War and the influence that it had on domestic political developments in each country. These two major events (the Second World War and the Cold War) had a negative influence on mutual perceptions. These perceptions continue to exist today. Ideational factors, therefore, weighed in against normalization, and combined with domestic factors they effectively killed off the *rapprochement* of 1996–2007. Had the international situation in East Asia in the 1990s deteriorated significantly, Japan and Russia might very well have normalized relations. For example, had China exhibited extremely aggressive behavior in the region, or had the United States undertaken a major strategic withdrawal from the region (due to world or domestic events), then it seems that leaders in Moscow and Tokyo might have been sufficiently alarmed to undertake a fundamental reassessment of their respective strategic situations. However, international exigencies and trends, as alarming as they were, were not enough to send the two nations into one another's arms. Over the course of the next few years and decades, given unsettling trends in Korea and elsewhere in the region, Japan and Russia just may agree that they cannot move forward without normalizing relations and becoming strategic partners. In the period 1996–2007, there was just not strong enough of an external stimulus to force the top leadership to act and make bold policy decisions.

Thus, by 2007 the two nations were unable to sign a peace treaty, end the technical state of war, and find a territorial solution, in spite of the positive momentum that was created by the international situation and by the motivations of the top leadership in both nations. The reasons for this failure repeated themselves from earlier in the twentieth century: in the end, domestic and ideational obstacles outweighed the exigencies of the international situation.

The major differences of 1996–2007

Having identified the similarities between the eras of *rapprochement* during the twentieth century, it also important to state that major differences did exist in the period 1996–2007. Although these differences were unable to positively affect bilateral relations to the extent that made normalization possible during the decade covered, they are nonetheless worth noting because they might come to play an increasingly important (and positive) role as the twenty-first century unfolds.

By the end of the 1990s, for the first time in 200 plus years of bilateral relations, neither Japan nor Russia viewed the other as a direct security threat. Throughout the first half of the twentieth century the two nations distrusted one another and fought four conflicts against one another. After the Second World War, Japan viewed the Soviet Union as the greatest threat to Japan's national security, while the Soviet Union saw Japan as potentially a major staging area for US forces in the event of an attack on the Russian Far East. By the mid-1990s this perception had been erased from the minds of almost all policymakers and military planners in both nations. Furthermore, a potential new security threat has arisen in the eyes of many in both nations – an increasingly powerful China. During the first half of the twentieth century China was in fact the opposite: it was a weak nation over which Japan and Russia (the Soviet Union) contested with one another for influence. In addition, as pointed out previously, the United States has in the last decade exerted a much more positive influence in the bilateral relationship, and has not obstructed positive relations between Japan and Russia as it had during the previous four decades. The Korea factor, as mentioned, has risen again after a period of nearly 100 years of dormancy. And as in the case of China and the United States, this issue adds a positive element to Japanese-Russian bilateral relations. Furthermore, for the first time since the Cold War, the initial push for *rapprochement* came not from Moscow but from Tokyo, as Japanese leaders seem to have grasped the implications of international change and began to think and act proactively.

Related to the international factors is the energy equation. This is also a new element that could portend a major improvement in bilateral relations. As mentioned, oil exports from the Russian Far East have reached Japan for the first time since before the Second World War. Whereas in the 1970s the Siberian energy projects were simply paper projects, today energy projects on Sakhalin are being realized. Additionally, there has been a significant push from Tokyo to connect Japan with oil and gas fields in Siberia (see Chapter 4). Were Russia to become a major energy supplier for Japan, this could have major positive repercussions for the bilateral relationship.

Also in line with international factors is the growing warmth between the defense establishments in Japan and Russia. As pointed out in Chapter 5, for the first time since the First World War, contacts between the two nations' defense establishments have proliferated. This demonstrates the strategic concerns on both sides that helped facilitate the *rapprochement* of the period 1996–2007. It will be important to monitor the international situation in Northeast Asia over the next ten years, because drastic structural shifts in the future could indeed bring Japan and Russia to the point where normalization is a necessary strategy.

Domestically, what sets the relationship apart from earlier eras of *rapprochement* is the fact that both Japan and Russia are now democracies, and the governments are answerable to the publics as they have never been before. Japan has been a democracy for 50 years, but the political shake-ups of the 1990s mean that politicians and political parties are now answerable to popular opinion. Meanwhile, Russia has undergone a complete transformation from a totalitarian system to a popularly elected government with a market economy. Public opinion in Russia is now a very important factor. This new element has not necessarily been good for bilateral relations, because any compromise between the two will entail some sort of sacrifice for one or both sides, and the publics have not yet demonstrated the willingness to make sacrifices.

Mutual perceptions have undergone some change over the past decade, but negative stereotypes still dominate, particularly in Japan. The perception of the security threat is gone, but mistrust and misunderstanding continue to dominate thinking in both nations. Before any normalization in relations in the twenty-first century can be contemplated one of two things must happen. First, the leadership of both nations must educate their respective populations on the merits of a meaningful *rapprochement* and normalization (and help to erase the negative images and stereotypes that leaders implanted in the minds of the public for nearly five decades). Or, what also might change the equation would be a transformation of international factors so large and compelling that the two nations would be virtually forced to cooperate. It would be preferable for education and time to bring Japan and Russia closer together in the future, rather than a critical shift in the international order in Northeast Asia.

In some respects the 1996–2007 period could be seen as a success: the improvement in relations is more far-reaching and fundamental than at any time in the past century. It can be said a foundation underlying the relationship is beginning to solidify. This era of *rapprochement* seems different from past eras of *rapprochement* because it seems based more on the recognition of mutual merit, and not just geopolitical expediency. But a serious transformation of the environment influenced by international, domestic, *and* ideational factors will be necessary if normalization is to be achieved.

Notes

Introduction

1 The territorial dispute involves what the Russian side calls the 'Southern Kuril Islands,' and what the Japanese side refers to as the 'Northern Territories.' The dispute comprises three islands and one small, uninhabited island group. From 1855 until 1945 Japan had possession of the islands of Etorofu (or *Iturup* in Russian), Kunashiri (*Kunashir*), Shikotan, and the Habomai group of islets. Upon the Soviet Union's entry into the Pacific War (at the behest of the United States and Great Britain) in August 1945, Red Army troops occupied the islands, and expelled over 17,000 Japanese inhabitants. Soviet and Russian citizens have lived there ever since.

2 In November 1997 in the Siberian city of Krasnoyarsk, Russian President Boris Yeltsin and Japanese Prime Minister Ryutaro Hashimoto agreed that the leaders of the two nations would do their utmost to sign a peace treaty by the year 2000. As the year 2000 approached, both sides switched the timing to the *end* of the year 2000.

3 In December 1995 Japanese Minister of State Masaki Nakayama stated that, "from a legal standpoint, Russia and Japan are still in a state of war." Quoted in Sergei Agafanov, "For Moscow and Tokyo, the War is Not Yet Over." *The Current Digest of the Post-Soviet Press*, vol. 49, Dec. 1995, p. 25. During the Korean War US officials were concerned that the Soviet Union would take advantage of this diplomatic loophole to attack Hokkaido in northern Japan. See Tsuyoshi Hasegawa, *The Northern Territories Dispute and Russo-Japanese Relations*, Berkeley, Calif.: University of California Press, 1998, p. 89.

4 The Russo-Japanese War was fought on land and sea in and around the Korean Peninsula and Northeast China. The war cost each side more than one hundred thousand dead, not to mention the thousands of lives of Korean and Chinese civilians. In one battle alone (Mukden) the combined casualties reached nearly 200,000, while combined casualties at the siege of Port Arthur amounted to nearly 100,000. See Meirion and Susie Harries, *Soldiers of the Sun: The Rise and Fall of the Imperial Japanese Army*, New York: Random House, 1991, pp. 74–93.

5 George Kennan, *Russia Leaves the War*, Princeton NJ: Princeton University Press, 1957, pp. 309–311. See also Frederick Dickinson, *War and National Reinvention: Japan and the Great War, 1914–1919*, Cambridge, Mass.: Harvard University Asia Center, 1999.

6 Throughout this work I will refer to Japanese names using given names first and family names second, as is the custom in the West.

7 Consult the Bibliography for an extensive list of these and other such articles.

8 Tsuyoshi Hasegawa, Hiroshi Kimura, and Gilbert Rozman have all published excellent English-language books on this topic, but they all date from the late 1990s, prior to the presidency of Vladimir Putin. See the Bibliography for a list of their publications.

9 See, for example, Kimie Hara, *Japanese-Soviet/Russian Relations since 1945: A*

Difficult Peace, London: Routledge, 1998, pp. 231–234. Hara feels that the territorial dispute still exists because the international system determined by the Yalta conference of 1945 and the San Francisco peace treaty of 1951 is still in place in East Asia. She argues that this Cold War legacy is what is hindering a full normalization of relations. Like Hara, Tsuyoshi Hasegawa places much of the blame for the territorial dispute and the current stagnation in relations squarely on the shoulders of the United States, and its Cold War strategy. See Hasegawa, *The Northern Territories Dispute and Russo-Japanese Relations*, p. 105.

10 See, for example, Hasegawa, *The Northern Territories Dispute and Russo-Japanese Relations*. In Russian, see Vladimir Yeremin, *Rossiya-Yaponiya: Territorial'naya Problema, Poisk Resheniya*, Moscow: Respublika, 1992.

11 The two biggest exceptions in this regard are two works published recently in Japan by two former MOFA officials who were deeply involved in the formulation of Japan's Russia policy. See Kazuhiko Tōgō, *Hoppo Ryōdo Kōshō Hiroku: Ushinawareta Gotabai no Kikai*, Tokyo: Shinchosha, 2007. Also, see Masaru Satō, *Kokka no Wana: Gaimushō no Rasupuchin to Yobarete*, Tokyo: Shinchosha, 2005. Unfortunately, for those analysts and historians who cannot read Japanese they are impenetrable.

12 Donald Hellmann, *Japanese Foreign Policy and Domestic Policy: The Peace Agreement with the Soviet Union*, Berkeley, Calif.: University of California Press, 1969.

13 Gilbert Rozman, *Japan's Response to the Gorbachev Era, 1985–1991: A Rising Superpower Views a Declining One*, Princeton: Princeton University Press, 1992.

14 For a brief period during the late 1960s Japan did become one of the Soviet Union's primary capitalist trading partners (along with West Germany), but the overall amount of trade was still a small fraction of the amount of overall trade for Japan. See P.D. Dolgorukov, "Torgovo-Ekonomicheskie Otnosheniya SSSR s Yaponiei." In Akademii Nauk SSSR, *SSSR-Yaponiya: K 50-letno ustanovleniya sovietsko-yaponskikh diplomaticheskikh otnoshenii (1925–1975)*, Moscow: Nauka, 1977, pp. 104–131. See also Kazuo Ogawa, "Economic Relations with Japan," in Rodger Swearingen, ed., *Siberia and the Soviet Far East: Strategic Dimensions in Multinational Perspective*, Palo Alto, Calif.: Hoover Institution Press, Stanford University, 1987, pp. 158–178. For First World War trade, see George Lensen, *Japanese Recognition of the USSR: Japanese-Soviet Relations, 1921–1930*, Tallahassee, Fla.: The Diplomatic Press, 1970, p. 322.

15 Taku Igarashi, "Genyuu no Senryaku Bichiku ni Ugokidashita Ajia," *Foresight Magazine*, May 1997, pp. 42–43; also, "Juuyu Ryushutsu ga Ukibori ni shita 'Ajia Oiru Rodo' no Fuan," *Foresight Magazine*, Feb. 1997, pp. 90–91; Shinjiro Mori, "Kasupi Kai no Nami ga Sawagu," *Asahi Shimbun*, Feb. 28, 1998, p. 5; Kaoru Sakurai, "Enerugi Riken Sodassen ni Yusaburareru Roshia Shin-naikaku," *Foresight Magazine*, May 1998, pp. 32–33. More recently, see Shoichi Ito, *Kiro ni tatsu Taiheiyo Paipurainu Kosō*, ERINA Report, vol. 72, Nov. 2006, pp. 23–33, also vol. 73, Jan. 2007, pp. 31–41.

16 Michael Armacost and Kenneth B. Pyle, "Japan and the Unification of Korea: Challenges for US Policy Coordination," *NBR Analysis*, vol. 10, no. 1 (Mar. 1999). Also, Joseph P. Ferguson, "Russia's Role on the Korean Peninsula and Great Power Relations in Northeast Asia," *NBR Analysis*, vol. 14, no. 1 (June 2003), pp. 33–50.

17 Cyril Black, Marius Jansen, Herbert Levine, Marion Levy, Henry Rosovsky, Gilbert Rozman, Henry Smith, and S. Frederick Starr, *The Modernization of Japan and Russia*, New York: The Free Press, 1975.

18 Japan did not actually recognize the People's Republic of China until 1978, but this late recognition was due more to American pressure than to Japanese strategic calculus. In fact, Japan's first post-war Prime Minister, Shigeru Yoshida, had advocated an immediate recognition of Communist China in 1949, as Great Britain had done. Japanese conservatives saw the Chinese market as a potential savior for the war-shattered Japanese economy. Unlike US leaders, they did not see an explicit threat in

Chinese communism. See John Dower, *Empire and Aftermath: Yoshida Shigeru and the Japanese Experience, 1878–1954*, Cambridge, Mass.: Council on East Asian Studies, Harvard University, 1988. Also, Howard Schonberger, *Aftermath of War: Americans and the Remaking of Japan, 1945–1952*, Kent, Ohio: The Kent State Press, 1989.

19 The Japan Communist Party and the Japan Socialist Party have consistently taken a harder line in relations with the Soviet Union -and Russia with regard to the territorial issue. Until recently the Japan Communist Party demanded the return of the entire Kuril Archipelago and the southern half of Sakhalin Island. See Anzen Hosho Mondai Kenkyuujo, *Kawaru Nichiro Kankei: Roshiajin kara no 88 no Shitsumon*, Tokyo: Anzen Hosho Mondai Kenkyuujo, 1999, pp. 192–195.

20 Robert Jervis, *Perception and Misperception in International Politics*, Princeton: Princeton University Press, 1976, p. 28.

21 The dropping of the atomic bombs on Hiroshima and Nagasaki also figure prominently in arguments that stress Japan's status as a victim in the Second World War. In fact, this is the act most commonly pointed to in decrying Japan's suffering during the war. However, Russia's 'treacherous' attack in the closing days of the war is a close second. Many Japanese think of this act as Americans think of the attack on Pearl Harbor. See Yoshi Numachi, *Haruka-na Shiberia: Sengo 50-nen no Shōgen (Vol. I–II)*, Sapporo: Hokkaido Shimbun Press, 1995. Also Ian Buruma, *The Wages of Guilt: Memories of War in Germany and Japan*, London: Jonathan Cape, 1994.

22 Former Duma member and Ambassador to the United States Aleksandr Lukin enunciated this idea of Russian cognitive dissonance most clearly. See Mingjie Yang, "Russia's Regional Security Role," in Koji Watanabe (ed.), *Engaging Russia in Asia Pacific*, Tokyo: Japan Center for International Exchange, 1999, p. 49.

23 Globalization is a term often used, or misused, without proper explanation. Generally, however, it can be seen as a combination of the twin forces of democratization and a free market economy. Spurring the forces are the tremendous advances in communications technology that marked the decade of the 1990s. These have all created a system of growing economic interdependence across the globe.

24 This idea was made clear to the author at a conference in Khabarovsk in the Russian Far East in September 1999 when Japanese and Russian participants spoke of the need to help one another integrate their economies into the new global economy dominated by information technology and the United States.

1 The patterns begin

1 It is a common misperception that Russia is a nation with little maritime tradition, but in fact Russian sailors did much in opening the northern Pacific Ocean to exploration. It should be remembered that Russia first colonized Alaska, northern California, and was among the first of the European nations to explore the Hawaiian Islands after Great Britain. Although Russia may have been geographically hemmed in by the Black and Baltic Seas in Europe, it has had a free outlet on the Pacific Ocean for almost 400 years. See Walter McDougall, *Let the Sea Make a Noise*, New York: Avon Books, 1993. Also see John Stephan, *Russia on the Pacific*, The Sanwa Lecture Series of the North Pacific Program at the Fletcher School of Law and Diplomacy, Winter 1989.

2 Peter Berton, *The Secret Russo-Japanese Alliance of 1916*, Ann Arbor, Mich., University Microfilms no. 16272, 1956, p. 265. See also George Lensen, *Japanese Recognition of the USSR*, pp. 363–371. Also, Lensen's *The Strange Neutrality: Soviet-Japanese Relations during the Second World War, 1941–1945*, Tallahassee, Fla.: The Diplomatic Press, 1972, pp. 186–220.

3 Kunadze interviews, Moscow 1999–2001. This author had frequent contact with Dr. Kunadze (who is considered one of Russia's top Japan experts) while in residence as a Fulbright Fellow at the Institute of World Economy and International Relations (IMEMO) of the Russian Academy of Sciences in Moscow from 1999 to 2001, where

Kunadze now serves as Deputy Director. Kunadze made this argument on numerous occasions.

4 Meirion and Susie Harries, *Soldiers of the Sun*, pp. 74–93. For the origins of the war consult Ian Nish, *The Origins of the Russo-Japanese War*, London: Longman Press, 1985.

5 Many scholars have argued that the alliance was aimed primarily at opposing the United States. More on this below.

6 Ernest Batson Price, *The Russo-Japanese Treaties of 1907–1916 Concerning Manchuria and Mongolia*. Baltimore: The Johns Hopkins Press, 1933, pp. 26–38.

7 *Ibid.*, pp. 39–58.

8 *Ibid.*, p. 45.

9 *Ibid.*, pp. 59–76. See also, David Dallin, *The Rise of Russia in Asia*, New Haven, Conn.: Yale University Press, 1949, pp. 123–133.

10 Price, *The Russo-Japanese Treaties of 1907–1916 Concerning Manchuria and Mongolia*, pp. 77–90. Also, Peter Berton, *The Secret Russo-Japanese Alliance of 1916*.

11 Chihiro Hosoya, "Japan's Policies Toward Russia," in James Morley, ed., *Japan's Foreign Policy, 1868–1941*, New York: Columbia University Press, 1974, p. 372. See also L.N. Kutakov, *Rossiya i Yaponiya*, Moscow: RAN, Vostochnaya Literatura, 1988, pp. 276–280.

12 Kutakov, *Rossiya i Yaponiya*, p. 315. On Japanese defense plan see Hosoya, "Japan's Policies Toward Russia," p. 375.

13 Price, *The Russo-Japanese Treaties of 1907–1916 Concerning Manchuria and Mongolia*, pp. 59–60.

14 Hosoya, "Japan's Policies Toward Russia," p. 378.

15 Dallin, *The Rise of Russia in Asia*, p. 61. Britain was concerned about Japanese designs in China, and was thus disinclined to support Japan in the Far East either against the United States or Russia (its newest ally).

16 Hikomatsu Kamikawa and Michiko Kimura, *Japanese-American Diplomatic Relations in the Meiji-Taisho Era*, Tokyo: Pan Pacific Press, 1959.

17 Walter LaFeber, *The Clash: US-Japanese Relations Throughout History*, New York: W.W. Norton and Company, 1997, pp. 95–96.

18 Kamikawa and Kimura, *Japanese-American Diplomatic Relations in the Meiji-Taisho Era*, p. 293.

19 Kutakov, *Rossiya i Yaponiya*, pp. 311–315.

20 LaFeber, *The Clash* , p. 96.

21 Kutakov, *Rossiya i Yaponiya*, p. 241.

22 Berton, *The Secret Russo-Japanese Alliance of 1916*. Japan also sent to Russia German guns captured from concessions in China that Japan had taken over in 1914. Frederick Dickinson, *War and National Reinvention*, p. 121.

23 Berton, *The Secret Russo-Japanese Alliance of 1916*.

24 Dickinson, *War and National Reinvention,* p. 145.

25 LaFeber, *The Clash* , pp. 45–98. See also, Dickinson, *War and National Reinvention*, pp. 31–83.

26 Kutakov, *Rossiya i Yaponiya*, p. 343.

27 *Ibid.*, pp. 276–277.

28 William C. Fuller, *Strategy and Power in Russia, 1600–1914*, New York: The Free Press, 1992, pp. 412–418, 423–425, 459. Bruce Parrott has pointed out that Russia feared an attack from both Europe *and* Asia at the same time.

29 Dickinson, *War and National Reinvention*, p. 157.

30 As quoted in Dickinson, *War and National Reinvention*, p. 179.

31 *Ibid.*, p. 227.

32 Barbara Heldt, "'Japanese' in Russian Literature: Transforming Identities," in J. Thomas Rimer, *A Hidden Fire: Russian and Japanese Cultural Encounters, 1868–1926*, Palo

Alto, Calif.: Stanford University Press and Woodrow Wilson Center, 1995.

33 Tetsuo Mochizuki, "Japanese Perceptions of Russian Literature in the Meiji and Taisho Eras," in J. Thomas Rimer, *A Hidden Fire: Russian and Japanese Cultural Encounters, 1868–1926*. Also see, N.I. Chegodar', "Russkaya i Sovietskaya Literatura v Yaponii v Posleoktyabrskii Period," in Akademii Nauk SSSR, *SSSR-Yaponiya: K 50-letno ustanovleniya sovietsko-yaponskikh diplomaticheskikh otnoshenii (1925–1975)*, Moscow: Nauka, 1977, pp. 208–224.

34 George Lensen, *Japanese Recognition of the USSR*, preface.

35 *Japonisumu no Nazo*, Asahi Gurafu, Tokyo: Bessatsu Bijutsu Tokushu, 1990, no. 1. See also, Barbara Heldt, "'Japanese' in Russian Literature: Transforming Identities."

36 George Lensen, *The Damned Inheritance: The Soviet Union and the Manchurian Crises, 1924–1935*, Tallahassee, Fla.: The Diplomatic Press, 1974, p. 450.

37 Quoted in J. Thomas Rimer, *A Hidden Fire*, p. 8.

38 Black, Jansen, Levine, Levy, Rosovsky, Rozman, Smith, and Starr, *The Modernization of Japan and Russia*.

39 LaFeber, *The Clash*, p. 78.

40 Kamikawa and Kimura, *Japanese-American Diplomatic Relations in the Meiji-Taisho Era*. For more on the Siberian Intervention see George Kennan, *The Decision to Intervene*, Princeton: Princeton University Press, 1958 and, *Russia Leaves the War*, Princeton: Princeton University Press, 1957. Also, James Morley, *The Japanese Thrust into Siberia, 1918*, New York: Columbia University Press, 1954.

41 Admittedly, a blockaded Russia had nowhere else to turn as an outlet.

42 John Fairbank, Edwin Reischauer, Albert Craig, *East Asia: The Modern Transformation*, Boston, Mass.: The Houghton Mifflin Company, 1965, pp. 513–529.

43 See James Morley, *The Japanese Thrust into Siberia, 1918*.

44 George Lensen, *Japanese Recognition of the USSR*.

45 Herbert Bix. *Hirohito and the Making of Modern Japan*, New York: HarperCollins Publishers, 2000, p. 705, note no. 59.

46 Lensen, *Japanese Recognition of the USSR*, pp. 8–9.

47 *Ibid.*, pp. 5–84.

48 Fairbank, Reischauer, Craig, *East Asia*, p. 680.

49 Lensen, *Japanese Recognition of the USSR*, pp. 85–139.

50 *Ibid.*, pp. 141–316. See also United States, Department of State, *Foreign Relations of the United States, The Conferences at Malta and Yalta, 1945*, Washington, DC: United States Government Printing Office, 1955, pp. 385–386.

51 Fairbanks, Reischauer, Craig, *East Asia*, pp. 676–681. Also, see Jonathan Spence, *The Search for Modern China*, New York: W.W. Norton and Company, 1990, pp. 305–310.

52 Lane, Harold. *The Japanese Exclusion Act, 1906–1924,* Unpublished Master's Thesis from the *George Lensen Collection* (Sapporo, Japan), Haverford University, 1929.

53 The Ministry of Foreign Affairs, *Nihon Gaik Bunsho (Documents on Japanese Foreign Policy)*, Tokyo: vol. I, 1924, pp. 286–315, 376–535, 664–665, 675–679, 759–761, 821–855.

54 Quoted in Patrick Mullins, *Japanese-Soviet Relations, 1925–1940*. Unpublished Master's Thesis from the *George Lensen Collection* (Sapporo, Japan), Tallahassee, Fla.: Florida State University, 1960.

55 Quoted in Jane Degras, *Soviet Documents on Foreign Policy (Vol. II, 1925–1932)*, Oxford: Oxford University Press, 1952–53, p. 9.

56 Xenia Eudin and Robert North, eds., *Soviet Russia and the East, 1920–1927: A Documentary Survey*, Palo Alto, Calif.: Stanford University Press, 1957, p. 254.

57 Lensen, *Japanese Recognition of the USSR*, pp. 196–202.

58 LaFeber, *The Clash*, pp. 148–149.

59 Lensen, *Japanese Recognition of the USSR*, p. 106.

60 Hosoya, "Japan's Policies Toward Russia," p. 393–394. See also Lensen, *Japanese*

Recognition of the USSR, p. 104.

61 Lensen, *Japanese Recognition of the USSR*, pp. 105–106.

62 For a look at Gotō's continuing efforts to improve relations see, Michio Yoshimura, "Gotō Shimpei Saigo no Hoso wo Megutte," *Gaikōshiryokanpo*, Ministry of Foreign Affairs of Japan,, no. 3 (Mar. 1990), pp. 50–66.

63 Adam Ulam, *Expansion and Coexistence: The History of Soviet Foreign Policy, 1917–67*, New York: Frederick Praeger Publishers, 1968.

64 George Kennan, *Russia and the West Under Lenin and Stalin*, Boston, Mass.: Little, Brown and Company, 1960, pp. 178–240.

65 Quoted in the *Japan Times*, Jan. 21, 1925, p. 1.

66 Lensen, *Japanese Recognition of the USSR*, pp. 343–344.

67 *Ibid.*, p. 98. This issue came up during discussions of who was to blame for the Nikolaevsk Massacre, and to whom an apology was owed.

68 *Ibid.*, pp. 137–142. The *Lenin* had on board over one million pounds of wheat, rice, and fish. Likewise in 1994 and 1995 the Russian government was reluctant to accept aid from the Japanese government in the wake of two horrible earthquakes on the islands of Shikotan and Sakhalin.

69 *Ibid.*, p. 344. For a description of the deterioration in bilateral relations after recognition see *ibid.*, pp. 343–361.

70 Hosoya, "Japan's Policies Toward Russia," pp. 395–396.

71 See Lensen, *The Damned Inheritance*, pp. 361–445.

72 Bix, *Hirohito and the Making of Modern Japan*, pp. 209–210.

73 Soviet leaders were also gravely concerned about a Japanese attack in the Far East. See Jonathan Haslam, *The Soviet Union and the Threat from the East, 1933–41: Moscow, Tokyo and the Prelude to the Pacific War*, Pittsburgh, Penn.: University of Pittsburgh Press, 1992.

74 Lensen, *The Damned Inheritance*, pp. 361–445.

75 Edmund Clubb, *China and Russia: The Great Game*, New York: Columbia University Press, 1971, p. 310. Also, see John Patton Davies Jr., *Dragon by the Tail: American, British, Japanese, and Russian Encounters with China and One Another*, New York: W.W. Norton & Company, 1972, pp. 192–193. Officers who participated in training Chinese soldiers included General Georgii Zhukov of the conqueror of Berlin fame, and General Vasilii Chuikov of Stalingrad fame. See also A.A. Arkad'yev, "Politika Yaponii v Otnoshenii SSSR Posle Nachala Vtoroi Mirovoi Voiny," in Akademii Nauk SSSR, *SSSR-Yaponiya*, p. 32.

76 L.N. Kutakov, "Bor'ba SSSR za Ustanovlenie i Razvitie Dobrososedskikh Otnoshenii s Yaponiei," in Akademii Nauk SSSR, *SSSR-Yaponiya*, pp. 7–30.

77 A. Lobanov-Rostovsky, *Russia and Asia*, Ann Arbor, Mich.: The George Wahr Publishing Company, 1965, p. 305.

78 Clubb, *China and Russia*, pp. 283–287.

79 The most complete account of these border wars can be found in Alvin Coox, *Nomonhan: Japan Against Russia, 1939*, Palo Alto, Calif.: Stanford University Press, 1985. Later at the Tokyo War Crimes Trial, and at another war crimes trial in Khabarovsk, Soviet prosecutors accused Japanese troops of experimenting with biological weapons against Red Army forces at Khalkhin-Gol/Nomonhan. See Bix, *Hirohito and the Making of Modern Japan*, pp. 351, 640.

80 See Lensen, *The Strange Neutrality*, p. xiv.

81 For a look of the effects of the purges on the Russian Far East, see John Stephan, *The Russian Far East: A History*, Palo Alto, Calif.: Stanford University Press, 1994, pp. 209–224.

82 Jonathan Haslam, *The Soviet Union and the Threat from the East, 1933–41*, pp. 135–136.

83 George Lensen, *The Strange Neutrality*, pp. 1–20.

84 *Ibid.*, p. 20.

85 John R. Deane, *The Strange Alliance: The Story of Our Efforts at Wartime Co-Operation with Russia*, New York: The Viking Press, 1947.
86 Lensen, *The Strange Neutrality*, pp. 258–267. The first such request was made on December 8, 1941, the morning after the Japanese attack on Pearl Harbor.
87 *Ibid.*, pp. 254–257.
88 Stephan, *The Russian Far East* , pp. 238–240.
89 Richard A. Russell, *Project Hula: Secret Soviet-American Cooperation in the War Against Japan*, Washington, DC: The Naval Historical Center, The US Navy in the Modern World Series, no. 4, Department of the Navy, 1997.
90 Lensen, *The Strange Neutrality*, pp. 105–155.
91 Haslam, *The Soviet Union and the Threat from the East, 1933–41*, pp. 136–137.
92 Soviet aid to the KMT was not insubstantial. See Clubb, *China and Russia* , and Davies, *Dragon by the Tail.*
93 Hosoya, "Japan's Policies Toward Russia," pp. 397–404. See also Haslam, *The Soviet Union and the Threat from the East, 1933–41*, pp. 135–150; and Bix, *Hirohito and the Making of Modern Japan,* pp. 350–357. Arkad'yev also gives a Soviet historian's view of which groups in Japan were most opposed to the Soviet Union during this time; see A.A. Arkad'yev, "Politika Yaponii v Otnoshenii SSSR Posle Nachala Vtoroi Mirovoi Voiny," pp. 38–40.
94 Robert Tucker, *Stalin in Power: The Revolution from Above, 1928–41 (vol. 2)*, New York: W.W. Norton and Company, 1990, pp. 257–258. Stalin was often noncommittal on foreign policy, and he encouraged subordinates to fight it out politically before he chose to go with one policy. In this way, if the policy failed, Stalin could blame it on the subordinate. Thanks to Bruce Parrott for pointing this out to me.
95 Lensen, *The Strange Neutrality,* pp. 15–20.
96 *Ibid.*, pp. 209–210.
97 Robert Whiting, *You Gotta Have Wa*, New York: Vintage Books, 1990, p. 45.
98 C.T. Mazhorov, "Sovietsko-Yaponskie Otnosheniya v Khode Vtoroi Mirovoi Voiny (1941–1945)," in Akademii Nauk SSSR, *SSSR-Yaponiya*, pp. 57–63. See also Lensen, *The Strange Neutrality*, pp. 130–155.
99 Bix, *Hirohito and the Making of Modern Japan*, pp. 494, 505–508.
100 Hosoya, "Japan's Policies Toward Russia," p. 404. See also Lensen, *The Strange Neutrality*, pp. 21–34.
101 Bix, *Hirohito and the Making of Modern Japan*, pp. 399–400.
102 United States, Department of State, *Foreign Relations of the United States, The Conferences at Cairo and Teheran, 1943*, Washington, DC: United States Government Printing Office, 1961, pp. 147, 489, 529, 618–619.
103 United States, Department of State, *Foreign Relations of the United States, The Conferences at Malta and Yalta,* pp. 361–400. See also, Lensen, *The Strange Neutrality*, pp. 258–267. Those concessions in China were the ones taken by Japan at the end of the Russo-Japanese War in 1905.
104 Coox, *Nomonhan*, pp. 1033–1074.

2 Cold War patterns

1 Aleksandr N. Panov, "Bor'ba Sovietskogo Soyuza za Mirnyi Demokraticheskii Put' Razvitiya Yaponii." in Akademii Nauk SSSR, *SSSR-Yaponiya*, pp. 65–82.
2 The Soviet delegation, led by a young official named Andrei Gromyko (who would later become Foreign Minister of the USSR), walked out of the signing ceremony in protest at the conditions. Soon thereafter the United States and Japan signed a security treaty, allowing the basing of US troops in Japan.
3 R.K. Jain, *The USSR and Japan, 1945–1980*, New Delhi: Radiant Publishers, 1981, p. 23.
4 Donald Hellmann, *Japanese Foreign Policy and Domestic Policy.*

5 *Ibid.*, pp. 31–32. Also, Motohide Saito, *The 'Highly Crucial' Decision making Model for Postwar Japan and Prime Minister Hatoyama's Policy Toward the USSR*, Ann Arbor, Mich.: University Microfilms, 1986, pp. 94–95.

6 By signing the San Francisco Treaty in 1951 Japan had renounced all claims to the Kuril Islands and Sakhalin, as well as to Korea and Taiwan. The treaty specified neither the sovereignty of these territories nor the exact geographical definition of the Kuril Islands. Japan claims that the four southernmost group of islands (the 'Northern Territories') are not and have never been geographically part of the Kuril archipelago. See Hasegawa, *The Northern Territories Dispute and Russo-Japanese Relations*. Also, see Hellman, *Japanese Foreign Policy and Domestic Policy*, and Saito, *The 'Highly Crucial' Decision making Model for Postwar Japan and Prime Minister Hatoyama's Policy Toward the USSR*. See also Rozman (ed.), *Japan and Russia: The Tortuous Path to Normalization, 1949–1999*, New York: St. Martin's Press, 2000.

7 Hellman, *Japanese Foreign Policy and Domestic Policy*, pp. 29–40. Hasegawa, *The Northern Territories Dispute and Russo-Japanese Relations*, pp. 106–124.

8 Savitri Vishwanathan, *Normalization of Japanese-Soviet Relations, 1945–1970*, Tallahassee, Fla.: The Diplomatic Press, 1973, pp. 78–82.

9 Hasegawa, *The Northern Territories Dispute and Russo-Japanese Relations*, p. 123.

10 *Ibid.*, pp. 124–127. Hara, *Japanese-Soviet/Russian Relations Since 1945: A Difficult Peace*, pp. 42–46. Saito, *The 'Highly Crucial' Decision making Model for Postwar Japan and Prime Minister Hatoyama's Policy Toward the USSR*, pp. 228–231.

11 For the Japanese viewpoint see Hasegawa, *The Northern Territories Dispute and Russo-Japanese Relations*; Hara, *Japanese-Soviet/Russian Relations Since 1945*; Saito, *The 'Highly Crucial' Decision making Model for Postwar Japan and Prime Minister Hatoyama's Policy Toward the USSR*. For the Russian viewpoint see, for example, Akakdemii Nauk SSSR, *SSSR-Yaponiya: K 50-letno ustanovleniya sovietsko-yaponskikh diplomaticheskikh otnoshenii (1925–1975)*. Also, Andrei Markov, *Rossiya i Yaponiya: V poiskakh soglasiya*, Moscow: RAN, Institut Dal'nego Vostoka, 1996, pp. 46–47.

12 United States, Department of State, *Foreign Relations of the United States, 1955–57, Volume XXIII, Part 1: Japan*, Washington, DC: United States Government Printing Office, 1991, pp. 310–313.

13 Bruce Elleman, Michael Nichols, and Matthew Ouimet, "A Historical Reevaluation of America's Role in the Kuril Islands Dispute," *Pacific Affairs*, vol. 71, no. 4 (Winter 1998–99), pp. 489–504. It should also be pointed out that the Japanese side was not above using the Soviet threat as a bargaining tool against the United States. The historian Howard Schonberger pointed out that in 1950 Japanese Prime Minister Yoshida attempted to bring pressure on the American government to negotiate an early peace treaty with Japan, before the Soviet Union did. He suggested that the Soviets might even offer to return the Kuril Islands and Sakhalin. Howard Schonberger, *Aftermath of War*, p. 244.

14 Vishwanathan, *Normalization of Japanese-Soviet Relations, 1945–1970*, p. 81.

15 Hasegawa, *The Northern Territories Dispute and Russo-Japanese Relations*, p. 135–139.

16 It is common knowledge that Soviet pilots flew pursuit missions against American fighter jets and bombers over northern Korea. It is less commonly known that Japanese minesweepers and coast guard cutters participated in naval operations off the coast of Korea, resulting in at least one death. See Walter LaFeber, *The Clash*, pp. 285–286.

17 Nikita Khrushchev, *Khrushchev Remembers: The Glasnost Tapes*, Boston, Mass.: Little, Brown and Company, 1990, pp. 83–90.

18 Saito, *The 'Highly Crucial' Decision making Model for Postwar Japan and Prime Minister Hatoyama's Policy Toward the USSR*, pp. 109–111. Hara, *Japanese-Soviet/Russian Relations Since 1945* , p. 39.

19 Dower, *Empire and Aftermath*, pp. 11, 387–388, 394–395, 401–414.

20 Vishwanathan, *Normalization of Japanese-Soviet Relations, 1945–1970*, pp. 61–65.
21 Edmund Clubb, *China and Russia*, pp. 228–253. See also, Edgar Snow, *Red Star over China*, New York: Grove Weidenfeld, 1968, pp. 161, 163, 423–424, 426–431.
22 "The Cold War in Asia," *The Cold War International History Project Bulletin*, Woodrow Wilson International Center for Scholars, issues 6–7 (Winter 1995/1996), pp. 4–29.
23 Khrushchev, *Khrushchev Remembers*, pp. 83–90.
24 Saito, *The 'Highly Crucial' Decision making Model for Postwar Japan and Prime Minister Hatoyama's Policy Toward the USSR*, pp. 105–117. Hellman, *Japanese Foreign Policy and Domestic Policy*, pp. 41–45. Hellman's work offers the best insight as to how Japan's domestic political situation influenced relations with the Soviet Union.
25 Hellman, *Japanese Foreign Policy and Domestic Policy*, p. 65.
26 *Ibid.*, pp. 140–141.
27 *Ibid.*, pp. 121–133.
28 Saito, *The 'Highly Crucial' Decision making Model for Postwar Japan and Prime Minister Hatoyama's Policy Toward the USSR*, pp. 282–283.
29 Khrushchev, *Khrushchev Remembers*, p. 85.
30 Aleksei Zagorsky, "Reconciliation in the Fifties: The Logic of Soviet Decision Making," in Rozman, *Japan and Russia: The Tortuous Path to Normalization, 1949–1999*, pp. 47–72.
31 James G. Richter, *Khrushchev's Double Bind: International Pressures and Domestic Coalition Politics*, Baltimore: Johns Hopkins University Press, 1994, pp. 54–55, 60–61, 70–71.
32 The Germans and the Poles might also be strong contenders for people that suffered most.
33 Vishwanathan, *Normalization of Japanese-Soviet Relations, 1945–1970*, p. 19.
34 Hellman, *Japanese Foreign Policy and Domestic Policy*, p. 32. For concern of socialist gains and communist subversion see also pp. 30, 48–49, 92–93, 95.
35 *Ibid.*, p. 49.
36 Stephan, *The Russian Far East*, pp. 246–249. Also, see Donald Gillin with Charles Etter, "Staying On: Japanese Soldiers and Civilians in China 1945–1949," *The Journal of Asian Studies*, vol. XLII, no. 3 (May 1983), pp. 497–518.
37 *The New York Times*, Oct. 29, 1994; *The Japan Times*, Nov. 19, 1997.
38 Hasegawa, *The Northern Territories Dispute and Russo-Japanese Relations*, pp. 72, 76. For an in-depth treatment of the Siberian POW issue see Yoshi Numachi, *Haruka-na Shiberia*. For one man's account see the interesting book by American-born *nisei* (second generation Japanese) Peter Sano Iwao, *1,000 Days in Siberia: The Odyssey of a Japanese-American POW*, Lincoln, Neb.: The University of Nebraska Press, 1997. Sano was in Japan when the war broke out in 1941 and in 1945 was drafted into the Imperial Japanese Army, and captured in Manchuria by Soviet troops at war's end.
39 Bix, *Hirohito and the Making of Modern Japan*, p. 511.
40 Saito, *The 'Highly Crucial' Decision making Model for Postwar Japan and Prime Minister Hatoyama's Policy Toward the USSR*, p. 216.
41 Khrushchev, *Khrushchev Remembers*, p. 86.
42 As cited in Zagorsky, "Reconciliation in the Fifties: The Logic of Soviet Decision Making,", p. 66.
43 See for example, Akademii Nauk SSSR, *SSSR-Yaponiya: K 50-letno ustanovleniya sovietsko-yaponskikh diplomaticheskikh otnoshenii (1925–1975)*.
44 Jain, *The USSR and Japan, 1945–1980*, p. 34.
45 Kazuo Ogawa, "Economic Relations with Japan," pp. 158–178.
46 Hasegawa, *The Northern Territories Dispute and Russo-Japanese Relations*, pp. 145–146.
47 LaFeber, *The Clash*, pp. 310–314.
48 Hasegawa, *The Northern Territories Dispute and Russo-Japanese Relations*, p. 141.

49 LaFeber, *The Clash*, pp. 352–358.
50 Ogawa, "Economic Relations with Japan," pp. 158–178.
51 *Ibid.*
52 Hasegawa, *The Northern Territories Dispute and Russo-Japanese Relations*, pp. 145–148.
53 Jain, *The USSR and Japan, 1945–1980*, p. 133.
54 *Ibid.*, p. 134.
55 Cited in Peter Berton's "Two Decades of Soviet Diplomacy and Andrei Gromyko," in Gilbert Rozman's *Japan and Russia: The Tortuous Path to Normalization, 1949-1999*, p. 76.
56 *Ibid.*, pp. 77–78. This was to be the first of many references to the now infamous East-Siberian-Pacific Ocean Pipeline that Japan and China have been lobbying for. Ito, *Kiro ni tatsu Taiheiyo Paipurainu Kosō.*
57 Hasegawa, *The Northern Territories Dispute and Russo-Japanese Relations*, p. 152.
58 Section on Tanaka visit to Moscow in Glaubitz, *Between Tokyo and Moscow: The History of an Uneasy Relationship, 1972 to the 1990s*, Honolulu: University of Hawaii Press, 1995. pp. 54–62. Also, see Tōgō, *Hoppo Ryōdo Kōshō Hiroku*, p. 93.
59 Walter Isaacson, *Kissinger: A Biography*, New York: Simon & Schuster, 1992, pp. 496–497. Raymond Garthoff, *Détente and Confrontation: American-Soviet Relations from Nixon to Reagan*, Washington, DC: The Brookings Institution, 1994, p. 404.
60 Attributed to Tadashi Aruga as cited in LaFeber, *The Clash*, p. 358.
61 The economic tensions resulted from Japan's ever-increasing trade surplus with the United States. In response, the Nixon administration forced a yen revaluation, imposed a surcharge on imports to the United States, and shut dollar-gold convertibility to all nations.
62 "Special Report: Japan and the Middle East," *Middle East Economic Digest,* Nov. 28, 1997.
63 Hara, *Japanese-Soviet/Russian Relations Since 1945*, p. 120.
64 Konstantin Preobrazhenskii, *KGB v Yaponii: Shpion, kotoryi lyubil Tokio*, Moscow: Tsenterpoligraf, 2000, pp. 56–89.
65 Gerald Curtis, "The Tyumen Oil Development Project and Japanese Foreign Policy Decision-Making," in Robert Scalapino, ed., *The Foreign Policy of Modern Japan*, Berkeley: University of California Press, 1977, p. 148.
66 *Ibid.*, pp. 154–160.
67 For a closer look at Tanaka and his legacy, see Jacob Schlesinger, *Shadow Shoguns: The Rise and Fall of Japan's Postwar Political Machine*, New York: Simon and Schuster, 1997.
68 LaFeber, *The Clash*, pp. 355–358.
69 Young Kim, *Japanese-Soviet Relations: Interaction of Politics, Economics, and National Security*, Washington, DC: The Center for Strategic and International Studies, 1974, pp. 72–75.
70 Curtis, "The Tyumen Oil Development Project and Japanese Foreign Policy Decision-Making," p. 160.
71 Martin Malia, *The Soviet Tragedy: A History of Socialism in Russia, 1917–1991*, New York: The Free Press, 1994, pp. 361–373.
72 Hara, *Japanese-Soviet/Russian Relations Since 1945*, p. 134.
73 Kim, *Japanese-Soviet Relations*, pp. 56–60.
74 Harry Gelman, *Russo-Japanese Relations and the Future of the US-Japanese Alliance*, Santa Monica, Calif.: RAND, 1993, pp. 6, 18.
75 Jain, *The USSR and Japan, 1945–1980*, pp. 87–105. Allen Whiting notes that not all fishing groups in Japan see eye to eye on fishery issues. See Whiting, *Siberian Development and East Asia Threat or Promise?*, Palo Alto, Calif.: Stanford University Press, 1981, p. 124. Also, based on interviews with Hiromasa Shinkura and Kenryō Yamaya, journalists from the *Hokkaido Shimbun*, Sapporo and Tokyo, Mar. 20, Apr. 9, 1997.

76 T.J. Pempel, *Regime Shift: Comparative Dynamics of the Japanese Political Economy*, Ithaca, NY: Cornell University Press, 1998, p. 46.
77 See Oleg Troyanovsky, "Soviet-Japanese Relations from 1967 to 1976." An unpublished paper presented at the conference *Japan and Russia: Postwar Relations, Mutual Influences, and Comparisons*, at Princeton University, September 1997. Troyanovsky was the Soviet ambassador to Japan from 1967 to 1976. Hasegawa concurs with this assessment; see Hasegawa, *The Northern Territories Dispute and Russo-Japanese Relations*, pp. 155–156. Troyanovsky also partially blamed the outbreak of the Yom Kippur War and the subsequent distraction for the failure of the 1973 summit to achieve any lasting results.
78 Jain, *The USSR and Japan, 1945–1980*, pp. 126–127. Also, Hasegawa, *The Northern Territories Dispute and Russo-Japanese Relations*, pp. 151–157.
79 Berton, "Two Decades of Soviet Diplomacy and Andrei Gromyko," p. 81.
80 Semyon Verbitsky, "Russian Perceptions of Japan," in James Goodby, Vladimir Ivanov, and Nobuo Shimotomai, eds., *"Northern Territories" and Beyond: Russian, Japanese, and American Perspectives*, Westport, Conn.: Praeger Publishers, 1995, pp. 64–65. Hasegawa, *The Northern Territories Dispute and Russo-Japanese Relations*, p. 145.
81 See for example, Georgii Kunadze, "Militarizm v Yaponii: Voprosy Metodologii Analiza," *Mirovaya Ekonomika i Mezhdunarodnye Otnosheniya*, no. 2 (1989), pp. 116–130. Kunadze's article analyzes the trend in Soviet writings on so-called Japanese militarism beginning in the 1970s. Also, see Myles Robertson, *Soviet Policy Towards Japan: An Analysis of Trends in the 1970s and 1980s*, Cambridge: Cambridge University Press, 1988, pp. 16–17. John Stephan also gives a brief listing of some of these articles. See Stephan's *The Kuril Islands: Russo-Japanese Frontier in the Pacific*, Oxford: Clarendon Press, 1974, pp. 206–207.
82 See for example, Tetsuya Kataoka, "Japan's Northern Threat," *Problems of Communism*, March–April, 1984, pp. 1–16. Also, see Glaubitz, *Between Tokyo and Moscow*, pp. 182–183.
83 Curtis, "The Tyumen Oil Development Project and Japanese Foreign Policy Decision-Making," pp. 154–171.
84 Kim, *Japanese-Soviet Relations*, pp. 68–69; Curtis, "The Tyumen Oil Development Project and Japanese Foreign Policy Decision-Making," p. 170.
85 Jain, *The USSR and Japan, 1945–1980*, pp. 120–121.
86 Hasegawa, *The Northern Territories Dispute and Russo-Japanese Relations*, p. 159.
87 Glaubitz, *Between Tokyo and Moscow*, pp. 181–198.
88 Hiroshi Kimura, *Distant Neighbors (Vol. I): Japanese-Russian Relations under Brezhnev and Andropov*, Armonk, NY: M.E. Sharpe, 2000, p. 211.
89 Glaubitz, *Between Tokyo and Moscow*, p. 137.
90 Berton, "Two Decades of Soviet Diplomacy and Andrei Gromyko," p. 79.
91 Kimura, *Distant Neighbors (Vol. I)*, pp. 144–148.
92 Tōgō, *Hoppō Ryōdo Kōshō Hiroku*, p. 98.
93 These included plans to construct a plant producing electro-magnetic steel plates, a timber resources project, exploration of resources off Sakhalin, coal projects in Yakutia, and the export of steel pipes to construct gas pipelines to Western Europe. Japan rethought its position on the last project and did sell the steel pipes. See Kimura, *Distant Neighbors (Vol. I)*, p. 227.
94 February 7 is the anniversary of the 1855 Shimoda Treaty that originally gave Japan control of the southernmost Kuril Islands, what the Japanese today refer to as the 'Northern Territories.' See Hasegawa, *The Northern Territories Dispute and Russo-Japanese Relations*, p. 169.
95 Hiroshi Kimura, "Japan-Soviet Political Relations from 1976–1983," p. 88.
96 Kimura dates the break to 1976. *Ibid.*, pp. 87–106.
97 Economic relations did continue to grow for a while, and ironically the worst years of political relations were the best years for economic relations. See Ogawa, "Economic

Relations with Japan," pp. 160–161. Also, Hasegawa, *The Northern Territories Dispute and Russo-Japanese Relations*, p. 166.

98 Statistics from the Japanese Ministry of Finance, cited in Okada Kunio, "Nikkei Kigyō Shinshutsu no Kōki wa?" in *Gaikō Forum*, no. 149 (Dec. 2000), pp. 56–61. Also, see Joachim Glaubitz, *Between Tokyo and Moscow*, pp. 282–283.

99 Kazuhiko Tōgō, *Nichirō Shinjidai he no Josō: Dakai no kagi wo Motomete*, Tokyo: Saimaru, 1993, pp. 42, 182.

100 Rozman, *Japan's Response to the Gorbachev Era, 1985–1991*, p. 151.

101 The 1956 joint declaration had stated that upon a signing of a peace treaty the Soviet Union would return the island of Shikotan and the island group of Habomai to Japan. See Nobuo Shimotomai, "Japan-Soviet Relations under Perestroika: Perceptions and Interaction between Two Capitals," in Gilbert Rozman, ed., *Japan and Russia: The Tortuous Path to Normalization, 1949–1999*, p. 112.

102 Hasegawa, *The Northern Territories Dispute and Russo-Japanese Relations*, p. 239.

103 *Ibid.*, pp. 246–249. See also Rozman, *Japan's Response to the Gorbachev Era, 1985–1991*, p. 151. Also, Tōgō, *Nichiro Shinjidai he no Josō*, pp. 42, 182.

104 Hasegawa, *The Northern Territories Dispute and Russo-Japanese Relations*, pp. 250–263.

105 Aleksandr Yakovlev, a member of the Politburo and Gorbachev's trusted lieutenant in the early days of his premiership, in an interview with the author, attested to the new thinking not only in the Soviet Ministry of Foreign Affairs and at IMEMO, but also in the International Department of the Central Committee of the Communist Party of the Soviet Union. Yakovlev interview, Moscow, March 2, 2000. One of Gorbachev's top aides, Anatoly Cherniaev, concurs with this assessment. See, A.S. Cherniaev, *Shest' let s Gorbachevym*, Moscow: Kultura, 1993. An English translation of the chapter Cherniaev wrote on Gorbachev's Japan policy can be found at the Woodrow Wilson International Center for Scholars' Cold War International History Project web site: <http://cwihp.si.edu/>. The document there is entitled "The Last Official Visit by M.S. Gorbachev as President of the USSR: The Road to Tokyo."

106 Both Rozman and Hasegawa date the change in thinking among Japanese to 1988. Hasegawa points in particular to the efforts of the new Soviet desk officer in the Japanese Ministry of Foreign Affairs in 1988, Kazuhiko Tōgō, to establish a new policy vis-à-vis the Soviet Union. See Hasegawa, *The Northern Territories Dispute and Russo-Japanese Relations*, pp. 285–289; Rozman, *Japan's Response to the Gorbachev Era 1985–1991*, p. 112. Nobuo Shimotomai dates the change in Japanese thinking to 1989, "Japan-Soviet Relations under Perestroika: Perceptions and Interaction between Two Capitals," p. 113.

107 Tōgō, *Nichiro Shinjidai he no Josō*, pp. 17–19, 24–25.

108 *Ibid.*, p. 60.

109 Dunlop, *The Rise of Russia and the Fall of the Soviet Empire*, Princeton: Princeton University Press, 1993. See also Archie Brown, *The Gorbachev Factor*, Oxford: Oxford University Press, 1997. Hasegawa describes Japan's diplomacy toward the Soviet Union during this period as being "immobile, inflexible, and lethargic." See Hasegawa, *The Northern Territories Dispute and Russo-Japanese Relations*, p. 329.

110 Schlesinger, *Shadow Shoguns*, pp. 202–251.

111 Yakovlev interview, Moscow, Mar. 2, 2000.

112 Tōgō, *Nichirō Shinjidai he no Josō*, p. 181. Also, Tōgō, *Hoppo Ryōdo Kōshō Hiroku*, pp. 126–129.

113 Sumio Edamura, *Teikoku Kaitai Zengo*, Tokyo: Toshishuppan, 1997, p. 62.

114 Hasegawa, *The Northern Territories Dispute and Russo-Japanese Relations*, pp. 364–368.

115 Andrei Kozyrev, *Preobrazhenie*, Moscow: Mezhdunarodnie Otnosheniya, 1995, p. 295. Also, Edamura, *Teikoku Kaitai Zengo*, p. 143. Gelman cites the figure $26 billion and gives a detailed account of how the aid would have been dispersed. See Harry

Gelman, *Russo-Japanese Relations and the Future of the US-Japanese Alliance*, pp. 21–23. Gelman also writes that the LDP, the *Keidanren*, and MITI were lined up with Ozawa against the Ministry of Foreign Affairs on this issue.

116 Vladimir Yeremin, *Rossiya-Yaponiya*, pp. 41–42. See also, Andrei Kozyrev, *Preobrazhenie*, p. 295. Also, Hasegawa, *The Northern Territories Dispute and Russo-Japanese Relations*, pp. 382–388. Artem Tarasov, a businessman who also had strong ties to Yeltsin, presented this report to the Soviet press.

117 This plan specifically called for: (1) Soviet recognition of the existence of the territorial dispute; (2) the establishment of a free economic zone on the disputed islands; (3) demilitarization of the islands; (4) the signing of a peace treaty; and (5) a settlement of the territorial issue by future generations. See Hiroshi Kimura, *Borisu Eritsin: Ichi Roshia Sejika no Kidō*, Tokyo: Maruzen Library, 1997, pp. iii–ix.

118 Hasegawa, *The Northern Territories Dispute and Russo-Japanese Relations*, pp. 389–407.

119 Edamura, *Teikoku Kaitai Zengo*, pp. 197–198, 205. Edamura, Japan's ambassador to Moscow, lamented Japan's slow reaction to the coup. Most Western governments had immediately denounced the coup, while Kaifu even briefly considered recognizing the new hard-line cabal. Also see Hasegawa, *The Northern Territories Dispute and Russo-Japanese Relations*, pp. 418–419.

120 Viktor Baranets, *Yel'tsin i ego Generaly*, Moscow: Kollektsiya 'Sovershenno Sekretno,' 1998, pp. 305–306. Glaubitz, *Between Tokyo and Moscow*, pp. 208–209, 212.

121 Tōgō, *Nichirō Shinjidai he no Josō*, p. 204.

122 Of the aid dispensed $1.8 billion was in the form of trade insurance for Japanese companies doing business in Russia (most of which was eventually used for energy projects), $500 million for humanitarian assistance distributed through loans by the Japan Bank for International Cooperation (JBIC), and $200 million in JBIC export credits (source: Ministry of Foreign Affairs, Japan's Assistance for the Russian Federation, April 2000). Some Russian critics claim that this aid was so tightly linked to Japanese institutions that it amounted to a mere "propaganda gesture." See Igor Latyshev, *Kto i kak prodaet Rossiiu: Khronika rossiisko-yaponskikh territorial'nykh torgov (1991–94 gody)*, Moscow: Paleia, 1994, p. 36.

123 Gelman, *Russo-Japanese Relations and the Future of the US-Japanese Alliance*, pp. 34–35.

124 Yeremin, *Rossiya-Yaponiya: Territorial'naya Problema, Poisk Resheniya*,, p. 130.

125 Latyshev, *Kto i kak prodaet Rossiiu*, pp. 39–54; Yeremin, *Rossiya-Yaponiya*, pp. 150–153. Also, Tōgō, *Nichiro Shinjidai he no Josō*, pp. 222–224; Glaubitz, *Between Tokyo and Moscow*, pp. 236–237.

126 Edamura, *Teikoku Kaitai Zengo*, pp. 234–235.

127 *Asahi Shimbun*, May 21, 2002. Apparently Hisashi Owada, Kunihiko Saitō, and Nagao Hyōdō of the Foreign Ministry indicated that they were amenable to such an agreement, provided Russia immediately recognized Japan's sovereignty over all four islands. Owada and Hyōdō were senior members of the Foreign Ministry's Russia school, and were seen as hard-liners in the debate over the Soviet Union/Russia. Their acquiescence in principle to Kozyrev's offer therefore was a sign of some softening on the Japanese side.

128 Edamura, *Teikoku Kaitai Zengo*, p. 252.

129 Ivan Tselitschev, "Russian Economic Reforms and Japan," in Goodby, Ivanov, and Shimotomai, eds., *"Northern Territories" and Beyond*, p. 185.

130 *Nihon Keizai Shimbun*, Jul.y 13, 1992. As cited in Gelman, *Russo-Japanese Relations and the Future of the US-Japanese Alliance*, p. 56. On European and US pressure, see also Gelman, *ibid.*, pp. 19–20, 42, 45–46, 56–57.

131 Hasegawa, *The Northern Territories Dispute and Russo-Japanese Relations*, pp. 442–445.

132 Latyshev, *Kto i kak prodaet Rossiiu*, pp. 88–101. Edamura, *Teikoku Kaitai Zengo*, p. 259–262. More on these hearings below.

133 Edamura, *Teikoku Kaitai Zengo*, p. 268–269, 282.
134 *Ibid.*, pp. 272–275. Hasegawa, *The Northern Territories Dispute and Russo-Japanese Relations*, pp. 462–466. Kimura, *Distant Neighbors (Vol. II): Japanese-Russian Relations under Gorbachev and Yeltsin,* Armonk, NY: M.E. Sharpe, 2000, pp. 129–158.
135 Latyshev, *Kto i kak prodaet Rossiiu*, p. 165.
136 Edamura, *Teikoku Kaitai Zengo*, p. 345. Kimura, *Distant Neighbors (Vol. II)*, pp. 164–165. Hasegawa, *The Northern Territories Dispute and Russo-Japanese Relations*, p. 483.
137 Edamura, *Teikoku Kaitai Zengo*, pp. 346–350. Kimura, *Distant Neighbors (Vol. II)* , pp. 166–171.
138 *Washington Post*, Oct. 17, 19, & 20, 1993. *Asahi Shimbun*, Oct. 17–18, 1993. Edamura, *Teikoku Kaitai Zengo*, pp. 352–353. In fact, the United States, Japan, and other Western nations had also been dumping low-level nuclear wastes in the oceans for years. This probably explains the muted reaction from the Japanese government at first. See Tsuneo Akaha, "Environmental challenge in the Russian Far East," in Tsuneo Akaha, ed., *Politics and Economics in the Russian Far East: Changing Ties with Asia-Pacific*, New York: Routledge, 1997, pp. 124, 132 (note 18).
139 *The Daily Japan Digest*, vol. IV, no. 188 (Oct. 21, 1993).
140 Edward Lincoln, *Japan's New Global Role*, Washington, DC: The Brookings Institution, 1993, pp. 182, 186–191. Also, Michael Green, *Japan's Reluctant Realism*, New York: Palgrave Press, 2001, p. 98.
141 *The Wall Street Journal*, Oct. 13, 1993.
142 For more on Sino-Soviet/Russian relations see O.B. Rakhmanin, *K Istorii Otnoshenii Rossii-SSSR s Kitaem v XX Veke*, Moscow: RAN, Institut Dal'nego Vostoka, 2000. Also, Boris Kulik, *Sovietsko-Kitaiskii Raskol: Prichiny i Posledstviya*, Moscow: RAN, Institut Dal'nego Vostoka, 2000. In English, see Sherman Garnett, ed., *Rapprochement or Rivalry?: Russia-China Relations in a Changing Asia*, Washington, DC: Carnegie Endowment for International Peace, 2000.
143 Glaubitz, *Between Tokyo and Moscow*, p. 229.
144 Gelman, *Russo-Japanese Relations and the Future of the US-Japanese Alliance*, pp. 80–98.
145 Raymond Garthoff, *The Great Transition: American-Soviet Relations and the End of the Cold War*, Washington, DC: The Brookings Institution, 1994, p. 468. Garthoff, a former high-ranking US diplomat, notes that while Bush's offer "earned Japanese gratitude, it was resented in the Soviet Union as American pressure at a time of Soviet weakness and vulnerability.... Reinforcing an overbearing Japanese stand while pressuring a weak but proud Soviet (Russian) leadership seeking a compromise was not the way to help resolve the problem." [pp. 666–667] Also, see Yeremin, *Rossiya-Yaponiya*, pp. 116–117. Yeremin is one Russian who feels that the United States has a role to play in settling the territorial dispute. See Latyshev, *Kto i kak prodaet Rossiiu*, p. 160.
146 Andrei Markov, *Rossiya i Yaponiya*, pp. 78, 93.
147 Tōgō, *Nichiro Shinjidai he no Josō*, p. 165.
148 In fairness to the Japanese, US leaders at times were unsure about the Soviets and often were slow in deciding about the extent of aid and assistance for Moscow (see Gelman, *Russo-Japanese Relations and the Future of the US-Japanese Alliance*, p. 46). Igor Tyshetskii, "The Gorbachev-Kaifu Summit: The View from Moscow," in Tsuyoshi Hasegawa, Jonathan Haslam, and Andrew Kuchins, eds., *Russia and Japan: An Unresolved Dilemma Between Distant Neighbors*, Berkeley: University of California Press, 1993, p. 95.
149 Tōgō, *Nichiro Shinjidai he no Josō*, pp. 42, 181, 185. Hasegawa, *The Northern Territories Dispute and Russo-Japanese Relations*, p. 411.
150 Leszek Buszynksi, "Russia and Northeast Asia: aspirations and reality," in *The Pacific Review*, vol. 13, no. 3 (2000), pp. 411–412.

151 Glaubitz, *Between Tokyo and Moscow*, pp. 209–210. The South Korean government would disburse only $1.47 billion before halting payments because of the inability of the Russian government to meet its payment schedule.

152 Kanemaru's links with Kim Il-sung and the *Chosen Sōren* in Japan were the source of much speculation. The *Far Eastern Economic Review* reported that there was suspicion within political circles in Japan that the *Chosen Sōren* had privy information concerning financial irregularities related to campaign funding for leading politicians in the LDP. Kanemaru was reported to have received financial contributions from DPRK-linked groups. This was reported to have included some of the unmarked gold bars found in his house when he was charged for tax evasion in 1992. See "Prepared for the Worst," *The Far Eastern Economic Review*, Feb. 10, 1994, pp. 10–11.

153 Glaubitz, *Between Tokyo and Moscow*, p. 210. See also, Gelman, *Russo-Japanese Relations and the Future of the US-Japanese Alliance*, p. 82.

154 Markov, *Rossiya i Yaponiya*, p. 59.

155 Tōgō, *Nichiro Shinjidai he no Josō*, p. 185.

156 Tsuyoshi Hasegawa, "The Gorbachev-Kaifu Summit: Domestic and Foreign Policy Linkages," in Hasegawa, Haslam, and Kuchins, eds., *Russia and Japan: An Unresolved Dilemma Between Distant Neighbors*, pp. 51, 53.

157 Owada, the father of the Japanese crown princess, enumerated some of his views in an article in 1993. See Hisashi Owada, "Kore ga Roshia shien no Ronri da," *Chuuō Kōron* (July 1993), pp. 30–38. This article gives the impression of a man open to policy changes, but such was evidently not the case.

158 Tsuyoshi Hasegawa, "Why Did Russia and Japan Fail to Achieve *Rapprochement* in 1991–1996?" in Rozman, ed., *Japan and Russia: The Tortuous Path to Normalization*, pp. 184–185, 200.

159 Markov, *Rossiya i Yaponiya*, p. 58. Also, Yeremin, *Rossiya-Yaponiya*, pp. 81–82.

160 Yeremin, *Rossiya-Yaponiya*, pp. 43–44. Also, see Akaha and Murakami, "Soviet/Russian-Japanese Economic Relations," in Hasegawa, Haslam, and Kuchins, eds., *Russia and Japan*, p. 172.

161 Akaha and Murakami, "Soviet/Russian-Japanese Economic Relations," in Hasegawa, Haslam, and Kuchins, eds., *Russia and Japan*, p. 166.

162 Gelman, *Russo-Japanese Relations and the Future of the US-Japanese Alliance*, p. 73.

163 Mike Mochizuki, "The Soviet/Russian Factor in Japanese Security Policy," in Hasegawa, Haslam, and Kuchins, eds., *Russia and Japan*, p. 152.

164 Vladimir Ivanov, "Russia and the United States-Japan Partnership," in Goodby, Ivanov, and Shimotomai, eds., *"Northern Territories" and Beyond*, pp. 275–276. See also, Glaubitz, *Between Tokyo and Moscow*, pp. 194–196.

165 Semyon Verbitsky, "Russian Perceptions of Japan," p. 67. Yeremin, *Rossiya-Yaponiya*, p. 99. Tōgō, *Nichiro Shinjidai he no Josō*, p. 218.

166 Brown, *The Gorbachev Factor*, pp. 269–285.

167 This according to a conversation Bessmertnykh later had with Japanese ambassador Edamura. Edamura, *Teikoku Kaitai Zengo*, pp. 150–151.

168 Yeremin, *Rossiya-Yaponiya*, pp. 46–63, 85. Tōgō, *Nichiro Shinjidai he no Josō*, pp. 166–167. Edamura, *Teikoku Kaitai Zengo*, pp. 150–151.

169 Much of the following section is based on a number of interviews the author had with Kunadze in Moscow between 1999 and 2001. Kunadze's views can also be found in several articles. See for example Kunadze and Konstantin Sarkisov, "Razmyshliaya o sovetssko-yaponskikh otnosheniiakh," *Mirovaya Ekonomika i Mezhdunarodnye Otnosheniya*, no. 5 (1989), pp. 83–93. G. Kunadze, "V Poiskakh Novogo Myshleniya: O Politike SSSR v Otnoshenii Yaponii," *Mirovaya Ekonomika i Mezhdunarodnye Otnosheniya*, no. 8 (1990), pp. 51–67. In English see G. Kunadze, "A Russian View of Russo-Japanese Relations," in Rozman, ed., *Japan and Russia: The Tortuous Path to Normalization*, pp. 159–173.

170 Baranets, *Yel'tsin i ego Generaly*, p. 285. Also, Kimura, *Distant Neighbors (Vol. II)*, pp. 137, 183.
171 See for example Latyshev, *Kto i kak prodaet Rossiiu*, pp. 39–45.
172 Robert Valliant, "The Political Dimension," in Akaha, ed., *Politics and Economics in the Russian Far East: Changing Ties with Asia-Pacific*, p. 16. See also, Semiyon Verbitsky, "Perceptions of Japan in the USSR During the Cold War and Perestroika," in *The Carl Beck Papers* No. 1503, University of Pittsburgh, Oct. 2000, p. 32.
173 Fedorov's actual title was the Chairman of the Executive Committee of the Sakhalin Oblast Soviet. He began calling himself the 'Governor of Sakhalin' and the self-styled moniker stuck. One of the books written by Fedorov is entitled, "Foreigners and us." See, V.P. Fedorov, *Inostrantsy i My*, Moscow: MP Russkoe Polye, 1992. For a shorter background of Fedorov, see John Stephan, "The Political and Economic Landscape of the Russian Far East," in Hasegawa, Haslam, and Kuchins, eds., *Russia and Japan*, p. 284.
174 Asahi Shimbun Weekly, *AERA*, Sept. 29, 1992. See also, Glaubitz, *Between Tokyo and Moscow*, p. 247.
175 Valliant, "The Political Dimension," p. 19.
176 Yeremin, *Rossiya-Yaponiya*, pp. 150–151.
177 Latyshev, *Kto i kak prodaet Rossiiu*, pp. 88–101. In English, see Herbert Ellison, "The Debate before the Summit," in Goodby, Ivanov, and Shimotomai, eds., *"Northern Territories" and Beyond*, pp. 93–101. Also, see *The Current Digest of the Post-Soviet Press*, vol. 44, no. 30 (Aug. 26, 1992), pp. 1–7.
178 Markov, *Rossiya i Yaponiya*, p. 64. Latyshev, *Kto i kak prodaet Rossiiu*, p. 40.
179 Latyshev, *Kto i kak prodaet Rossiiu*, p. 46.
180 Baranets, *Yel'tsin i ego Generaly*, pp. 272–307.
181 See Hasegawa, *The Northern Territories Dispute and Russo-Japanese Relations*. Also, Rozman, *Japan's Response to the Gorbachev Era 1985–1991*, and Kimura, *Distant Neighbors (Vol. II)*.
182 Rozman, *Japan's Response to the Gorbachev Era 1985–1991*.
183 *Ibid.*, p. 14.
184 Ivan Tselishev, " 'Upravlenie po-yaponski' za predelami Yaponii," *Mirovaya Ekonomika i Mezhdunarodnye Otnosheniya*, Aug. 1991, pp. 58–79. In English, see Kimura, *Distant Neighbors (Vol. II)*, pp. 55–76.
185 LaFeber, *The Clash*, p. 388.
186 Igor Latyshev, *Kto i kak prodaet Rossiiu*. See Evgenii Prokhorov and Leonid Shevchuk, "O territorial'nykh pretenziiakh Yaponii k SSSR," *Mezhdunarodnaya Zhizn'*, no. 1 (1989), pp. 47–52, for a similar view.
187 See Kunadze and Sarkisov, "Razmyshliaya o sovetssko-yaponskikh otnosheniiakh."
188 Aleksei Bogaturov, "Diplomatiia pered vyborom," *Novoe Vremya*, no. 32 (1989), pp. 20–23. Aleksei Abramov and Boris Makeev, "Kuril'skii Bar'er," *Novoe Vremya*, nos. 40–41 (Oct. 1992) pp. 16–18, 24–25. Vladimir Yeremin, *Rossiya-Yaponiya*. Yeremin gives an excellent summary of the debate in the Soviet and Russian press during this period (pp. 46-63).
189 Oleg Bondarenko, *Neizvestnye Kurily: Ser'eznye razmyshleniia o statuse Kuril'skikh ostrovov*, Moscow: VTI Deita Press, 1992. Bondarenko in the end calls for reversion to Japan, but he maintains a critical line toward both Japan and Russia in much of his book.
190 Glaubitz, *Between Tokyo and Moscow*, pp. 212, 238.
191 Stepan Pesh, "Zapretnyi paltus: Yaponskie brakonery stanut rybolevami posle uregolirovaniia spora o severnykh territoriiakh," *Novoe Vremya*, no. 35 (September 1994), pp. 24–25.
192 Kimura, *Distant Neighbors (Vol. II)*, p. 199.
193 Hasegawa, *The Northern Territories Dispute and Russo-Japanese Relations*, pp. 491–493.

194 In an interview in Moscow at the end of May Yeltsin said that Russia did not need Japanese assistance and suggested that Japan would only demand back territory in return. Tōgō, *Hoppo Ryōdo Kōshō Hiroku*, p. 190.

3 Another *rapprochement*

1 At least one observer feels that the 1990s were quite different from earlier periods of Japanese-Soviet relations. The historian C.W. Braddick states that the 1990s had a "distinctly mercurial quality." In fact, Japanese-Russian relations throughout the entire twentieth century have been mercurial. See, Braddick's "The Waiting Game: Japan-Russia Relations," in Takashi Inoguchi and Purnendra Jain, eds., *Japanese Foreign Policy Today*, New York: Palgrave, 2000, p. 216.
2 This quote is attributed to Gorō Saitō and cited in Richard Samuels and Eric Heginbotham, "Mercantile Realism and Japanese Foreign Policy," *International Security*, vol. 22, no. 4 (Spring 1998), pp. 181, note 26. Tamba's statement was confided to the author by a US participant of the meeting in San Francisco.
3 Hisahiko Okazaki, "Roshia kyowakoku no kassika wa Nihon ni totte son ka toku ka," *Shukan Bunshun*, no. 33 (Sept. 5, 1991), pp. 169-170. As cited in Mike Mochizuki, "The Soviet/Russian Factor in Japanese Security Policy," in Hasegawa, Haslam, and Kuchins, eds., *Russia and Japan: An Unresolved Dilemma Between Distant Neighbors*, p. 142.
4 Yoichi Nishimura, "Yurashia Gaikō no Butaiura," *Sekai*, January 1998, p. 143. Tōgō, *Hoppo Ryōdo Kōshō Hiroku*, pp. 191–92.
5 *Itar-Tass*, Mar. 20, 1996.
6 Hasegawa, *The Northern Territories Dispute and Russo-Japanese Relations*, pp. 500–501.
7 *Radio Free Europe/Radio Liberty Daily Digest*, June 30, 1996. Henceforth to be cited as *RFE/RL*.
8 *RIA Novosti*, Nov. 12, 1996.
9 *Sankei Shimbun*, Nov. 12, 1996.
10 Yoichi Nishimura, "Yurashia Gaikō no Butaiura," p. 142.
11 Koichi Watanabe, "Hōkai Suru Hoppō Yon Shima," *Sekai*, Jan. 1997, pp. 116–126. Also, see Taku Igarashi, "Juuyu Ryuushutsu ga Ukibori ni shita 'Ajia Oiru Rodo' no Fuan," pp. 90–91.
12 Sergei Sevastyanov, "Russian Reforms: Implications for Security Policy and the Status of the Military in the Russian Far East." *NBR Analysis*, vol. 11, no. 4 (Dec. 2000), p. 11. Also, see "Baturin Bewildered by Redeployment Plans of US Troops in Japan," *Jamestown Monitor*, Jan. 29, 1997; also, "Sakhalin Governor Protests Plans to Move US Marines to Hokkaido," *Jamestown Monitor*, May 9, 2000.
13 Scott Parrish, "Cancelled Visit Sparks Diplomatic Spat with Tokyo," *Jamestown Monitor*, Mar. 7, 1997.
14 Panov, however, was not involved in the clean-up operation, nor did the Russian government offer to help. In fact, it spent much of the time deflecting criticism that the boat was not seaworthy and should not have been transporting oil.
15 The groundwork for this acceptance was set up by Bill Clinton, who personally phoned Hashimoto on March 30, 1997 and urged the Japanese government to accept Russian attendance as a gesture to Moscow in return for Russia's acceptance to swallow NATO expansion. Tōgō, *Hoppo Ryōdo Kōshō Hiroku*, pp. 221–222.
16 *Japan Daily Digest*, May 19, 1997.
17 Nishimura, "Yurashia Gaikō no Butaiura," pp. 138–147. Also, see Shigeki Hakamada, "Aratana Nichirō Kankei Kochaku e no Teigen," *Foresight Magazine*, August 1997, pp. 6–9. Reportedly, when Yeltsin told Hashimoto that Russia would no longer target its nuclear weapons on Japan, the Japanese Prime Minister half-jokingly responded that he didn't know Japan was targeted in the first place.
18 "Address by Prime Minister Ryutaro Hashimoto to the Japan Association of Corporate

Executives," translation issued by the Embassy of Japan, Washington, July 1997, pp. 5–6.

19 Nishimura, "Yurashia Gaikō no Butaiura," p. 145. Also, see Shinjiro Mori, "Nichirō Shuunō 'Kawana Kaidan' no Butaiura," *Sekai*, June 1998, p. 130.

20 Kazuhiko Tōgō actually places the genesis of the 'two-track' formula even earlier to 1988. Tōgō, *Hoppo Ryōdo Kōshō Hiroku*, pp.134–135.

21 *Asahi Shimbun*, May 19, 2002.

22 *Itar-Tass*, Sept. 11, 1997.

23 Nishimura, "Yurashia Gaikō no Butaiura," p. 143.

24 Boris Yel'stin, *Prezidentskii Marafon*, Moscow: AST, 2000, p. 135.

25 "Kurasunoyarusuku ni okeru Nichi-Rō Shunō Kaidan (Gaiyō to Hyōka)," report issued by the Embassy of Japan, Washington, DC, November 1997. The Hashimoto-Yeltsin Plan called for: (1) an initiative for joint investment; (2) Russia's integration into international economic organizations; (3) expanding assistance for Russian reforms; (4) Japanese training of young Russian business managers; (5) a strengthening of the energy dialogue; and (6) cooperation toward the peaceful uses of nuclear energy.

26 Based on interviews with Mikhail Galuzin, Ministry of Foreign Affairs, Moscow, February 23 and November 22, 2000. Galuzin was Yeltsin's interpreter at Krasnoyarsk. In the interviews he admitted that he was surprised by Yeltsin's statement, but he said that it was perfectly within the president's rights and that sometimes top-level initiatives were necessary to advance relations, pointing to Nixon's 1972 China visit as an example. At the time of the interviews Galuzin was Deputy Director of the Second Asia Division (Japan and Korea) at the Ministry of Foreign Affairs.

27 See, for example, "Russia dismisses press speculation on future of northern islands," *BBC Summary of World Broadcasts*, Nov. 6, 1997.

28 Yel'stin, *Prezidentskii Marafon*, pp. 135–138. Yeltsin misinterpreted Japanese intentions: both Hashimoto and Vice Deputy Foreign Minister Minoru Tamba made it clear on numerous occasions that the Japanese approach to Russia in 1997 was about geopolitics and energy. See, for example, Minoru Tamba, "Kurasunoyarusuku Kaidan kara Eritsin Daitōryō no Hōnichi he," *Gaikō Forum*, no. 117 (Apr. 1998), pp. 12–17.

29 Nobuo Miyamoto, "Nichirō Kankei ha Naze Nichibei no Tsugi-ni Juuyō-ka," *Chuuō Kōron*, Feb. 1999, p. 246.

30 Okada, "Nikkei Kigyō Shinshutsu no Kōki ha?" pp. 56–61.

31 *Itar-Tass*, May 21 and June 10, 1996.

32 *Itar-Tass*, Nov. 20, 1996.

33 "Bōeichō ga tsuini Roshia-sei Sentōki wo Kōnyuu ka." *Foresight Magazine*, June 1996, p. 22.

34 *Asahi Shimbun,* Mar. 30, May 23, 1998.

35 S. Enders Wimbush, "Time for a New Northeast Asian Security Order," *The Wall Street Journal* (Asia edition), Oct. 23, 1996.

36 "MSDF ships pass through sensitive channel," *The Daily Yomiuri*, Oct. 3, 1997. In fact, later the Russian Foreign Ministry expressed its dismay that the Japanese side had so publicized the voyage. A Russian diplomat stated, "Without having any claims on the Japanese from a legal point of view, we think that the reports were probably not quite timely, and that the very passage should have been considered more profoundly," *RFE/RL*, Oct. 10, 1997.

37 The influential Japanese fishing lobby was very happy about this agreement. On the heels of the Krasnoyarsk summit some in Japan suggested that this could be the first step toward complete Japanese sovereignty over the disputed islands. For details on the agreement, see *Jamestown Monitor*, Jan. 6, 1998. This information is based partly on interviews with correspondents from the *Hokkaido Shimbun* in Sapporo, Tokyo, and Washington, March, April, and December 1997.

38 Mori, "Nichirō Shunō 'Kawana Kaidan' no Butaiura," p. 130. *The Financial Times*, Feb. 24, 1998.

39 Shigeki Hakamada, "Roshia no Seihen to Eritsin Rainichi," *Chuuō Kōron*, Apr. 1998, p. 233.
40 Mori, "Nichirō Shunō 'Kawana Kaidan' no Butaiura," pp. 126–127.
41 See Hiroshi Kimura, *Distant Neighbors (Vol. II)*, pp. 218–219.
42 Moscow Mayak Radio Network, as cited in *Foreign Broadcast Information Service* (hereafter *FBIS*), FBIS-SOV-98-127, May 7, 1998.
43 *Yomiuri Shimbun*, Nov. 14, 1998. See also, Michael Green, *Japan's Reluctant Realism*, p. 156.
44 *Asahi Shimbun*, Nov. 14, 1998.
45 "Russia Said to Reject Japanese Proposal on Kuril Islands," *Jamestown Monitor*, Dec. 9, 1998.
46 *Yomiuri Shimbun*, Dec. 2, 1998.
47 Tetsuo Sugano, "Russia's Economy and Development of the Far East," in Koji Watanabe, editor, *Engaging Russian in Asia Pacific*, Tokyo: Japan Center for International Exchange, 1999, p. 85.
48 Hakamada is the daughter of Mutsuo Hakamada, a member of the Japanese Communist Party who in 1944 was drafted into the Japanese Imperial Army, sent to Manchuria, and imprisoned as a POW in Siberia after the Second World War. Because he cooperated with Soviet authorities in helping to indoctrinate Japanese POWs, he was derisively nicknamed the 'Emperor of Siberia' by fellow Japanese internees. Hakamada was granted asylum in the Soviet Union after he was declared *persona non grata* by the Japanese government. In Moscow he married a Russian woman and raised a family. Interestingly, Irina Hakamada's half brother Shigeki Hakamada (from a Japanese mother to whom the elder Hakamada was married before he was drafted into the army in 1944) is now one of Japan's top Russia experts. This information is based on conversations with Shigeki Hakamada and Tomohisa Sakanaka in Tokyo, 1996–97. See also, Stephan, *The Russian Far East*, p. 248.
49 Ichiyanagi interview, Vladivostok, Aug. 23, 1998.
50 "Oil Drilling to Be Resumed in Sakhalin," *Itar Tass*, Apr. 13, 2000.
51 Based on interviews with Takeshi Kondo and Ryoichi Nozaki of the Itochu Corporation, and Koji Hitachi of the Nissho Iwai Corporation, Sept. 1–2, 1998, Tokyo.
52 *Asahi Shimbun*, Apr. 6, 1998.
53 "Northeast Asia Energy and Environmental Cooperation," report written and published by *IMEMO* (Moscow) and *NIRA* (Tokyo), 1999, p. 47. Also, see *Itar-Tass*, Dec. 1, 1998.
54 "Rough Sailing for Russian Naval Exercises?" *Jamestown Monitor*, July 28, 1998.
55 "Special Economic Zone for the Disputed Kuril Islands?" *Jamestown Monitor*, Jan. 19, 1999.
56 Russian leaders, like Chinese leaders, were put off by the vague geographical definition of the sphere of Japanese defense responsibility in the Far East under the aegis of the US-Japan security treaty. "Moscow Comments Warily on Japanese-US Defense Pact," *Jamestown Monitor*, Feb. 12, 1999.
57 "Maslyukov Back Empty-Handed After Tokyo Talks," *Jamestown Monitor*, Mar. 17, 1999.
58 "Some Japanese Favour 'Intermediary Treaty' with Russia," *Itar-Tass*, Aug. 25, 1999.
59 Braddick, "The Waiting Game: Japan-Russia Relations," p. 220. See also, Kimura, *Distant Neighbors (Vol. II)*, p. 259.
60 "Japan 'Unfreezes' $1.5 Billion Credit to Russia," *Itar-Tass*, Aug. 31, 1999.
61 *Nezavisimaya Gazeta*, Nov. 12, 1999; *Izvestia*, Dec. 16, 1999.
62 *Izvestia*, Nov. 10, 1999. Also, Matsuda interview, Moscow, December 1999.
63 This statement was apparently issued after a bitter struggle within the Japanese government and within the Foreign Ministry over the issue of Chechnya. More on this issue below.

64 *Izvestia*, Nov. 25, 1999.
65 *Izvestia*, Dec. 18, 1999.
66 *Izvestia*, Nov. 17, 1999; *Nezavisimaya Gazeta*, Nov. 11–12, 1999; *Moscow Times*, Nov. 17, 1999.
67 *Sankei Shimbun*, Dec. 21, 1999; *Asahi Shimbun*, Dec. 22, 1999.
68 *Reuters*, Mar. 12, 1999; *Itar-Tass*, Mar. 15, 1999; May 14, 1999.
69 *Itar-Tass*, Apr. 17, 1999.
70 "Roshia gensen 50-ki kaitai he zaisei shien," *Yomiuri Shimbun*, Apr. 21, 1999. "Roshia kakukaitai ni 2 oku doru shinki shien," *Yomiuri Shimbun*, June 16, 1999.
71 James Clay Moltz, "Russian Nuclear Regionalism: Emerging Local Influences over Far Eastern Facilities," *NBR Analysis*, vol. 11, no. 4 (Dec. 2000), p.8. Also, see Sergei Sevastyanov, "Russian Reforms.".
72 *Nucleonics Week*, Jan. 22, 1998, p. 13.
73 For a look at a Japanese perspective of Russia's nuclear 'graveyard', see Yoichi Nishimura, *Purometeusu no Hakaba* [Prometheus' Graveyard], Tokyo: Asahi Shimbun Press, 1997.
74 *Yomiuri Shimbun*, Apr. 4, 1999.
75 *Nezavisimaya Gazeta*; *Moskovskie Novosti*, Feb. 29, Mar. 6, 2000.
76 *Izvestia*, Feb. 10, 2000; *Segodnya*, Feb. 11, 2000; *Nezavisimaya Gazeta*, Feb. 12, 2000.
77 *Nezavisimaya Gazeta* Feb. 12, 16, 2000. See also, *Moscow Tribune*, Feb. 25, 2000.
78 See, for example, *Nezavisimaya Gazeta*, Apr. 20, 2000.
79 Tōgō, *Hoppo Ryōdo Kōshō Hiroku*, p. 203.
80 *Itar-Tass*, Apr. 27, 2000.
81 *Yomiuri Shimbun*, Aug. 9, 2000.
82 For more on the divide within the influential group of Russian experts in Japan and the Nonaka statement see Tōgō, *Hoppo Ryōdo Kōshō Hiroku*, pp. 314–317.
83 *Sankei Shimbun*, July 28, 2000; *Yomiuri Shimbun*, Aug. 4, 2000. Also, see Haruki Wada, "Sukyandaru to Gaikō," *Sekai*, May 2002, pp. 71–77.
84 *Nezavisimaya Gazeta*, Aug. 31, 2000.
85 *Napsnet Daily Report*, Sept. 5, 2000, as quoted in Peggy Falkenheim Meyer, "Is Japan's New Eurasian Diplomacy a Failure?" A paper presented to the Workshop *Russian National Security: Perceptions, Policies, and Prospects*, Army War College, Carlisle Barracks, PA, Dec. 4–6, 2000, p. 14.
86 Tōgō, *Hoppo Ryōdo Kōshō Hiroku*, pp. 318–322.
87 For extensive coverage of the 2000 Tokyo summit see the December 2000, no. 149, issue of the Japanese journal *Gaikō Forum*, pp. 12–47.
88 Kunadze interview, Moscow, Oct. 16, 2000; Sarkisov interview, Tokyo, Sept. 29, 2000. Also see Hiroshi Kimura, "Islands Apart," *Look Japan*, February 2001, pp. 14–15.
89 Interviews with Kazuhiko Tōgō and Jirō Kodera, Ministry of Foreign Affairs, Tokyo, Sept. 29, 2000. Also Hakamada, Kimura, Shimotomai, and Suetsugu interviews in Tokyo, Sept. 27–30, 2000.
90 *Reuters*, Nov. 2, 2000.
91 *Kommersant'*, Nov. 16, 21, 2000.
92 See for example, Takerō Nagashi, "Shippai ni Owatta Tai-ro Seisaku Kudeta," *Sekai Shuuho*, Apr. 2, 2002, pp. 6–9; also, Kazuya Kitagawa, "Puchin Seiken ni Ryōdo Mondai Sakiokuri no Kōjitsu," *Sekai Shuuho*, Apr. 2, 2002, pp. 10–12. Tōgō, *Hoppo Ryōdo Kōshō Hiroku*, pp. 338–340.
93 *Yomiuri Shimbun*, Jan. 27, 2001.
94 *Sankei Shimbun*, Dec. 27–29, 2000. These articles were meant to be addressed to Japanese conservatives that would be opposed to any compromise, hence the decision to publish them in the *Sankei Shimbun*. A detailed rebuttal was published in the *Sankei Shimbun* on Feb. 3, 2001. The author of the Mr. X articles was Kazuhiko Tōgō, Director-General of the Bureau European and Oceanic Affairs in the Foreign Ministry.

In his memoirs Tōgō stated that he felt the Irkutsk summit was the most important breakthrough in Japanese-Russian relations since the end of the Second World War. Tōgō, *Hoppo Ryōdo Kōshō Hiroku*, pp. 347–371.

95 Zaitsev interview, Moscow, Mar. 30, 2001.

96 See Takao Toshikawa, "'Muneo Paji' de Miushinatta Gaikō-Gaimusho Kaikaku," *Sekai*, May 2002, pp. 78–87. Also, Shigeki Hakamada, "Irukutsuku no Gensō," *Sekai Shuuhō*, May, 2001, pp. 6–9. Nonaka was one of a small group of LDP politicians that had placed Mori in the prime minister's office after Obuchi fell into a coma.

97 *Yomiuri Shimbun,* Apr. 8, 2001. Also, see Shigeki Hakamada, "Irukutsuku no Gensō," pp. 6–9; also, *Sankei Shimbun*, Mar. 26, 2001.

98 Georgii Kunadze, "Ostrova Nevezeniya," *Novoe Vremya*, no. 14 (Apr. 2001), pp. 24–25. *Nezavisimaya Gazeta*, Mar. 24, 2001; *Segodnya*, Mar. 26, 2001. Also for polling results on perceptions toward Japan after the summit, see *Itar-Tass*, Apr. 9, 2001. Interviews with Kunadze, Pavlyatneko, Zaitsev, Moscow, Mar. 25–31, 2001.

99 "Japan May Push Harder for Return of Four Disputed Islands," *Jamestown Monitor*, May 1, 2001.

100 *Asahi Shimbun*, May 10–11, 2001.

101 *Izvestia*, May 19, 2001; *Itar-Tass,* May 21, 27, 2001; *Yomiuri Shimbun*, June 21, 26, 28, 2001. Tanaka was the daughter of Kakuei Tanaka, himself a very controversial politician. She apparently felt that the 1973 summit was an 'achievement,' although the only thing of note seems to have been the argument about whether Brezhnev was clearing his throat or saying "Da." See Chapter 2.

102 *Izvestia*, *Nezavisimaya Gazeta*, *Vremya Novostei*, May 15, 2001.

103 *Yomiuri Shimbun*, May 17, 2001.

104 *Yomiuri Shimbun*, July 4, Aug. 3, 17, 2001; *Sankei Shimbun*, June 21, July 4, Aug. 1, 3, 9, 2001; *Nikkei Shimbun*, June 25, Aug. 11, 2001. The moderate *Asahi Shimbun*, predictably, simply asked the Japanese government to "remain cool" (Aug. 9, 2001).

105 "Tokyo Objects to Extended Fishing Rights off Disputed Islands," *Jamestown Monitor*, Aug. 9, 2001.

106 *Asahi Shimbun,* Oct. 6, 2001.

107 *Asahi Shimbun, Yomiuri Shimbun*, Oct. 22, 2001.

108 Kunio Okada, "The Japanese Economic Presence in the Russian Far East," in Thornton and Ziegler, eds. *Russia's Far East*, p. 439, note 12.

109 *Moscow Times*, Mar. 16, 2000.

110 *Itar-Tass*, July 26, 2000.

111 *Izvestia*, Sept. 7, 2000; *Kommersant'*, July 22, Sept. 2, 2000; *Moscow Times*, July 25, 2000. On oil shipments, see *Asahi Shimbun*, May 31, 2001.

112 *Yomiuri Shimbun*, *Reuters*, June 14, 2001.

113 *Itar-Tass*, as cited in Stratfor.com Global Intelligence Update, Sept. 15, 2000.

114 *Kommersant Daily*, Nov. 23, 2001. As cited in Elizabeth Wishnick, "One Asia Policy or Two?: Moscow and the Russian Far East Debate Russia's Engagement in Asia," *NBR Analysis*, vol. 13, no. 1 (Jan. 2002), p. 47.

115 *Izvestia*, June 9, 2001.

116 "Russian Far East: Rotten to the Core," *The Economist*, Oct. 18, 1997.

117 *Izvestia*, Aug. 31, 2000. Also, "The Wild, Wild East." *Forbes Magazine*, Dec. 10, 2001.

118 Kunio Okada, "Nikkei Kigyō Shinshutsu no Kōki ha?" p. 58.

119 "Moscow Blasts US, Japan for Regional ABM System Plans," *Moscow Interfax* Feb. 24, 2000, as cited in FBIS-EAS-2000-0224.

120 *Interfax*, Apr. 26, 2000, as cited in FBIS-SOV-2000-0426.

121 *Asahi Shimbun*, Sept. 10, 2000.

122 Shigeki Hakamada, "Roshia ni Gokai wo Ataeru you na Nihon-teki 'Kikubari Gaikō' wo Aratameyo," *Nihon no Ronten, 2001*, pp. 130–135.

123 *Itar-Tass*, Jan. 23, 2001.

124 *Itar-Tass*, Apr. 3, 2001
125 *Reuters*, Apr. 11–13, 2001.
126 *Yomiuri Shimbun*, Aug. 24, 2001.
127 *Asahi Shimbun*, *Nikkei Shimbun*, *Yomiuri Shimbun*, Jan. 19, 2002.
128 *Vremya Novostei*, Jan. 28, 2002.
129 Shigeki Hakamada, "Roshia ni Gokai wo Ataeru you na Nihon-teki 'Kikubari Gaikō' wo Aratameyo,' pp. 130–135.
130 "Russian Foreign Minister in Tokyo," *Jamestown Monitor*, Feb. 5, 2002.
131 *Asahi Shimbun*, *Yomiuri Shimbun*, Feb. 15, 2002.
132 *Asahi Shimbun*, *Yomiuri Shimbun*, *Nikkei Shimbun*, Feb. 21–22, 2002. A damning piece of evidence against Suzuki's testimony was produced when it was revealed that the two local firms were unable to carry out the order and had to subcontract out much of the work.
133 *Asahi Shimbun*, Mar. 2, 2002; *Mainichi Shimbun*, Feb. 28, 2002, *Yomiuri Shimbun*, March 7, 2002. Also interview with Yoichi Nishimura, Washington, DC, Apr. 20, 2002.
134 *Asahi Shimbun*, *Yomiuri Shimbun*, *Nikkei Shimbun*, Mar. 11–13, 2002.
135 *Vremya Novostei*, Mar. 19, 2002. *Yomiuri Shimbun*, Apr. 11, 2002. In English see, "Russian-Japanese Talks Suffer Setback," *Jamestown Monitor*, Mar. 20, 2002.
136 For more on the so-called Suzuki scandal and the 'massacre' of the MOFA's Russia School see Masaru Satō's book *Kokka no Wana*. Satō wrote the book while in prison, and his explanation, though perhaps fair, needs to be read knowing that he undoubtedly wished to protect his own reputation. Two of the other major players in the Japanese Foreign Ministry at the time, Minoru Tamba and Kazuhiko Tōgō, also wrote memoirs of the time, but they avoid detailed explanation of the events.
137 *Blue Book of the Japanese Foreign Ministry*, 2004–2005. See the URL: <http://www.mofa.go.jp/mofaj/area/russia/kodo_0301.html> (accessed July 21, 2007).
138 *Japan Times*, Jan. 15, 2003.
139 *Yomiuri Shimbun*, Jan. 14, 2003.
140 *Japan Times*, Jan. 10, 2003.
141 See for example, Lyle Goldstein and Vitaly Kozyrev, "China, Japan and the Scramble for Siberia," *Survival*, vol. 48, no. 1 (Spring 2006), pp. 163–178.
142 See the interview with Japan's Ambassador to Moscow Issei Nomura in the Russian journal *Novoe Vremya*, January 2005.
143 Tsuyoshi Kitagawa, "Strategicheskii Vybor V.V. Putina: Kitai ili Yaponia," Moscow Carnegie Center, Mar. 15, 2005. See the URL: <http://www.carnegie.ru/ru/print/71530-print.htm> (accessed July 22, 2007).
144 *Wall Street Journal*, Dec. 2–3, 2002.
145 *Financial Times*, Jan. 14, 2003.
146 "Moscow Testing China's Patience on Oil Exports," *RFE/RL*, Feb. 26, 2003.
147 Isabel Gorst, "Russian Pipeline Strategies: Business Versus Politics," paper prepared for the James A. Baker Institute for Public Policy of Rice University, Oct. 2004, p. 16. See the URL: <http://www.rice.edu/energy/publications/docs/PEC_Gorst_10_2004.pdf> (accessed July 20, 2007).
148 Nodari Simonia, "Russian Energy Policy in East Siberia and the Far East," paper prepared for the James A. Baker Institute for Public Policy of Rice University, October 2004, pp. 7–9. See the URL: <http://www.rice.edu/energy/publications/docs/PEC_SimoniaFinal_10_2004.pdf> (accessed July 20, 2007).
149 Shoichi Itō, "Kiro ni Tatsu Taiheiyō Paipurain Kōsō," ERINA Report, vol. 72, Nov. 2006, p. 24.
150 Simonia, "Russian Energy Policy in East Siberia and the Far East,", p. 22.
151 Itō, "Kiro ni Tatsu Taiheiyō Paipurain Kōsō," p. 24.
152 Gorst, "Russian Pipeline Strategies: Business Versus Politics," p. 1.
153 Simonia, "Russian Energy Policy in East Siberia and the Far East,", p. 1.

154 Gorst, "Russian Pipeline Strategies: Business Versus Politics," pp. 14–15.
155 *Asia Times*, Dec. 9, 2003.
156 Simonia, "Russian Energy Policy in East Siberia and the Far East," p. 21.
157 Simonia, "Russian Energy Policy in East Siberia and the Far East," p. 23.
158 *Ibid.*, p. 24.
159 "Japan Favored Over China in Pipeline Duel," *Moscow Times*, Feb. 25, 2004.
160 *Mainichi Shimbun*, Nov. 15, 2004.
161 *BBC News*, Oct. 16, 27, 2004.
162 *Kommersant'*, Nov. 15, 2004.
163 *Asahi Shimbun*, Nov. 16, 2004.
164 *Kommersant'*, Nov. 15, 2004.
165 *Nezavisimaya Gazeta*, Nov. 15, 2004.
166 Takayuki Nakazawa, "Puchin Hōnichi to Hoppō Ryōdo Mondai," *Sekai Shuuhō*, Apr. 5, 2005.
167 *Kommersant'*, Nov. 15; *Nezavisimaya Gazeta*, November 15; *Vremya Novostei*, Nov. 16, 2004.
168 *Yomiuri Shimbun*, Nov. 23–24, 2005.
169 *Asahi Shimbun*, Nov. 23, 2004.
170 Bilateral trade figures in 2004 reached $8.8 billion; an all time high. See, "Tokyo hopes for progress on territorial problem this year," Interfax, Apr. 15, 2005.
171 *Nikkei Shimbun*, Jan. 1, 2005.
172 *Ibid.*, Dec. 15, 2004.
173 Vladimir Ivanov, "Russia and Japan Beyond 2005," in *ERINA Report*, vol. 66, Nov. 2005, p. 11. Since Mitsubishi and Mitsui are junior partners in the Sakhalin-2 project, the investment figures are officially listed as being from Shell (thus Great Britain and Shell).
174 *Yomiuri Shimbun*, Nov. 23, 2004.
175 RFE/RL Newsline, May 25, 2005. See the URL: <http://www.hri.org/news/balkans/rferl/2005/05-05-25.rferl.html> (accessed July 20, 2007); also *AP press Reports*, Steve Gutterman, May 23, 2005.
176 James Brooke, "Japan and Russia, With an Eye on China, Bury the Sword," *New York Times*, Feb. 13, 2005. Also, Ivanov, "Russia and Japan Beyond 2005," p. 10.
177 Sergei Blagov, "Russia's Gazprom Eyes East Asian Markets," *Eurasia Daily Monitor*, vol. 2, issue 51, Mar. 15, 2005.
178 *Sankei Shimbun*, Feb. 2, 2005.
179 *Kommersant'*, Mar. 9; *Nezavisimaya Gazeta*, Mar. 10, 2005.
180 *Asahi Shimbun*, Feb. 10, 2005.
181 *Ekspert*, Apr. 4, 2005.
182 Interview by author with Nodari Simonia, Moscow, Mar. 18, 2005.
183 *Asahi Shimbun*, Feb. 20, 2005.
184 Rail shipments of oil to China from Russia increased by over 20 percent in 2005 compared to 2004. Vladimir Ivanov, "Russia's Energy Politics: Focusing on New Markets in Asia," in *ERINA Report*, vol. 67, Jan. 2006, p. 12.
185 *Kommersant'*, Nov. 3, 2004.
186 *Sankei Shimbun*, Dec. 11, 2004.
187 Sergei Blagov, "Russian Energy Minister Visits Tokyo," *Eurasia Daily Monitor*, vol. 2, issue 82, (Apr. 27, 2005).
188 John Helmer, "Putin opts for China-First oil plan, Japan and India relegated," *Russia Journal*, July 12, 2005. Also, *Japan Times*, July 10, 2005.
189 Sergei Blagov, "Russia Sheds No Tears Over Peace Treaty with Japan," *Eurasia Daily Monitor*, vol. 2, issue 193 (Oct. 18, 2005).
190 *Moscow Times*, Feb. 2, 2005. Also, Nakazawa, "Puchin Hōnichi to Hoppō Ryōdo Mondai," p. 17.
191 *Interfax*, *BBC*, July 29, 2005.

192 Sergei Blagov, "Russian Personnel Changes to Affect Far East Region," *Eurasia Daily Monitor*, vol. 2, issue 216, Nov. 18, 2005.
193 Pavel Felgenhauer, "Secrets of Japanese Cuisine," *Novaya Gazeta*, Nov. 21, 2005.
194 *Yomiuri Shimbun*, Nov. 22, 2005.
195 *Asahi Shimbun*, Nov. 23, 2005.
196 *Mainichi Shimbun*, Feb. 19, 2006.
197 *RIA Novosti*, Jan. 24, 2006.
198 Sergei Blagov, "Moscow Looks To Asia's Growing Demand," *Eurasia Daily Monitor*, vol. 3, issue 40 (Feb. 28, 2006).
199 <Http://www.minprom.gov.ru/activity/energy/appearance/17/print> (accessed July 20, 2007).
200 Ivanov, "Russia's Energy Politics: Focusing on New Markets in Asia," p. 12.
201 Itō, "Kiro ni Tatsu Taiheiyō Paipurainu Kosō," pp. 32–33. It must also be pointed out that the northern route above Lake Baikal may also make more sense economically because it passes closer to oil and gas fields in northern Irkutsk Oblast (Verkhnechonskoye) and Southern Yakutia (Tarakan).
202 *Nihon Keizai Shimbun*, Jan. 14, 2006.
203 *Kyodo News Service*, July 16, 2006.
204 *Asahi Shimbun*, *Sankei Shimbun*, *Tokyo Shimbun*, *Yomiuri Shimbun*, Aug. 17–18, 2006.
205 *Yomiuri Shimbun*, Aug. 21, 2006; *Sankei Shimbun*, Aug. 23, 2006.
206 *Asahi Shimbun*, Sept. 18, 2006.
207 *Vremya Novostei*, Sept. 27, 2006.
208 *Kommersant'*, Sept. 20, 2006; *Nezavisimaya Gazeta*, Sept. 21, 2006.
209 *Wall Street Journal*, Apr. 19, 2007. The exchange was below market value if one considers the $20 billion developmental costs of the project. Gazprom should have paid more than $10 billion.
210 *Kommersant'*, Sept. 20, 2006.
211 *RIA-Novosti*, Sept. 21, 2006.
212 *Moscow Times*, June 25, 2007.
213 *ITAR-TASS*, Nov. 18, 2006.
214 *Yomiuri Shimbun*, Dec. 18 and 21, 2006.
215 *Rossiskaya Gazeta*, Feb. 27, 2007. On nuclear cooperation see *Moscow Times*, Feb. 28, 2007.
216 *Yomiuri Shimbun*, May 4–5, 2007. *Kommersant'*, May 4–5, 2007.
217 *Yomiuri Shimbun*, June 4, 2007. *Nihon Keizai Shimbun*, June 4, 2007.
218 Gabe Collins, "Fueling the Dragon: China-bound Pipelines are Russia's Most Realistic Asian Energy Option," *Geopolitics of Energy*, vol. 28, no. 9 (Sept. 2006), p. 13.
219 Michael Richardson, "Russia plays oil and gas as political instrument," *The Jakarta Post*, July 9, 2007. Cost estimates in 2007 were just under $500 million for the Daqing spur, *RIA Novosti*, July 10, 2007.
220 *RIA Novosti*, July 12, 2007.
221 Sergei Blagov, "Moscow Considers Enormous Investments in Eastern Russia's Gas Sector," *Eurasia Daily Monitor*, vol. 4, issue 119 (June 19, 2007).
222 "Pacific Pipeline Delayed until 2015," *Moscow Times*, July 20, 2007.

4 The international context

1 See, for example, Nobuo Miyamoto, "Nichirō Kankei ha Naze Nichibei no Tsugi-ni Juuyō-ka." Also, see Michael Green, *Japan's Reluctant Realism*, pp. 145–166. For an argument outlining why a Japanese-Russian *rapprochement* in the 1990s was important to balance against China's rise see Rajan Menon, "Japan-Russia Relations and Northeast Asian Security," *Survival*, vol. 38, no. 2 (Summer 1996), pp. 59–78.
2 China is the one nation on the UN Security Council opposing a permanent seat for Japan. See Green, *Japan's Reluctant Realism*, p. 205.

3 Not coincidentally, this is about the time that China's economic rise began to garner international attention. For changes in Japan's China policy in the mid-1990s see Michael Green and Benjamin Self, "Japan's Changing China Policy: From Commercial Liberalism to Reluctant Realism," *Survival*, vol. 38, no. 2 (Summer 1996), pp. 35–58.
4 The suspended grant assistance was but a small amount (roughly $75 million), but ODA to China in the early 1990s amounted to nearly $1 billion annually. See Green, *Japan's Reluctant Realism*, p. 78. Also, see Peggy Falkenheim Meyer, "Sino-Japanese Relations: The Economic Security Nexus," in Tsuneo Akaha, editor, *Politics and Economics in Northeast Asia: Nationalism and Regionalism in Contention*, New York: St. Martin's Press, 1999, p. 144.
5 Green and Self, "Japan's Changing China Policy," pp. 36–37.
6 The Senkaku (or Diaoyutai, in Chinese) Islands are located east of the Japanese island of Okinawa and are administratively part of Okinawa prefecture. China also claims the islands, as does Taiwan. Occasional Chinese air flights near the islands cause Japanese warplanes to scramble from bases in Okinawa. A Chinese oil research ship was reported in Japanese-claimed waters in the fall of 1995, and five more Chinese sonar ships were spotted in contested waters in April 1996. Some geologists have speculated that geological formations in the East China Sea resemble formations found in the oil-rich North Sea. Both China and Japan would want to exploit any reserves that might be found. See the *New York Times*, May 19, 1996.
7 "Defense Agency Voices 'Concern' Over PRC Military Buildup," *FBIS-EAS-96-140*, July 19, 1996.
8 See for example Yoichi Funabashi, "Engejimento, Antei, Baransu: Ajia Taiheiyo no 21 Seiki Senryaku," *Sekai*, January 1997, and "Nichibeichuu Shin-Jidai ni Wa-Kan-Yo no Kyoyō to Shuuyō wo," *Foresight Magazine*, January 1997, pp. 6-10. Also, see Jitsuro Terashima's "Nicchuubei Toraianguru Kuraishisu wo dō Seigyo suru ka," *Chuuō Kōron*, August 1996; and Takubo Tadae's "Nichibei Kankei ga Shinpai da," *Seiron*, March 1997. Though these authors do not all necessarily support the 'zero-sum' game theory, they do describe the fears in Japan of such a phenomenon. Takubo calls on Japan's leadership to reassess relations with Russia in light of the improving US-Chinese relations, pp. 48–49.
9 *Sankei Shimbun*, June 25, 1998. Also, see *Beijing Zhongguo Xinwen She* as cited in *FBIS-CHI-98-220*, Aug. 8, 1998.
10 *Asahi Shimbun*, Aug. 3, 1998.
11 *Xinhua News*, July 16, 2007. <http://news.xinhuanet.com/english/2007-07/16/content_6385169.htm>
12 Green, *Japan's Reluctant Realism*, p. 99.
13 *Financial Times*, June 19, 1997; *Agence France Presse*, Sept. 28, 1997.
14 For a look at Japanese concerns about Chinese energy consumption in the mid-1990s see Yoshihisa Murayama, "2010-nen 'Shigen Yuunyuu Daikoku' Chuugoku no Kyōi," *Foresight Magazine*, November 1996, pp. 50–53. Also, Taku Igarashi "Juuyu Ryuushutsu ga Ukibori ni shita 'Ajia Oiru Rodo' no Fuan.".
15 Green, *Japan's Reluctant Realism*, p. 101.
16 "Yaponskoe Predlozhenie," *Vremya Novostei*, Jan. 10, 2003.
17 "Yaponiya Opozdala," *Izvestia*, May 5, 2003.
18 *Sankei Shimbun*, Feb. 6, 2003.
19 "Your pipe or mine?" *The Economist*, Sept. 27, 2003, p. 40.
20 "Scramble for natural resources," *Yomiuri Shimbun*, June 6, 8, 9, 2007.
21 Tatsumi Okabe, "Nicchuu Kankei no Kako to Shōrai," *Gaikō Forum*, no. 151 (Feb. 2001), pp. 13–15.
22 Gilbert Rozman, "Russia's calculus and Japan's Foreign Policy in Pacific-Asia," in Takashi Inoguchi, editor, *Japan's Asian Policy*, New York: Palgrave, 2002.
23 Green, *Japan's Reluctant Realism*, pp. 96–98.

24 *FBIS-EAS-96-128*, June 11, 1996.
25 Chen Lineng, "The Japanese Self Defense Forces Are Marching Toward the Twenty First Century," *FBIS-CHI-96-085*, Feb. 8, 1996.
26 *Wall Street Journal*, Sept. 8, 1997.
27 Both quotes in *Foresight Magazine*, July 1998, p. 21, as cited in *FBIS-EAS-98-191*, July 10, 1998.
28 Vasilii Golovnin, "Heokochennyi Poyedinok Bogatyrya i Samuraya," *Novoe Vremya*, no. 1–2 (1997), p. 35.
29 For an interesting Russian viewpoint of Japan's concerns about the Sino-Russian 'strategic partnership' see Valerii Kistanov, "Yaponiya i Perspektiva Rossiisko-Kitaiskogo Strategicheskogo Partnyorstva," *Problemy Dal'nego Vostoka*, no. 2 (1997), pp. 48–56.
30 Hakamada, "Aratana Nichirō Kankei," pp. 7–8.
31 This according to Japanese MOFA officials, as cited in Green, *Japan's Reluctant Realism*, p. 104.
32 Nobuo Miyamoto, "Nichirō Kankei ha Naze Nichibei no Tsugi-ni Juuyō-ka," pp. 248–250. Miyamoto ends up his essay with the statement that, "to actively place Russia in the position of being a positive presence in our diplomatic and security picture is without precedent in Japan's diplomatic history. It is a new requirement for a new age. We should bring our wisdom to bear and continue to respond in a creative manner." [p. 251]. In fact there was a precedent in 1907–16 and the mid-1920s. See Chapter 1.
33 Tamba interview, Moscow, Dec. 7, 2000; Nishimura interview, Tokyo, Sept. 1, 1998; Shinoda interview, Tokyo, Sept. 1, 1998.
34 Report published by the *Sankei Shimbun*, Jan. 7, 1996, as cited in Harry Gelman, "The Changing Asian Arena," in Garnett, ed., *Rapprochement or Rivalry?*, p. 420.
35 *Asahi Shimbun*, May 19, 2002.
36 *Ibid.*, Aug. 3, 1998.
37 A.D. Voskresenskii, "Kitai vo Vneshnepoliticheskoi Strategii Rossii," in A.V. Kortunov, *Vneshaya Politika i Besopasnost' Sovremennoi Rossii*, vol. I, no. 2, Moscow: Moskovskii Obshestvennyi Nauchnyi Fond, 1999, pp. 146–147.
38 For a background of Sino-Russian relations in the 1990s, see Garnett, ed., *Rapprochement or Rivalry?*
39 Sergey Kortunov, "Rossiya isshet Soyuznikov," *Mezhdunarodnaya Zhizn'*, no. 5 (1996), p. 19.
40 Analyses in the American press often tend to view Sino-Russian relations in one of two ways; either they see the 'strategic partnership' as all but a military alliance aimed at the United States, or they scoff at the notion that Russia and China could be partners. For the former view see, for example, Paula J. Dobriansky, "Be Wary When the Bear Sides With a Dragon," *Los Angeles Times*, Sept. 18, 2000. Also, Martin Sieff, "Russia-China pact echoes 1939," *UPI*, July 16, 2001. For the latter view see, for example, Colin McMahon, "Russia, China to sign friendship accord; 2 nations wary of the West, but suspicions of each other still run deep," *Chicago Tribune*, March 7, 2001. Also, John Daniszewski, "Far East Void Eats at Russia," *Los Angeles Times*, July 19, 2001.
41 *Nezavisimaya Gazeta*, Feb. 19, 1998.
42 Official web site of the Russian President, <http://www.kremlin.ru/text/appears/2000/07/28796.shtml> (accessed July 23, 2007).
43 "Russia cracking down on illegal migrants," *International Herald Tribune*, Jan. 15, 2007.
44 See O.B. Rakhmanin, *K Istorii Otnoshenii Rossii-SSSR s Kitaem v XX veke*; Boris Kulik, *Sovietsko-Kitaiskii Raskol*; Sherman Garnett, ed., *Rapprochement or Rivalry?*
45 Viktor Larin, "Rossiya i Kitai ha Poroge Tret'ego Tysyachletiya: Kto-zhe bydet otstaivat' nashi natsional'nye interesy?" in A.V. Kortunov, *Vneshaya Politika i Besopasnost' Sovremennoi Rossii*, pp. 159–176. In English, see Vladimir Portyakov, "Are the Chinese Coming? Migration Processes in Russia's Far East," *International*

Affairs (Moscow), vol. 42, no. 1 (Jan.–Feb. 1996), pp. 132–140. Although neither Larin nor Portyakov personally voice grave concern about Chinese migration, they do cite various alarmist sources.

46 Andrew Kuchins, "Russia and great power security in Asia," in Gennady Chufrin, editor, *Russia and Asia: The Emerging Security Agenda*, SIPRI: Oxford University Press, 1999, pp. 440–441 (also quote).

47 Kunadze interview, Moscow, Feb. 4, 2000. Kunadze feels that Putin had recognized the danger of leaning too closely toward China in late 1999, and had begun to rectify this in early 2000. Also, see Georgii Kunadze, "Sindrom Kuz'kinoi Materi," *Novoe Vremya*, no. 37 (Sept. 1999), pp. 17–18.

48 "More Arbatov on National Security Approach." *FBIS-SOV-98-216*, Aug. 4, 1998.

49 For two fine, balanced views of Russia's 'China dilemma,' see V.S. Myasnikov, "Rossia i Kitai: Perspektivy Partnytorstva v ATR v XXI V," in A. Kortunov *Vneshaya Politika i Besopasnost' Sovremennoi Rossii*, pp. 177–204. Also, Vladimir Lukin, "Strategicheskoe Partnyorstvo Rossii i Kitaya – predskazuemaya real'nost'," *Problemy Dal'nego Vostoka*, no. 3 (1997), pp. 39–44.

50 For a look at Putin's Asia policy see, Hiroshi Kimura, "Puchin ha Tainichi Seisaku wo Dono-yōni Kettei suru no-ka," *Gaikō Forum*, no. 149 (Dec. 2000), pp. 12–19.

51 To follow current events in Sino-Russian and US-Russian relations see the Center for Strategic and International Studies (CSIS) Pacific Forum electronic journal *Comparative Connections*: <http://www.csis.org/pacfor/ccejournal.html> (accessed July 20, 2007).

52 Aleksei Bogaturov, *Velikie Derzhavy ha Tikhom Okeane: Istoriya i Teoriya Mezhdunarodnykh Otoshenii v Vostochnoi Azii Posle Vtoroi Mirovoi Voiny*, Moscow: RAN, Institut SShA i Kanady, 1997, pp. 301, 307.

53 Interviews with Vladimir Malakhov and Sergey Sherstyuk, Vladivostok, Aug. 28, 1998; Sergey Kastyornov and Vitaly Yelizariev, Yuzhno-Sakhalinsk, Sept. 26, 2000. At a 1999 conference a delegation of officials from the government of Khabarovskii Krai, including Governor Viktor Ishayev, called for a strong Japanese economic presence to help the economy and to balance against others, including China and the United States. The author attended this conference in Khabarovsk, Sept. 20–22, 1999.

54 Lukin interview, Moscow, Mar. 3, 2000; Galuzin interview, Moscow, Feb. 23, 2000; Pushkov interview, Moscow, Nov. 22, 2000. Losyukov talk at IMEMO, Moscow, Oct. 29, 2000.

55 Expressed numerous times to the author in a number of interviews in Moscow. Blagovolin, Kunadze, Zaitsev interviews, Moscow, 1999–2001.

56 Rozman, "Japan and Russia: Great Power Ambitions and Domestic Capacities," in Rozman, ed., *Japan and Russia*, p. 381.

57 Ryutaro Hashimoto, "Roshia ha Kanzen na Paatonaashippu wo," *Gaikō Forum*, no. 149, December 2000, pp. 30–35. Also, Minoru Tamba, "Fukugan-teki, Juuso-teki Nichiro Kankei Kochiku no Tame ni." *Gaikō Forum*, no. 149 (Dec. 2000), pp. 36–41.

58 "Highlights of US Foreign Trade," The Department of Commerce: Bureau of Economic Analysis, 1995. Also, "Asia-Pacific Economic Update," Honolulu: United States Pacific Command, Summer 1995, p. vi.

59 Green, *Japan's Reluctant Realism*, pp. 6, 23–24.

60 Yoichi Funabashi, *Alliance Adrift*, New York: Council on Foreign Relations Press, 2000.

61 Mikio Fujimura, "Hoppō Ryōdo Mondai de Nihon Shiji kara Kōtai-suru Beikoku," *Foresight Magazine*, July 1997, pp. 34–35.

62 Michael Green, "The Forgotten Player," *The National Interest*, Summer 2000, p. 47. For a comprehensive look at the troubled alliance in the mid-1990s see Funabashi, *Alliance Adrift*.

63 Sergei Kortunov, "Rossiya isshet Soyuznikov," p. 22.

64 Green, *Japan's Reluctant Realism*, p. 146.

65 The author was included in the planning for some of these meetings.

66 Tsuyoshi Hasegawa, "Russo-Japanese relations and the security of North-East Asia in the twenty-first century," in Gennady Chufrin, editor, *Russia and Asia: The Emerging Security Agenda*, p. 336. Hasegawa frequently alludes to the role of the United States in preventing the normalization of Japanese-Russian relations.

67 Krupyanko, *Yaponia 90-kh: V Poiskakkh Modeli Otnoshenii s Novoi Rossiei*, Moscow: RAN, Vostochnaya Literatura, 1997, pp. 6, 21–24, 43, 53. Krupyanko, however, also argues that worsening US-Russian relations are good for Japan, as are worsening relations between China and Russia.

68 "East Asia Braces as American Influence Fades," *Defense News*, Mar. 19, 2007.

69 Gilbert Rozman, "Russia's Resurgence in Northeast Asia: Views from the Region," *Russian Analytical Digest*, no. 25, July 17, 2007.

70 See Aleksei Arbatov, "Russia's Foreign Policy Alternatives." Lukin interview, Moscow, March 3, 2000.

71 Gilbert Rozman, "Sino-Russian Relations: Mutual Assessments and Predictions," in Garnett, ed., *Rapprochement or Rivalry?*, p. 166.

72 Hasegawa, "Russo-Japanese relations and the security of North-East Asia in the twenty-first century," pp. 327–329.

73 Yeltsin, *Prezidentskii Marafon*, p. 148.

74 Hiroshi Kimura, Shaojun Li, Il-Dong Koh, "Frontiers are the Razor's Edge: Russia's borders with its eastern neighbors," in Gilbert Rozman, Mikhail Nosov, and Koji Watanabe, editors, *Russia and East Asia: The twenty first Century Security Environment*, Armonk, NY: East West Institute, M.E. Sharpe, 1999, p. 151.

75 For a Japanese view of quadrilateral relations, see Nobuo Miyamoto, "Nichi-Bei-Chuu-Ro Shijuusō no Fukyōwaon," *Chuuō Kōron*, February 1998, pp. 138–149. Gilbert Rozman is one of the few American scholars to take on the complicated task of evaluating quadrilateral relations between China, Japan, Russia, and the United States. See, for example, "The Strategic Quadrangle and the Northeast Asian Region" in Rouben Azizian, ed., *Strategic and Economic Dynamics of Northeast Asia: global, regional, and New Zealand perspectives*, Centre for Strategic Studies, Wellington, New Zealand, 1999, pp. 8–21, or any one of Rozman's other works in the Bibliography dealing with this issue.

76 Chinese leaders, similarly, are unlikely to strongly oppose NATO expansion to the point that it would damage Sino-Western ties. Some analysts argue that apart from opposition to US 'hegemony' there is a lack of a common strategic vision between Beijing and Moscow. See Sherman Garnett, "A Limited Partnership," in Garnett, ed., *Rapprochement or Rivalry?*, pp. 11–13.

77 In late 2001, Gen Nakatani, Director General of the Japan Defense Agency, proposed that Japan and Russia cooperate in mine-clearing operations in Afghanistan, *Itar-Tass*, Dec. 18, 2001.

78 Rozman, "Russia's Resurgence in Northeast Asia."

79 Gelman, "The Changing Asian Arena," pp. 422–423.

80 The South Korean government had promised the Soviet Union $3 billion in economic credits in 1990. South Korea had paid out approximately $1.47 billion of this when it halted payments because of the inability of the Russian government to repay interest on the loans. The Asian economic 'flu' that hit Seoul in 1997 effectively killed off any hopes of future payments for the time being. See *The New York Times*, July 3, 1996, as cited in Gelman, "The Changing Asian Arena," p. 408.

81 Funabashi, *Alliance Adrift*, pp. 85–87.

82 Green, *Japan's Reluctant Realism*, pp. 124–126.

83 James Clay Moltz, "Russian Nuclear Regionalism," pp. 8–9.

84 The perception is also that a unified Korea, minus US troops, would more than likely lean toward China. The 1995 JDA report mentions this. See Gelman, "The Changing Asian Arena," pp. 413, 420. Also, Green, *Japan's Reluctant Realism*, pp. 142–143.

85 *Korea Herald*, Aug. 20, 1997, as cited in Gelman, "The Changing Asian Arena," p. 429, note 16.
86 Gilbert Rozman, "Japan and Korea: should the US be worried about their new spat in 2001?" *The Pacific Review*, vol. 15, no. 1 (2002), pp. 1–28.
87 Green, *Reluctant Realism, passim.*
88 Vladimir Li, "Russia's Interests in the Versatile Structure of Security in Northeast Asia and the Korean Issue," paper presented at the Research Institute for International Affairs (Seoul) at the conference *The New Discourses on Peace Regime in Northeast Asia and Korea: Contending Views and New Alternatives*, Nov. 22–23, 1996.
89 Georgii Kunadze, "Politika Rossii v Otnoshenii KNDR," *Mirovaya Ekonomika i Mezhdunarodnye Otnosheniya*, no. 12 (1999), p. 39.
90 Leszek Buszynski, "Russia and Northeast Asia: aspirations and reality," pp. 411–414. During 1996–97 Russia had transferred to South Korea weapons systems (including tanks) worth approximately $200 million in exchange for debt relief. See Alexander Fedorovsky, "Russian Policy and Interests in the Korean Peninsula," in Chufrin, ed., *Russia and Asia*, p. 397.
91 Zaitsev interview, Moscow, June 21, 2000.
92 "Russia Loses Its Ticket to Asia," *STRATFOR Review*, Aug. 16, 2000.
93 *The Moscow Times*, Aug. 23, 2002. Also, see *Izvestia*, Oct. 29, 2002.
94 Hiroshi Kimura, "Puchin no Tai-Nichi Seisaku (3)," *Journal of World Affairs* (Tokyo), 4/2003, pp. 82–83.
95 Sang-Woo Rhee, "Russia and the new balance of power in East Asia: implications for stability on the Korean Peninsula," in Chufrin, ed., *Russia and Asia*, p. 408.
96 *RIA Novosti*, Nov. 19, 2006.
97 For more on the strategic perceptions of Japan and Russia concerning the Korean Peninsula, see Rajan Menon and Charles E. Ziegler, "The Balance of Power and US Foreign Policy Interests in the Russian Far East," *NBR Analysis*, vol. 11, no. 5 (Dec. 2000), p. 6; also, Gilbert Rozman, "The Strategic Quadrangle and the Northeast Asian Region," p. 13.
98 Buszynski, "Russia and Northeast Asia," p. 413.
99 *Izvestia*, Feb. 10, 2000; *Segodnya* Feb. 11, 2000; *Nezavisimaya Gazeta* Feb. 12, 2000.
100 For more on the 2002–03 crisis and the Russian role therein, see Joseph Ferguson, "Russia's Role on the Korean Peninsula and Great Power Relations in Northeast Asia," pp. 33–50.
101 "Moscow Asks Tokyo to Extend Economic Aid to DPRK," in *Seoul Yonhap* as cited in *FBIS-EAS-2000-0217*, Feb. 17, 2000. One wonders whether Russia hoped to get Pyongyang to use this Japanese assistance to repay North Korea's debt to Russia, which amounts to roughly $3 billion.
102 "Far Eastern Commercial Bank Participates in Transfer of DPRK Funds," *Vladivostok Times*, June 19, 2007.
103 *The Russia Journal*, Nov. 4–10, 2000.
104 Kent Calder, "Japan's Energy Angst and the Caspian Great Game." *NBR Analysis*, vol. 12, no. 1, (Mar. 2001), p. 1.
105 For Japanese energy insecurities and Middle East dependence see, Yōshi Tachiyama, "Mega-Chuutō wo Misueru Arata-na Shisa wo," *Foresight Magazine*, Sept. 1997, pp. 6–9. In English, see Calder, "Japan's Energy Angst and the Caspian Great Game."
106 Yasuo Tanabe, "Kasupi Kai Sekiyu Shigen no Seiji-Keizai Gaku," *Gaikō Forum,* 8/9 (1998), pp. 64–70.
107 *Tokyo Shimbun*, Mar. 10, 1998.
108 Ahmed Rashid, "Power Play," *The Far Eastern Economic Review*, 10 Apr. 1997, pp. 22–24.
109 "Beichuurō ni Nirami Dokuji Senryaku: Nihon Kasupi Kai Shigen Arasoi Sannyu e," *Asahi Shimbun*, Mar. 3, 1998.

110 Kunihiko Miyake, quoted from a speech delivered at the Middle East Institute, Washington, DC, March 13, 1998.

111 Kent Calder argues that Japan's diplomatic initiatives in Central Asia are part of a design to gain leverage over Russia to use in negotiations over the status of the disputed islands. Kent Calder, "Japan's Energy Angst and the Caspian Great Game," p. 2.

112 Shigeki Hakamada, "Akino-shi no Nokoshita Yume," *Gaikō Forum*, no. 10 (1998), pp. 12–14.

113 See Gaye Christoffersen, "China's Intentions for Russian and Central Asian Oil and Gas." *NBR Analysis*, vol. 9, no. 2 (Mar. 1998).

114 For a chronological look at US-Russian relations in Central Asia see the Center for Strategic and International Studies (CSIS) Pacific Forum electronic journal *Comparative Connections*: <http://www.csis.org/pacfor/ccejournal.html> (accessed July 20, 2007).

115 See for example, *Nezavisimaya Gazeta*, March 2, 2002; in English, *The Russia Journal*, Oct. 12–18, 2001.

5 The domestic political context

1 For various accounts of the political dynamics of Japan in the 1990s see, for example, Gerald Curtis, *The Logic of Japanese Politics: Leaders, Institutions, and the Limits of Change*, New York: Columbia University Press, 1999. Also, Richard Katz, *Japan The System That Soured: The Rise and Fall of the Japanese Economic Miracle*, Armonk, NY: M.E. Sharpe, 1998. Also, Pempel, *Regime Shift*.

2 Green, *Japan's Reluctant Realism*, p. 58.

3 Hasegawa, *The Northern Territories Dispute*, *passim*. Rozman, ed., *Japan and Russia*, *passim*.

4 Nishimura, "Eurasia Gaikō no Butaiura," p. 144.

5 In 2004 Tamba published his memoirs upon returning from his ambassadorship in Russia. They give a good overview of the evolution of Japan's Soviet and Russia policy. Minoru Tamba, *Nichiro Gaikō Hiwa*, Tokyo: Chuokoron-Shinsha, 2004.

6 See, Ryutaro Hashimoto, "Roshia ha Kanzen na Paatonaashippu wo," pp. 30–35; Minoru Tamba, "Fukugan-teki, Juuso-teki Nichiro Kankei Kochiku no Tame ni."; Also, Tamba, *Nichiro Gaikō Hiwa*, pp. 11–66.

7 Tōgō published his memoirs in 2007: *Hoppō Ryōdo Kōshō Hiroku: Ushinawareta Gotabi no Kikai*.

8 Nishimura, "Yurashia Gaikō no Butaiura," p. 141.

9 The reassignments were partly rotational and partly due to political infighting in the Ministry during the Suzuki scandal. This will be addressed below.

10 Tsutomu Saitō, "Satō, Rasupuchin to Yobareta Otoko," *Bungei Shunju*, May 2002, pp. 114–121. Also, Takao Toshikawa, "'Muneo Paji' de Miushinatta Gaikō-Gaimusho Kaikaku," pp. 85–86.

11 *Asahi Shimbun*, May 24, 2002. Also, interview with Yoichi Nishimura, Washington, DC, May 30, 2002.

12 Sumio Edamura, "Nichiro Kōshō no 'Wakugumi' wo Tatenaose," *Chuuō Kōron*, May 2002, pp. 68–73. Also, Hiroshi Hasegawa, "Suzuki Muneo Nigen Gaikō no Aya," *Aera*, Mar. 5, 2001, pp. 29–31.

13 *Asahi Shimbun*, May 28, 2002.

14 Takao Toshikawa, " 'Muneo Paji' de Miushinatta Gaikō-Gaimusho Kaikaku," p. 84.

15 Gilbert Rozman, "A Chance for a Breakthrough in Russo-Japanese Relations: Will the Logic of Great Power Relations Prevail?" *Pacific Review*, vol. 15, no. 3 (2002), pp. 325–357.

16 See Chapter 6 for the breakdown in public opinion polls of this time.

17 After serving time in prison, Satō published an insider's account of the political scandal that erupted. See, Satō, *Kokka no Wana*.

18 "Japan's Assistance Programs for Russia," Tokyo: The Ministry of Foreign Affairs of Japan, 1998.
19 Based on interviews with Takeshi Kondo and Ryoichi Nozaki of the Itochu Corporation, and Koji Hitachi of the Nissho Iwai Corporation, Sept. 1–2, 1998, Tokyo. See also, Nishimura, "Yurashia Gaikō no Butaiura," p. 141. Also, Mori, "Nichirō Shunō 'Kawana Kaidan' no Butaiura," pp. 130–131. METI and ANRE have also been involved in the dismantling of nuclear weapons and in the commissioning of a radioactive waste facility in the Russian Far East.
20 "The Merits and Demerits of the Sakhalin Gas Pipeline: Drawn into a Muddy Stream of Political and Bureaucratic Communities," *Tokyo Enerugi Foramu*, April 1998. As cited in FBIS-EAS-98-180, June 29, 1998. It was pointed out that JNOC has a 50 percent stake in the consortium SODECO, which is a participant in the Sakhalin-1 project.
21 Green, *Japan's Reluctant Realism*, p. 60.
22 In 1999 the Ex-Im Bank merged with several other institutions and has since been known as the Japan Bank for International Cooperation.
23 Interview with Yoshiaki Nishimura, Tokyo, Jan. 18, 2000.
24 Mikhail Krupyanko, *Yaponia 90-kh*, pp. 72–75.
25 For example, the JDA and MITI fought over the implementation of Japanese investment in the Baikal-Amur Mainline Railway (BAM) in 1972–73. In the end the JDA won out. See Chapter 2.
26 See, for example, Vasilii Golovnin, "Russkaya Svetlana vkhodit v elitu 'Aum Sinrikyo," *Izvestia*, Feb. 23, 2000.
27 "Mafiya i More," *Nezavisimaya Gazeta*, Feb. 27, 2003.
28 See, for example, Galina Vitkovskaya, "Lawlessness, Environmental Damage, and Other New Threats in the Russian Far East," in Rozman, Nosov and Watanabe, eds., *Russia and East Asia: The twenty first Century Security Environment*, pp. 179–199. Also, see "The Death of Sushi?" in the *Far Eastern Economic Review*, Aug. 15, 2002.
29 *Nezavisimaya Gazeta*, May 22, 2002.
30 Kunio Okada, "The Japanese Economic Presence in the Russian Far East," in Judith Thornton and Charles Ziegler, editors, *Russia's Far East*, pp. 432–433, and note 22, p. 439.
31 See, for example, Vasilii Golovnin, "Idiotov daet dobro," *Izvestia*, Apr. 15, 2002.
32 Michael Robert Hickok, "Japan's Gambit: An Asian View of Eurasia," *Problems of Post-Communism*, vol. 47, no. 3 (May/June 2000), pp. 36–37.
33 Ryutaro Hashimoto, "Roshia ha Kanzen na Paatonaashippu wo."
34 *Tokyo Ekonomisuto*, May 19, 1998, pp. 72–74, as cited in FBIS-EAS-98-142, May 22, 1998. Also, *Bungei Shunju*, January 1998, pp. 224–228, as cited in FBIS-EAS-98-054, Feb. 23, 1998.
35 Hickok, "Japan's Gambit: An Asian View of Eurasia," p. 38.
36 *Asahi Shimbun*, May 24, 2002. Also, Takerō Nagashi, "Shippai ni Owatta Tai-ro Seisaku Kudeta."
37 Suetsugu interview, Tokyo, Sept. 28, 2000.
38 Haruki Wada, "Sukyandaru to Gaikō,"pp. 75–76.
39 Nagashi, "Shippai ni Owatta Tai-ro Seisaku Kudeta," p. 9. Also, Suetsugu interview, Tokyo, Sept. 28, 2000.
40 See, for example, Gerald Curtis, *The Japanese Way of Politics*, New York: Columbia University Press, 1988, pp. 114–116.
41 "The Merits and Demerits of the Sakhalin Gas Pipeline: Drawn into a Muddy Stream of Political and Bureaucratic Communities," *Tokyo Enerugi Foramu*, April, 1998. As cited in FBIS-EAS-98-180, June 29, 1998.
42 Green, *Japan's Reluctant Realism*, p. 49.
43 Ichiyanagi interview, Vladivostok, Aug. 23, 1998; Hitachi interviews Tokyo and Moscow 1998–2000. Also, see Gilbert Rozman, "Backdoor Japan: The Search for a

Way Out via Regionalism and Decentralization," *Journal of Japanese Studies*, 25(1) (1999), pp. 3–31.

44 Judith Thornton, "Sakhalin Energy: Problems and Prospects," in Thornton and Ziegler, eds., *Russia's Far East: A Region at Risk*, p. 169.

45 Kunio Okada, "The Japanese Economic Presence in the Russian Far East," in Thornton and Ziegler, eds., *Russia's Far East: A Region at Risk*, p. 430.

46 "Mitsui, Mitsubishi to sell Sakhalin shares," *Yomiuri Shimbun*, Sept. 22, 2006.

47 Investment in Sakhalin-1 is expected to reach $12 billion, and in Sakhalin-2 $10 billion. See Judith Thornton, "Sakhalin Energy: Problems and Prospects," pp. 172–174.

48 Okada Kunio, "The Japanese Economic Presence in the Russian Far East," p. 439, note 12.

49 Hitachi and Kondo interviews with the author, Tokyo, Aug. 28–29, 1998.

50 Green, *Japan's Reluctant Realism*, pp. 65–66.

51 Thornton, "Sakhalin Energy: Problems and Prospects," pp. 166–167.

52 *Kommersant'*, Aug. 7, 2006.

53 Guy-Pierre Chomette, "Enjeu Stratégique ou Conflit du Passé?" *Le Monde Diplomatique*, Sept. 2001, pp. 10–11. See also, *Asahi Shimbun*, May 25, 2002.

54 Okada, "The Japanese Economic Presence in the Russian Far East," pp. 427–429.

55 *Ibid.*, pp. 423–424.

56 For a Japanese view on the damage dome to Russia's reputation by the problems encountered in economic interaction see Akihiro Iwashita, "U Yaponii i Rossii malo obshevo," *Izvestia*, Jan. 17, 2001.

57 Krupyanko, *Yaponia 90-kh*, p. 68.

58 Based largely on numerous interviews with Shigeki Hakamada, Hiroshi Kimura, Tomohisa Sakanaka, and Ichirō Suetsugu, Tokyo 1996–2000. Also, see Gilbert Rozman, *Japan's Response to the Gorbachev Era, 1985–1991,* pp. 34–38. Many heartfelt thanks to the late Tomohisa Sakanaka for introducing me to Suetsugu in 1997.

59 Yakovlev interview, Mar. 2, 2000.

60 For an excellent review of the positions of Japan's academics, experts and journalists on relations with the Soviet Union, see Rozman, *Japan's Response to the Gorbachev Era, 1985–1991*. Most of the people cited by Rozman are still active participants today in the Russia debate.

61 Ken'ichi Itō, Matake Kamiya, Hiroshi Nakanishi, Kunihiko Yamaoka, "Japan's Initiatives towards US, China and Russia," Tokyo: Japan Forum on International Relations, Apr. 19, 1999. For the English-language version see <http://www.jfir.or.jp/e-jf-pr-18/pr18-body.html> (accessed Dec. 20, 1999).

62 See, for example, Guy-Pierre Chomette, "Enjeu Stratégique ou Conflit du Passé?" *Le Monde Diplomatique*, September 2001, pp. 10–11. Also, *Asahi Shimbun*, May 25, 2002.

63 *Asahi Shimbun*, May 25, 2002. Also, for more on the *Uyoku*, see Green, *Japan's Reluctant Realism*, p. 68.

64 For an excellent summary of Russia's complex political situation during this period see Richard Sakwa, *Russian Politics and Society*, New York: Routledge, 1996.

65 Based on numerous interviews with Sergei Blagovolin, Aleksei Bogaturov, Mikhail Nosov, and Valerii Zaitsev, Moscow, 1999–2001.

66 Many Russians insist that Yeltsin was duped by the Japanese into making such a statement. See Chapter 3 for details. Mikhail Galuzin of the Russian Foreign Ministry served as Yelstin's interpreter at the Krasnoyarsk summit and claims that it was Yeltsin that came up with the idea. Galuzin interview, Moscow, Feb. 23, 2000.

67 Yeltsin, *Presidentskii Marafon*, p. 137.

68 *Ibid.*

69 Vyacheslav Amirov, "Russia in the Asia-Pacific area: challenges and opportunities," in Chufrin, ed., *Russia and Asia: The Emerging Security Agenda*, p. 276.

70 Yeltsin, *Presidentskii Marafon*, p. 148.
71 Much of the information on Primakov came from numerous interviews with Georgii Kunadze, Moscow, 1999–2001. For the Primakov quote see Green, *Japan's Reluctant Realism*, p. 315, note 41. Also, see Chapter 4 of this study for more detail on Sino-Russian relations and Japan's reaction to it.
72 Liberals, such as Yegor Gaidar, were also concerned with China's rise. Amirov, "Russia in the Asia-Pacific area," p. 284.
73 *Itar-Tass*, as cited in FBIS-SOV-97-274, Oct. 1, 1997. Also, see Kimura, *Distant Neighbors (Vol. II)*, p. 311, note 37.
74 Nodari Simonia, "Domestic developments in Russia," in Chufrin, ed., *Russia and Asia: The Emerging Security Agenda*, Oxford: SIPRI/Oxford University Press, 1999, p. 59–60.
75 Gilbert Rozman, "Sino-Russian Relations: Mutual Assessments and Predictions," in Garnett, ed., *Rapprochement or Rivalry?*, pp. 158–159.
76 Kaoru Sakurai, "Enerugi Riken Sodassen ni Yusaburareru Roshia Shin-naikaku," *Foresight Magazine* (May 1998), pp. 32–33.
77 *Izvestia*, Sept. 7, 2000; *Kommersant'*, July 22, Sept. 2, 2000; *Moscow Times*, July 25, 2000.
78 Kimura, *Distant Neighbors (Vol. II)*, p. 226.
79 Nobuo Miyamoto, "Nichirō Kankei ha Naze Nichibei no Tsugi-ni Juuyō-ka," p. 246.
80 Blagovolin interview, Moscow, Dec. 6, 2000.
81 See Chapter 6 for numbers on public opinion surveys.
82 Suetsugu interview, Tokyo, Sept. 28, 2000.
83 Shigeki Hakamada, "Irukutsuku no Gensō," p. 8.
84 Blagovolin interview, Moscow, Dec. 6, 2000.
85 Takayuki Nakazawa, "Ryōdō Mondai kaiketsu no Kagi to naru no ha 'Tsuyoi Roshia'," *Sekai Shuuho*, Nov. 28, 2000, pp. 26–29.
86 See *Obshchaya Gazeta*, No. 31, Aug. 2, 2001 and *Vremya Novostei*, July 13, 2001.
87 Kimura, Shimotomai, Suetsugu interviews, Tokyo, Sept. 28, 2000.
88 The same was said about Yeltsin in 1996–97, when presidential advisors Anatolii Chubais and Aleksandr Korzhakov battled for influence over the ailing President. Chubais won out (temporarily) and Korzhakov was fired. See, "Media Focus on the Behind-the-scenes Kremlin Power Struggle," *Jamestown Monitor*, Nov. 29, 2001.
89 Hakamada and other Japanese observers insist that the Japanese side (through Muneo Suzuki and Sergei Ivanov) gave Putin the idea of bringing up the applicability of the 1956 Joint Declaration. See Shigeki Hakamada, "Irukutsuku no Gensō." Hakamada also points out that Putin would not necessarily return any of the four islands, even were a peace treaty signed specifying Russia's obligation to do so.
90 In Kozyrev's memoirs he barely even includes Japan in his chapter entitled "*Rossia idyot na Vostok* [Russia goes to the East]." He includes a section on Japan later on in the book when he discusses domestic politics. Kozyrev, *Preobrazhenie*, pp. 243–244, 293–301.
91 Simonia, "Russian Energy Policy in East Siberia and the Far East," p. 28.
92 Ivanov, "Russia's Energy Politics," p. 12.
93 Bogaturov interview, Moscow, Dec. 16, 1999.
94 Nosov interview, Moscow, Dec. 3, 1999.
95 Interview with Oleg Serov, Washington, DC, Jan. 16, 1998.
96 Baranets, *Yel'tsin i ego Generaly*, pp. 272–307.
97 Simonia, "Domestic developments in Russia," pp. 61–71.
98 Kuchins, "Russia and great power security in Asia," pp. 440–441.
99 *Asahi Shimbun,* Mar. 30, May 23, 1998.
100 Green, *Japan's Reluctant Realism*, p. 315, note 41.
101 Vasilii Golovnin, "Utrennii Perepolokh," *Izvestia*, Feb. 15, 2001.
102 "Rough Sailing for Russian Naval Exercises?" *Jamestown Monitor*, July 28, 1998.

103 *Nikkei Shimbun*, Aug. 31, 2000.
104 Lukin interview, Moscow March 3, 2000. Also, Kunadze interview Feb. 23, 2000. Also, see "No Peace Treaty Breakthrough to Be Expected from Summit," *Itar-Tass*, Mar. 15, 2001.
105 Kaoru Sakurai, "Enerugi Riken Sodassen ni Yusaburareru Roshia Shin-naikaku," pp. 32–33.
106 Itō, "Kiro ni tatsu Taiheiyō Paipurainu Kōsō," p. 30.
107 *Kommersant'*, July 29, 2006.
108 *Nezavisimaya gazeta*, Feb. 28, 2006.
109 *Moscow Times*, July 19, 2007. See <http://www.mofa.go.jp/announce/event/2007/7/1174282_852.html> (accessed July 22, 2007).
110 In 1991 after the fall of the Soviet Union the Ministry of Fisheries became the Russian Committee of Fisheries, a department within the Ministry of Agriculture. See Tony Allison, "The Crisis of the Region's Fishing Industry: Sources, Prospects, and the Role of Foreign Interests," in Thornton and Ziegler, eds., *Russia's Far East: A Region at Risk*, p. 141.
111 Yurii Sinel'nik, "Shto My Kurilam, Shto Kurily Nam?" *Nezavisimaya Gazeta*, Aug. 31, 2000.
112 Allison, "The Crisis of the Region's Fishing Industry," pp. 150–151, 154.
113 Okada, "The Japanese Economic Presence in the Russian Far East," p. 432.
114 Vyacheslav Zilanov, "Kuril'skoe Rybolovstvo i Diplomatiya," *Nezavisimaya Gazeta*, Jan. 20, 2001.
115 Much of this section based on numerous interviews with Georgii Kunadze, himself a veteran of these hearings. The author was also provided with a transcript of hearings by an anonymous participant in Duma hearings in the fall of 2000. See also, Oleg Bondarenko, *Neizvestnye Kurily*, Moscow: VTI-Data Press, 1992.
116 Kimura, *Distant Neighbors (Vol. II)*, p. 173.
117 "Zhirinovsky Suggests Returning Kurile Islands to Japan," *Jamestown Monitor*, June 1, 2002.
118 Kimura *Distant Neighbors (Vol. II)*, p. 174.
119 Lukin interview, Moscow, Mar. 3, 2000.
120 Aleksei Arbatov, "Russia's Foreign Policy Alternatives," *International Security*, vol. 18, no. 2 (Fall 1993), p. 37.
121 Arbatov, "Military Reform in Russia: dilemmas, obstacles, and prospects," *International Security*, vol. 22, no. 4 (Spring 1998), pp. 91–92.
122 *Nezavisimaya Gazeta*, as cited in BBC Summary of World Broadcasts, Nov. 7, 1998.
123 Kastyornov, Vishnevsky, and Yelizariev interviews, Yuzhno-Sakhalinsk, Sept. 26, 2000. Also, see Nakazawa, "Puchin Hōnichi to Hoppō Ryōdo Mondai," pp. 14–17.
124 Vishnevsky interview, Yuzhno-Sakhalinsk, Sept. 26, 2000.
125 Nakazawa, "Puchin Hōnichi to Hoppō Ryōdo Mondai," pp. 14–17.
126 From a series of articles published on the 'Northern Territories' in the *Sankei Shimbun*, Aug. 24–28, 1999. Also, Nakazawa, "Puchin Hōnichi to Hoppō Ryōdo Mondai," pp. 14–17.
127 *Ibid.* Also, "Russians Grow Restless and Desperate on Disputed Kuril Islands" *AP*, Jan. 3, 1999.
128 Nakazawa, "Puchin Hōnichi to Hoppō Ryōdo Mondai," pp. 14–17.
129 See, for example, "Changing Perception of the People Living in the Northern Territories," *Sentaku*, December 1998, pp. 14–16, as cited in FBIS-SOV-1999-0223, Dec. 1, 1998.
130 The author met these two on several occasions while traveling in the Russian Far East between 1998 and 2000, and continues to correspond with them.
131 Kunihiko Matsushima, "Nihon Kai Juuyu Ryuushutsu Jiko no Haigo ni Kyokutô Roshia no Keizai Hatan," *Foresight Magazine*, Mar. 1997, pp. 36–37. Also, Koichi Watanabe, "Hokai Suru Hoppô Yon Shima."

132 Gilbert Rozman, "The Crisis of the Russian Far East: Who is to blame?" *Problems of Post-Communism*, vol. 44, no. 5 (Sept./Oct. 1997), p. 4.
133 Vladivostok Mayor Viktor Cherepkov, a Chubais ally, was however reinstated, a slight victory for Chubais. See Michael McFaul, "The Far Eastern Challenge to Russian Federalism," in Garnett, ed., *Rapprochement or Rivalry?*, pp. 325–331.
134 Peter Kirkow, "Regionalism Warlordism in Russia: The Case of Primorskii Krai." *Europe-Asia Studies*, vol. 47, no. 6 (1995), pp. 934–935.
135 "Russia-Chinese Border Agreement Under Fire," *Itar-Tass*, as cited in *RFE/RL* Dec. 17, 1996.
136 *Itar-Tass*, Aug. 23, 1999.
137 *Yomiuri Shimbun*, May 17, 2001.
138 *Itar-Tass*, Dec. 3–4, 2000.
139 *Reuters*, Dec. 8, 2000.
140 *Nezavisimaya Gazeta*, May 22, 2002.
141 *Asahi Shimbun*, Aug. 21, 2001.
142 For a good summary of the VPK and its influence on foreign policy, see Simonia, "Domestic developments in Russia," pp. 61–71.
143 See, for example, *Itar-Tass*, Nov. 29, 2000.
144 *Moskovskie Novosti*, Feb. 29-Mar. 6, 2000.
145 Author's interview with Jonas Bernstein, formerly columnist of the *Moscow Times* and for the *Jamestown Foundation*. Mar. 21, 2000.
146 Many thanks to an anonymous staff member at IMEMO for providing this author with a copy of the proposal presented to the Kremlin.
147 Based on interviews with Yoichi Nishimura and Viktor Pavlyatenko, Moscow and Washington, DC, 2000–02.

6 The ideational context

1 See Robert Jervis, *Perception and Misperception in International Politics*. Also, see Ernest May, *"Lessons" of the Past: The Use and Misuse of History in American Foreign Policy*, New York: Oxford University Press, 1973.
2 For comparison's sake, take as contrasting examples relations between Canada and the United States, or France and Spain, where there is a freer flow of people and ideas between the nations.
3 *Yomiuri Shimbun*, Nov. 1, 2003. Also see, Anzen Hoshō Mondai Kenkyuujo, *Kawaru Nichirō Kankei*, pp. 36–39. This book gives samples from a number of polls taken both in Japan and in Russia. In English for an excellent summary of opinion polls taken in Japan see Hasegawa, "Japanese Perceptions of the Soviet Union and Russia in the Postwar Period," in Rozman, editor, *Japan and Russia*, pp. 281–320.
4 Alexander Panov, "From Stalin to Brezhnev," in Goodby, Ivanov, and Shimotomai, editors, *"Northern Territories" and Beyond*, p. 27. Panov ascribes the disjunction in Japanese-Russian historical memory to *negative* factors. Hasegawa Tsuyoshi (*The Northern Territories Dispute and Russo-Japanese Relations*, p. 70) ascribes it to *selective* factors. Perhaps a better term to refer to historical memory as it pertains to Japanese-Russian relations is '*subjective*,' because the two sides simply see it differently.
5 Rozman, "Russia's calculus and Japan's Foreign Policy in Pacific-Asia.".
6 Peter Katzenstein and Nobuo Okawara, "Japan's National Security: Structures, Norms, and Politics," *International Security*, vol. 17, no. 4 (Spring 1993), p. 98. Thomas Berger, "From Sword to Chrysanthemum: Japan's Culture of Anti-militarism," *International Security*, vol. 17, no. 4, Spring 1993, pp. 135, 139. Much of the feeling of the 'victim mentality' is directed toward the United States and the atomic bombings.
7 Yoshi Numachi, *Haruka-na Shiberia.*
8 Hasegawa, *The Northern Territories Dispute and Russo-Japanese Relations,* p. 71.

9 John Dower, *Embracing Defeat: Japan in the Wake of World War II*, New York: W.W. Norton and Company, 1999, pp. 482–484, 491, 496–497; quote taken from p. 513.

10 Tuomas Forsberg, "Explaining Territorial Disputes: From Power Politics to Normative Reasons," *Journal of Peace Research*, vol. 33, no. 4 (1996), pp. 433–449. Ironically, as described previously Russia has acceded to Chinese demands that some of the disputed territory in the Far East be returned but this border issue has a much longer history and has little to do with the Second World War. Nevertheless, Russians have been quite distraught over the loss of territory to the Ukraine, Belarus, and Kazakhstan, as Bruce Parrott has pointed out.

11 *The New York Times*, Feb. 7, 1992, as quoted in Akaha and Murakami, "Soviet/Russian-Japanese Economic Relations," in Hasegawa, Haslam, and Kuchins, editors, *Russia and Japan: An Unresolved Dilemma Between Distant Neighbors*, p. 181.

12 Hara, *Japanese-Soviet/Russian Relations since 1945*, pp. 194–195, 234.

13 Krupyanko, *Yaponia 90-kh*, pp. 88, 127–128.

14 Susumu Takahashi, "Toward a New Era of Russian-Japanese Relations," in Goodby, Ivanov, and Shimotomai, eds, *"Northern Territories" and Beyond*, pp. 204–205. One Japanese analyst was quoted as saying, "The nature of Russia has not changed basically from Imperial times," *Yomiuri Shimbun* Feb. 19, 2001.

15 Takahashi, "Toward a New Era of Russian-Japanese Relations," p. 208. Also, see Hasegawa, "Japanese Perceptions of the Soviet Union and Russia in the Postwar Period,", pp. 281–320. Hasegawa's work incorporates numerous opinion polls and demonstrates the consistently negative image of Russia in Japan during the 1980s and 1990s.

16 See, for example, Daniil Nizamutdinov, "Russia's friends and enemies: Who are they?" *RIA Novosti*, June 9, 2005. Also, *Yomiuri Shimbun*, June 23, 2005.

17 Japanese leaders in favor of normalization could have pointed to security fears surrounding the rise of China and the strategic ambiguity of the United States, but admittedly this would have been difficult to do diplomatically. Tamba, Tōgō, and others, however, could have pointed to the environmental dangers of allowing the Russian Far East to fall apart and to the energy concerns in Japan that could be partially alleviated with Russian assistance.

18 See Rozman, *Japan's Response to the Gorbachev Years, passim.*

19 See, for example, Kunihiko Matsushima, "Nihon Kai Juuyu Ryuushutsu Jiko no Haigo ni Kyokutō Roshia no Keizai Hatan," pp. 36–37. Also, Koichi Watanabe, "Hokai Suru Hoppō Yon Shima," pp. 116–126.

20 Tsuyoshi Hasegawa, "Japanese Misperceptions of the Soviet Union During the Gorbachev Era," *The Carl Beck Papers*, no. 1503, University of Pittsburgh, Oct. 2000, p. 62.

21 Kamo, "Japan, Russia, and Nuclear Nonproliferation," in Goodby, Ivanov, and Shimotomai, editors, *"Northern Territories" and Beyond,* pp. 258-259.

22 *Asahi Shimbun*, Nov. 7, 1998. For more recent data, see Daniil Nizamutdinov, "Russia's friends and enemies: Who are they?" *RIA Novosti*, June 9, 2005.

23 "Russian Attitudes Towards Japan," *VTsIOM Survey*, as cited in *Russian Analytical Digest*, No. 25, July 17, 2007.

24 Georgii Kunadze, "Militarizm v Yaponii: Voprosy Metodologii Analiza," 116–130.

25 Semyon Verbitsky, "Perceptions of Japan in the USSR during the Cold War and Perestroika," pp. 16–17.

26 Ivan Tselishev, "'Upravlenie po-yaponski' za predelami Yaponii," pp. 58-79. In English, see Kimura, *Japanese-Russian Relations under Gorbachev and Yeltsin*, pp. 55–76.

27 See, for example, Black, Jansen, *et al.*, *The Modernization of Japan and Russia.*

28 See, for example, Kenneth Pyle, *The Japanese Question*, Washington, DC: American Enterprise Press, 1992. Also, see Edward Lincoln, *Japan's New Global Role.*

29 See, for example, the series of articles in the *Japan Times* on Japan's security issues,

Aug. 23–30, 2001. Also, see the *Yomiuri Shimbun* Feb. 19, 2001.

30 *New York Times*, June 9, 2002. This article pointed out groups in the political mainstream, and the public support they have garnered in re-examining this once taboo issue.

31 In an interview with *Izvestia* in July 2002, Russian Foreign Minister Igor Ivanov asserted that Russia's threats no longer lie in the West, but to the South and the East. "The threat for Russia hides in the Caucasus Mountains region and its Asian border." He went on to say that, "one of the main threats we have seen has not been the United States or NATO, but Afghanistan," *Izvestia*, July 10, 2002.

32 This idea was expressed by Vladimir Lukin, former Deputy Speaker of the Russian Duma and Ambassador to the United States. Cited in Yang Mingjie, "Russia's Regional Security Role."

33 Krupyanko, *Yaponia 90-kh*, pp. 124–125.

34 *Versiya*, No. 19 May 29–June 4, 2001.

35 Aleksandr Dugin, *Osnovy Geopolitiki*, Moscow: Arktogeya, 1997. Also, see the web site: http://arctogaia.com/public/osnovygeo/

36 Mari Kuraishi Horne, "The Northern Territories Dispute: Source or Symptom?" *Journal of Northeast Asian Studies*, Winter 1989, p. 62.

37 See, for example, Anzen Hoshō Mondai Kenkyuujo, *Kawaru Nichirō Kankei*.

38 Quote taken from Rozman, "Russia and Japan: Mutual Misperceptions, 1992–1999," *The Carl Beck Papers*, no. 1503, University of Pittsburgh, October 2000, p. 84.

39 There is a good section on the forces of globalization and how they affect Japan and Russia in the book published by Suetsugu Ichiro's Council on National Security Problems. In Japanese, see Anzen Hoshō Mondai Kenkyuujo, *Kawaru Nichirō Kankei*, pp. 98–100. In Russian see, *Veki na Puti k Zaklyucheniyu Mirnogo Dogovora mezhdu Yaponiei I Rossiei: 88 Voprosov ot Grazhdan Rossii*, Council on National Security Problems, Tokyo, Moscow: Maternik, 2000, pp. 94–96.

40 Blair Ruble, "Bringing America In," in Rimer, ed., *A Hidden Fire: Russian and Japanese Cultural Encounters, 1868–1926*, pp. 234–235.

41 Very few Russian and Japanese films or television programs are shown in either country because of the dominance of the US film and television industry. The same applies to popular music.

42 Richard Katz, *Japan The System That Soured*, pp. 47–54.

43 Former Chairman of the US Federal Reserve Alan Greenspan is one person who is concerned about the possibility of Japan's economy collapsing. See "Greenspan: Stuck Between Japan and a Hard Place," *STRATFOR* analysis, March 16, 2001.

44 Green, *Japan's Reluctant Realism*, pp. 30–31.

45 For a look at Russian views of globalization see the article by Dmitri Glinski-Vassiliev at the web site: http://oppozitsiya.narod.ru/N34_Glinski.htm. In a speech at the Brookings Institution on Nov. 7, 2001, Russia analyst Clifford Gaddy also pointed out Russia's fears of globalization.

46 See the essay by Tadashi Anno, "*Nihonjinron* and *Russkaia Ideia*: Transformations of Japanese and Russian Nationalism in the Postwar Era and Beyond," in Rozman, editor, *Japan and Russia*, pp. 337–358.

47 Only in 1999 did the Japanese government tacitly recognize the Soviet/Russian annexation of Sakhalin Island by opening a consulate in Yuzhno-Sakhalinsk. Some groups in Japan still insist on the Russian return of southern Sakhalin and the entire Kuril archipelago.

48 Rozman, editor, *Japan and Russia*, p. 6.

49 Attributed to Harold Hinton as cited in Hiroshi Kimura, "Japanese Perceptions of Russia," p. 56. Also, see *The Journal of Japanese Studies*, 26(1) (2000), pp. 270–274. In this edition American historian Stephen Kotkin, in a review of Hasegawa's book *The Northern Territories Dispute and Russo-Japanese Relations*, argues that both George Lensen and John Stephan have successfully refuted the argument that Japanese-Russian relations have been nothing but acrimonious since the beginning. Hasegawa

(and Hinton) are more accurate in this author's estimation. Even when Russia/the Soviet Union and Japan reached a *rapprochement* it was more often than not a shallow bettering of relations more relative than actual, barely masking contempt and mistrust (see Chapter 1).

50 Hiroshi Kimura, "Japanese Perceptions of Russia," p. 57.

Conclusion

1 The excellent study by George Lensen on this period is titled *The Strange Neutrality*.

2 See, for example, Peter Berton, *The Secret Russo-Japanese Alliance of 1916*, p. 265. See also George Lensen, *Japanese Recognition of the USSR*, pp. 363–371. Also, Lensen's *The Strange Neutrality*, pp. 186–220.

Bibliography

Government documents

Agency of Natural Resources and Energy, Ministry of International Trade and Industry. "Enerugi: Mirai kara no Keishō." Tokyo, 1996.

Embassy of Japan, Washington, DC. "Kurasunoyarusuku ni okeru Nichirō Shunō Kaidan (Gaiyō to Hyōka)." [The Japanese-Russian Summit in Krasnoyarsk: a General Outline and Assessment]. November 1997.

———. Embassy of Japan, Moscow. *Yapono-Rossiskie Otnosheniya: Tokiiskaya Deklaratsiya o Yapono-Rossiskikh Otnosheniyakh – Moskovskaya Deklaratsiya ob Ustanovlenii Sozdatel'nogo Partnyorstva Mezhdu Yaponiei i Rossiskoi Federatsiei.* 1998.

———. *Vystuplenie Prem'er-Ministra Yaponii R. Khasimoto v Obshectbe Ekonomicheskikh yedinomyshlennikov (Keizai Doyuukai) 24 July 1997.*

Japanese Defense Agency. *Bōei Handobukku* [Defense Handbook], Tokyo, 1996 edition.

Ministry of Foreign Affairs of Japan. *Gaikōshiryokanpo (Journal of the Diplomatic Record Office).* Tokyo, March 1990, 1991, 1995, 1996.

———. *Japan's Northern Territories.* Tokyo, 1996.

———. *Nihon Gaikō Bunshō (Documents on Japanese Foreign Policy).* Tokyo, vol. I, 1924, pp. 286–315, 376–535, 664–665, 675–679, 759–761, 821–855.

———. *Wareware no Hoppō Ryōdo.* Tokyo, 1998.

———. *Yaponiya i Rossiya: V Interesakh Podlinnogo Vzaimoponimaniya.* Moscow, 2000.

Ministry of Foreign Affairs of Japan/Ministry of Foreign Affairs of the Russian Federation. *Nichirō-kan Ryōdo Mondai no Rekishi in Kansuru Kyodo Sakusei Shiryo-shu – Sovmestnii Sbornik Dokumentov po Istorii Territorial'nogo Pazmezhevaniya mezhdu Rossiei i Yaponiei. (Joint Compendium of Documents Pertaining to the History of the Territorial Demarcation Between Japan and Russia)*, 1992.

———. United States, Department of Energy/Energy Information Association. *International Energy Outlook, 1997.* Washington, 1997.

United States, Department of State. *Foreign Relations of the United States, The Conferences at Cairo and Teheran, 1943.* Washington, DC: United States Government Printing Office, 1961.

———. *Foreign Relations of the United States, The Conferences at Malta and Yalta, 1945.* Washington, DC: United States Government Printing Office, 1955.

———. *Foreign Relations of the United States, 1955–57, Volume XXIII, Part 1: Japan.* Washington, DC: United States Government Printing Office, 1991.

English language sources (articles, books, memoirs, monographs)

Abend, Hallett. *Japan Unmasked*. New York: Ives Washburn, Inc., 1943.

Agafanov, Sergei, "For Moscow and Tokyo, the War is Not Yet Over." *The Current Digest of the Post-Soviet Press*, vol. 49 (December 1995).

Akaha Tsuneo, editor. *Politics and Economics in Northeast Asia: Nationalism and Regionalism in Contention*. New York: St. Martin's Press, 1999.

– – –, editor. *Politics and Economics in the Russian Far East: Changing Ties with Asia-Pacific*. New York: Routledge, 1997.

– – – Murakami, Takashi, "Soviet/Russian-Japanese Economic Relations." In Tsuyoshi Hasegawa, Jonathan Haslam, and Andrew Kuchins, editors, *Russia and Japan: An Unresolved Dilemma between Distant Neighbors*. Berkeley, Calif.: University of California Press, 1993.

Allison, Tony. "The Crisis of the Region's Fishing Industry: Sources, Prospects, and the Role of Foreign Interests." In Judith Thornton and Charles Ziegler, editors, *Russia's Far East: A Region at Risk*. Seattle, Wash.: The National Bureau of Asian Research, University of Washington Press, 2002.

Anno, Tadashi. "*Nihonjinron* and *Russkaia Ideia*: Transformations of Japanese and Russian Nationalism in the Postwar Era and Beyond." In Gilbert Rozman, editor, *Japan and Russia: The Tortuous Path to Normalization, 1949–1999*. New York: St. Martin's Press, 2000, pp. 337–358.

Arbatov, Aleksei. "Military Reform in Russia: dilemmas, obstacles, and prospects." *International Security*, vol. 22, no. 4 (Spring 1998), pp. 91–92.

– – –. "Russia's Foreign Policy Alternatives." *International Security*, vol. 18, no. 2 (Fall 1993), pp. 5–43.

Armacost, Michael and Kenneth B. Pyle. "Japan and the Unification of Korea: Challenges for US Policy Coordination." *NBR Analysis*, vol. 10, no. 1 (March 1999).

Ausland, Anders. *How Russia Became a Market Economy*. Washington, DC: The Brookings Institution, 1995.

Barnett, Doak. *China and the Great Powers in East Asia*. Washington, DC: The Brookings Institution, DC, 1977.

Barnhart, Michael. *Japan Prepares for Total War: The Search for Economic Security, 1919–1941*. Ithaca, NY: Cornell University Press, 1987.

– – –. *Japan and the World Since 1868*. New York: St. Martin's Press, 1995.

Beasley, William. *Japanese Imperialism, 1894–1945*. Oxford: Clarendon Press, 1987.

Bellah, Robert. *Tokugawa Religion: The Cultural Roots of Modern Japan*. New York: The Free Press, 1985.

Beloff, Max. *The Foreign Policy of Soviet Russia, 1929–1941 (Vol. II, 1936–1941)*. Oxford: Oxford University Press, 1949.

– – –. *Soviet Policy in the Far East, 1944–51*. Oxford: Oxford University Press, 1953.

Benedict, Ruth. *The Chrysanthemum and the Sword: Patterns of Japanese Culture*. Boston, Mass.: Houghton Mifflin Company, 1989.

Berger, Thomas. "From Sword to Chrysanthemum: Japan's Culture of Anti-militarism." *International Security*, vol. 17, no. 4 (Spring 1993), pp. 119–150.

– – –. "Set for Stability?: Prospects for Conflict and Cooperation in East Asia." Unpublished paper written at Johns Hopkins University, 1999.

Bergsten, Fred and Marcus Noland. *Reconcilable Differences? United States-Japan Economic Friction*. New York: The Institute for International Economics, 1993.

Berton, Peter. *The Secret Russo-Japanese Alliance of 1916*. Ann Arbor, Mich.: University Microfilms No. 16272, 1956.

———. "The Russo-Japanese Alliance of 1916." *Pew Case Studies in International Affairs*, Case 326, Georgetown University, 1988.

———. "Two Decades of Soviet Diplomacy and Andrei Gromyko." In Gilbert Rozman, editor, *Japan and Russia: The Tortuous Path to Normalization, 1949–1999*. New York: St. Martin's Press, 2000.

Betts, Richard. "Wealth, Power and Instability: East Asia and the United States after the Cold War." *International Security*, vol. 18, no. 3 (Winter 1993–94), pp. 34–77.

Bix, Herbert. *Hirohito and the Making of Modern Japan*. New York: HarperCollins Publishers, 2000.

Black, Cyril, Marius Jansen, Herbert Levine, Marion Levy, Henry Rosovsky, Gilbert Rozman, Henry Smith, and S. Frederick Starr. *The Modernization of Japan and Russia*. New York: The Free Press, 1975.

Blagov, Sergei. "Moscow Considers Enormous Investments in Eastern Russia's Gas Sector." *Eurasia Daily Monitor*, vol. 4, issue 119 (June 19, 2007).

———. "Moscow Looks to Asia's Growing Demand." *Eurasia Daily Monitor*, vol. 3, issue 40 (February 28, 2006).

———. "Russia Sheds No Tears over Peace Treaty with Japan." *Eurasia Daily Monitor*, vol. 2, issue 193 (October 18, 2005).

———. "Russian Energy Minister Visits Tokyo." *Eurasia Daily Monitor*, vol. 2, issue 82 (April 27, 2005).

———. "Russian Personnel Changes to Affect Far East Region." *Eurasia Daily Monitor*, vol. 2, issue 216 (November 18, 2005).

———. "Russia's Gazprom Eyes East Asian Markets." *Eurasia Daily Monitor*, vol. 2, issue 51 (March 15, 2005).

Blank, Stephen and Alvin Rubinstein, editors. *Imperial Decline: Russia's Changing Role in Asia*. Durham, NC: Duke University Press, 1997.

Braddick, C.W. "The Waiting Game: Japan-Russia Relations." In Inoguchi Takashi and Purnendra Jain, editors, *Japanese Foreign Policy Today*. New York: Palgrave, 2000, pp. 209–225.

Bremmer, Ian and Ray Taras, editors. *New States, New Politics: Building the Post-Soviet Nations*. Cambridge: Cambridge University Press, 1997.

Brown, Archie. *The Gorbachev Factor*. Oxford: Oxford University Press, 1997.

Brzezinski, Zbigniew. *The Grand Chessboard: American Primacy and its Geostrategic Imperatives*. New York: HarperCollins Publishers, 1997.

———. *The Soviet Bloc: Unity and Conflict*. Cambridge, Mass.: Harvard University Press, 1971.

Bull, Hedley and Adam Watson, editors. *The Expansion of International Society*. Oxford: Oxford University Press, 1984.

Buruma, Ian. *The Wages of Guilt: Memories of War in Germany and Japan*. London: Jonathan Cape, 1994.

Buszynski, Leszek. "Russia and Northeast Asia: aspirations and reality." *The Pacific Review*, vol. 13, no. 3 (2000), pp. 399–420.

Calder, Kent. *Crisis and Compensation*. Princeton: Princeton University Press, 1988.

———. *Pacific Defense: Arms, Energy, and America's Future in Asia*. New York: William Morrow, 1996.

———. "Japan's Energy Angst and the Caspian Great Game." *NBR Analysis*, vol. 12, no. 1, (March 2001).

Chang, Felix. "Chinese Energy and Asian Security." *Orbis*, vol. 45, no. 2 (Spring 2001), pp. 211–240.

Chapman, J.W.M., Reinhardt Drift and Ian Gowe. *Japan's Search for Comprehensive Security: Defence, Diplomacy, and Dependence*. London: Francis Pinter, 1983.

Checkel, Jeffrey. *Ideas and International Political Change: Soviet/Russian Behavior and the End of the Cold War*. New Haven, Conn.: Yale University Press, 1997.

Chi Su. "Sino-Soviet Relations of the 1980s: From Confrontation to Conciliation." In Samuel Kim, editor, *China and the World*. Boulder, Col.: Westview Press, 1989.

Christensen, Thomas. "China, the US-Japan Alliance, and the Security Dilemma in East Asia." *International Security*, vol. 23, no. 4 (Spring 1999), pp. 49–80.

Christoffersen, Gaye. "China's Intentions for Russian and Central Asian Oil and Gas." *NBR Analysis*, vol. 9, no. 2 (March 1998).

Chufrin, Gennady, editor. *Russia and Asia: The Emerging Security Agenda*. Oxford: SIPRI/ Oxford University Press, 1999.

Clubb, Edmund. *China and Russia: The Great Game*. New York: Columbia University Press, 1971.

Cohen, Warren. *America's Response to China: A History of Sino-American Relations*. New York: Columbia University Press, 1990.

———, editor. *Pacific Passage: The Study of American-East Asian Relations on the Eve of the 21st Century*. New York: Columbia University Press, 1996.

Collins, Gabe. "Fueling the Dragon: China-bound Pipelines are Russia's Most Realistic Asian Energy Option," *Geopolitics of Energy*, vol. 28, no. 9 (September 2006).

Coox, Alvin. *Nomonhan: Japan Against Russia, 1939*. Palo Alto, Calif.: Stanford University Press, 1985.

——— and Hata Ikuhiko. "The Japanese-Soviet Confrontation, 1935–1939." In James Morley, *Deterrence Diplomacy*. New York: Columbia University Press, 1976.

Cossa, Ralph. "The Major Powers in Northeast Asian Security." *McNair Paper 51*, Institute for National Strategic Studies, Washington, DC, 1996.

Crowley, James. *Japan's Quest for Autonomy: National Security and Foreign Policy, 1930–38*. Princeton: Princeton University Press, 1966.

———. "New Deal for Japan and Asia: One Road to Pearl Harbor." In James Crowley, editor. *Modern East Asia: Essays in Interpretation*. New York: Harcourt, Brace and World, 1970.

———, editor. *Modern East Asia: Essays in Interpretation*. New York: Harcourt, Brace and World, 1970.

Curtis, Gerald. *The Japanese Way of Politics*. New York: Columbia University Press, 1988.

———. *The Logic of Japanese Politics: Leaders, Institutions, and the Limits of Change*. New York: Columbia University Press, 1999.

———. "The Tyumen Oil Development Project and Japanese Foreign Policy Decision-Making." In Robert Scalapino, editor, *The Foreign Policy of Modern Japan*. Berkeley: University of California Press, 1977.

———, editor. *Japan's Foreign Policy After the Cold War: Coping with Change*. Armonk, NY: M.E. Sharpe, 1993.

Dallin, David. *The Rise of Russia in Asia*. New Haven, Conn.: Yale University Press, 1949.

Davies, John Patton Jr. *Dragon by the Tail: American, British, Japanese, and Russian Encounters with China and One Another*. New York: W.W. Norton & Company, 1972.

Dawisha, Karen and Bruce Parrot, editors. *Democratic Changes and Authoritarian Reactions in Russia, Ukraine, Belarus, and Moldova*. Cambridge: Cambridge University Press, 1997.

———. *Russia and the New States of Eurasia: The Politics of Upheaval*. Cambridge: Cambridge University Press, 1994.

Deane, John R. *The Strange Alliance: The Story of Our Efforts at Wartime Co-Operation with Russia*. New York: The Viking Press, 1947.

Degras, Jane. *Soviet Documents on Foreign Policy (Vol. II, 1925–1932; Vol. III, 1933–1941)*. Oxford: Oxford University Press, 1952–53.

Der Derian, James. *International Theory: Critical Investigations*. New York: New York University Press, 1995.

Deudney, Daniel and G. John Ikenberry. "The International Sources of Soviet Change." *International Security*, vol. 16, no. 3 (Winter 1991/92), pp. 74–118.

Dickinson, Frederick. *War and National Reinvention: Japan and the Great War, 1914–1919*. Cambridge, Mass: Harvard University Asia Center, 1999.

Dower, John. *Embracing Defeat: Japan in the Wake of World War II*. New York: W.W. Norton and Company, 1999.

———. *Empire and Aftermath: Yoshida Shigeru and the Japanese Experience, 1878–1954*. Cambridge, Mass: Council on East Asian Studies, Harvard University, 1988.

Dunlop, John. *The Rise of Russia and the Fall of the Soviet Empire*. Princeton: Princeton University Press, 1993.

Duus, Peter, Ramon Meyers and Mark Peattie, editors. *Japan's Informal Empire in China, 1895–1937*. Princeton: Princeton University Press, 1989.

Elleman, Bruce, Michael Nichols, and Matthew Ouimet. "A Historical Reevaluation of America's Role in the Kuril Islands Dispute." *Pacific Affairs*, vol. 71, no. 4 (Winter 1998–99), pp. 489–504.

Ellings, Richard and Aaron Friedberg, editors. *Strategic Asia: Power and Purpose, 2001–02*. Seattle, Wash.: The National Bureau of Asian Research, 2001.

Ellison, Herbert and Bruce Acker. "The New Russia and Asia: 1991–1995." *NBR Analysis*, vol. 7, no. 1 (June 1996).

Ellison, Herbert. *The Soviet Union and Northeast Asia*. New York: Monograph of the Asia Society, published by University Press of America, 1989.

———. "The Debate before the Summit." In James Goodby, Vladimir Ivanov, and Nobuo Shimotomai, editors, *"Northern Territories" and Beyond: Russian, Japanese, and American Perspectives*. Westport, Conn.: Praeger Publishers, 1995.

English Robert. *Russia and the Idea of the West: Gorbachev, Intellectuals, and the End of the Cold War*. New York: Columbia University Press, 2000.

Eudin, Xenia and Robert North, editors. *Soviet Russia and the East, 1920–1927: A Documentary Survey*. Palo Alto, Calif.: Stanford University Press, 1957.

Fairbank, John K. *The United States and China*. Cambridge, Mass.: Harvard University Press, 1980.

———, Edwin Reischauer, and Albert Craig. *East Asia: The Modern Transformation*. Boston, Mass.: Houghton Mifflin Company, 1965.

Fedorovsky, Alexander. "Russian Policy and Interests in the Korean Peninsula." in Gennady Chufrin, editor, *Russia and Asia: The Emerging Security Agenda*. Oxford: SIPRI/Oxford University Press, 1999.

Ferguson, Joseph. "Japan's attempts to formulate a Silk Road strategy." Johns Hopkins SAIS, Central Asia-Caucasus Institute *The Cyber-Caravan*, vol. 1, no. 3 (February 14, 1999), pp. 2–4.

———. "*Nordpolitik*: Japan's New Russia Policy." *Journal of Public and International Affairs*, Woodrow Wilson School, Princeton University, 1998.

———. "Russia's Role on the Korean Peninsula and Great Power Relations in Northeast

Asia," *NBR Analysis*, vol. 14, no. 1 (June 2003), pp. 33–50.

Forsberg, Tuomas. "Explaining Territorial Disputes: From Power Politics to Normative Reasons." *Journal of Peace Research*, vol. 33, no. 4 (1996), pp. 433–449.

Friedberg, Aaron. "Ripe for Rivalry: Prospects for Peace in a Multipolar Asia." *International Security*, vol. 18, no. 3 (Winter 1993–94), pp. 5–33.

———. "Warring States: Theoretical Models of Asia Pacific Security." *Harvard International Review*, vol. XVIII, no. 2 (Spring 1996), pp. 12–15.

Fuller, William C. *Strategy and Power in Russia, 1600–1914*. New York: The Free Press, 1992.

Funabashi Yoichi. *Alliance Adrift*. New York: Council on Foreign Relations Press, 2000.

———. *Asia-Pacific Fusion: Japan's Role in APEC*. Washington, DC: The Institute for International Economics, 1995.

———, editor. *Japan's International Agenda*. New York: New York University Press, 1994.

Garnett, Sherman, editor. *Rapprochement or Rivalry?: Russia-China Relations in a Changing Asia*. Washington, DC: Carnegie Endowment for International Peace, 2000.

Garthoff, Raymond. *Détente and Confrontation: American-Soviet Relations from Nixon to Reagan*. Washington, DC: The Brookings Institution, 1994.

———. *The Great Transition: American-Soviet Relations and the End of the Cold War*. Washington, DC: The Brookings Institution, 1994.

Garver, John. *Foreign Relations of the People's Republic of China*. Englewood Cliffs, NJ: Prentice Hall, 1993.

Gayn, Marc. *Japan Diary*. Rutland, Vermont: Charles E. Tuttle Company, 1981.

Gelman, Harry. *Russo-Japanese Relations and the Future of the US-Japanese Alliance*. Santa Monica, Calif.: RAND, 1993.

———. "The Changing Asian Arena," in Sherman Garnett, editor, *Rapprochement or Rivalry?: Russia-China Relations in a Changing Asia*. Washington, DC: Carnegie Endowment for International Peace, 2000.

Gibney, Frank, editor. *Unlocking the Bureaucrat's Kingdom: Deregulating the Japanese Economy*. Washington, DC: The Brookings Institution Press, 1998.

Gillin, Donald with Charles Etter. "Staying On: Japanese Soldiers and Civilians in China 1945–1949." *The Journal of Asian Studies*, vol. XLII, no. 3 (May 1983), pp. 497–518.

Glaubitz, Joachim. *Between Tokyo and Moscow: The History of an Uneasy Relationship, 1972 to the 1990s*. Honolulu: University of Hawaii Press, 1995.

Gleason, Abbot. *Totalitarianism: The Inner History of the Cold War*. Oxford: Oxford University Press, 1995.

Goldstein, Lyle and Vitaly Kozyrev. "China, Japan and the Scramble for Siberia," *Survival*, vol. 48, no. 1 (Spring 2006), pp. 163–178.

Goodby, James, Vladimir Ivanov, and Nobuo Shimotomai, editors. *"Northern Territories" and Beyond: Russian, Japanese, and American Perspectives*. Westport, Conn.: Praeger Publishers, 1995.

Gorst, Isabel. "Russian Pipeline Strategies: Business Versus Politics," paper prepared for the James A. Baker Institute for Public Policy of Rice University, October 2004, p. 16. http://www.rice.edu/energy/publications/docs/PEC_Gorst_10_2004.pdf

Green, Michael. *Japan's Reluctant Realism*. New York: Palgrave Press, 2001.

———. "The Forgotten Player." *The National Interest*, (Summer 2000), pp. 42–49.

——— and Benjamin Self. "Japan's Changing China Policy: From Commercial Liberalism to Reluctant Realism." *Survival*, vol. 38, no. 2 (Summer 1996), pp. 35–58.

Hakamada Shigeki. "Building a New Japanese-Russian Relationship." An unpublished paper delivered at the International Symposium, Nara, Japan, 1996.

Hara Kimie. *Japanese-Soviet/Russian Relations since 1945: A Difficult Peace*. London: Routledge, 1998.

———. "Rethinking the 'Cold War' in the Asia-Pacific." *The Pacific Review*, vol. 12, no. 4 (1999), pp. 515–536.

Harada Chikahito. "Russia and Northeast Asia." *Adelphi Paper 310*, International Institute for Strategic Studies, Oxford University Press, 1997.

Harding, Harry. *The Fragile Relationship: the United States and China since 1972*. Washington, DC: The Brookings Institution, 1992.

Harries, Meirion and Susie Harries. *Soldiers of the Sun: The Rise and Fall of the Imperial Japanese Army*. New York: Random House, 1991.

Hasegawa Tsuyoshi. *The Northern Territories Dispute and Russo-Japanese Relations* (Vol. I–II). Berkeley, Calif.: University of California Press, 1998.

———. "Japanese Misperceptions of the Soviet Union During the Gorbachev Era." *The Carl Beck Papers*, No. 1503, University of Pittsburgh, October 2000.

———. "Japanese Perceptions of the Soviet Union and Russia in the Postwar Period." In Gilbert Rozman, editor, *Japan and Russia: The Tortuous Path to Normalization, 1949–1999*. New York: St. Martin's Press, 2000, pp. 281–320.

———. "The Gorbachev-Kaifu Summit: Domestic and Foreign Policy Linkages." In Tsuyoshi Hasegawa, Jonathan Haslam, and Andrew Kuchins, editors, *Russia and Japan: An Unresolved Dilemma Between Distant Neighbors*. Berkeley, Calif.: University of California Press, 1993.

———. "Russo-Japanese relations and the security of North-East Asia in the twenty-first century." In Gennady Chufrin, editor, *Russia and Asia: The Emerging Security Agenda*. Oxford: SIPRI/Oxford University Press, 1999.

———. "Why Did Russia and Japan Fail to Achieve *Rapprochement* in 1991–1996?" In Gilbert Rozman, editor, *Japan and Russia: The Tortuous Path to Normalization, 1949–1999*. New York: St. Martin's Press, 2000.

———. Jonathan Haslam, and Andrew Kuchins, editors. *Russia and Japan: An Unresolved Dilemma Between Distant Neighbors*. Berkeley, Calif.: University of California Press, 1993.

Haslam, Jonathan. *The Soviet Union and the Threat from the East, 1933–41: Moscow, Tokyo and the Prelude to the Pacific War*. Pittsburgh, Penn.: University of Pittsburgh Press, 1992.

Hauner, Milan. *What is Asia to Us?: Russia's Asian Heartland Yesterday and Today*. Winchester, Mass.: Unwin Hyman, Inc., 1990.

Heldt, Barbara. " 'Japanese' in Russian Literature: Transforming Identities," in J. Thomas Rimer. *A Hidden Fire: Russian and Japanese Cultural Encounters, 1868–1926*. Palo Alto, Calif.: Stanford University Press and Woodrow Wilson Center, 1995.

Hellmann, Donald. *Japanese Foreign Policy and Domestic Policy: The Peace Agreement with the Soviet Union*. Berkeley, Calif.: University of California Press, 1969.

Helmer, John. "Putin opts for China-First oil plan, Japan and India relegated." *Russia Journal*, July 12, 2005.

Hickok, Michael Robert. "Japan's Gambit: An Asian View of Eurasia." *Problems of Post-Communism*, vol. 47, no. 3 (May/June 2000), pp. 36–47.

Hill, Fiona. " 'A Disagreement Between Allies': The United Kingdom, the United States, and the Soviet-Japanese Territorial Dispute, 1945–56." *The Journal of Northeast Asian Studies* (Fall 1995), pp. 3–49.

Hoffman, Erik and Frederic Fleron, Jr. *The Conduct of Soviet Foreign Policy*. New York: Aldine Press, 1980.

Holloway, David. "State, Society and the Military under Gorbachev." *International Security*, vol. 14, no. 3 (Winter 1989/90), pp. 5–24.

Horne, Mari Kuraishi. "The Northern Territories Dispute: Source or Symptom?" *Journal of Northeast Asian Studies* (Winter 1989), pp. 60–76.

Hosoya Chihiro. "Japan's Policies Toward Russia." In James Morley, editor, *Japan's Foreign Policy, 1868–1941*. New York: Columbia University Press, 1974.

Hough, Jerry and Merle Fainsod. *How the Soviet Union is Governed*. Cambridge, Mass.: Harvard University Press, 1979.

Hsu, Immanuel. *The Rise of Modern China*. Oxford: Oxford University Press, 1990.

Inoguchi Takashi and Purnendra Jain, editors. *Japanese Foreign Policy Today*. New York: Palgrave, 2000.

Inouye Yuichi. "The Russo-Japanese Entente and Railway Diplomacy." Ministry of Foreign Affairs of Japan. *Gaikōshiryokanpo* (Journal of the Diplomatic Record Office), March 1991, pp. 121–30.

Institute of International Economy and International Relations (Moscow) and the National Institute for Research Advancement (Tokyo). *Northeast Asia Energy and Environmental Cooperation*. Moscow-Tokyo, 1999.

Iriye, Akira. *Pacific Estrangement: Japan and American Expansion, 1897–1911*. Cambridge, Mass.: Harvard University Press, 1972.

———. "Imperialism in East Asia." In James Crowley, editor, *Modern East Asia: Essays in Interpretation*. New York: Harcourt, Brace and World, 1970.

Isaacson, Walter. *Kissinger: A Biography*. New York: Simon & Schuster, 1992.

Itō Kenichi, Kamiya Matake, Nakanishi Hiroshi, Yamaoka Kunihiko. "Japan's Initiatives towards US, China and Russia." *Japan Forum on International Relations*, Tokyo, April 19, 1999.

Ivanov, Vladimir. "Russia and Japan Beyond 2005." *ERINA Report*, vol. 66 (November 2005).

———. "Russia's Energy Politics: Focusing on New Markets in Asia." *ERINA Report*, vol. 76 (January 2006), pp. 6–14.

———. "Russia and the United States-Japan Partnership." In James Goodby, Vladimir Ivanov, and Nobuo Shimotomai, editors, *"Northern Territories" and Beyond: Russian, Japanese, and American Perspectives*. Westport, Conn.: Praeger Publishers, 1995.

Iwao, Peter Sano. *1,000 Days in Siberia: The Odyssey of a Japanese-American POW*. Lincoln, Neb.: University of Nebraska Press, 1997.

Iwata Kazumasa. "Rule Maker of World Trade: Japan's Trade Strategy and the World System." In Yoichi Funabashi, editor, *Japan's International Agenda*. New York: New York University Press, 1994.

Jacob, Jo Dee Catlin. *Beyond the Hoppo Ryodo: Japanese, Soviet, American Relations in the 1990s*. Washington, DC: American Enterprise Institute, 1991.

Jain, Purnendra. "Emerging Foreign Policy Actors: Subnational Governments and Nongovernmental Organizations." In Inoguchi Takashi and Purnendra Jain, editors, *Japanese Foreign Policy Today*, New York: Palgrave, 2000.

Jain, R.K. *The USSR and Japan, 1945–1980*. New Delhi: Radiant Publishers, 1981.

Jansen, Marius. *Japan and China: From War to Peace, 1894–1972*. Chicago, Ill.: Rand McNally, 1975.

"Japan and the Middle East." *The Middle East Economic Digest*. November 28, 1997.

Jervis, Robert. *Perception and Misperception in International Politics*. Princeton: Princeton University Press, 1976.

Johnson, Chalmers. *MITI and the Japanese Economic Miracle: The Growth of Industrial*

Policy 1925–75. Palo Alto, Calif.: Stanford University Press, 1982.

———, editor. *Change in Communist Systems*. Palo Alto, Calif.: Stanford University Press, 1970.

Johnston, Alastair Iain. *Cultural Realism: Strategic Culture and Grand Strategy in Chinese History*. Princeton: Princeton University, 1995.

Kaiser, Robert. *The Geography of Nationalism in Russia and the USSR*. Princeton: Princeton University Press, 1994.

Kamikawa Hikomatsu and Kimura Michiko. *Japanese-American Diplomatic Relations in the Meiji-Taisho Era*. Tokyo: Pan Pacific Press, 1959.

Kamo, Takehiro, "Japan, Russia, and Nuclear Nonproliferation." in James Goodby, Vladimir Ivanov, and Nobuo Shimotomai, editors, *"Northern Territories" and Beyond: Russian, Japanese, and American Perspectives*. Westport, Conn.: Praeger Publishers, 1995.

Kataoka Tetsuya. "Japan's Northern Threat." *Problems of Communism*, March–April, 1984.

Katz, Richard. *Japan The System That Soured: The Rise and Fall of the Japanese Economic Miracle*. Armonk, NY: M.E. Sharpe, 1998.

Katzenstein, Peter and Okawara Nobuo. "Japan's National Security: Structures, Norms, and Politics." *International Security*, vol. 17, no. 4 (Spring 1993), pp. 84–118.

Keene, Donald. *The Japanese Discovery of Europe, 1720–1830*. Palo Alto, Calif.: Stanford University Press, 1969.

Kennan, George F. *The Decision to Intervene*. Princeton, NJ: Princeton University Press, 1958.

———. *Russia and the West Under Lenin and Stalin*. Boston, Mass.: Little, Brown and Company, 1960.

———. *Russia Leaves the War*. Princeton, NJ: Princeton University Press 1957.

Keohane, Robert and Lisa Martin, "The Promise of Institutionalist Theory." *International Security*, vol. 20, no. 1 (Summer 1995), pp. 39–51.

Khrushchev, Nikita. *Khrushchev Remembers: The Glasnost Tapes*. Boston, Mass.: Little, Brown and Company, 1990.

Kim, Samuel, editor. *China and the World*. Boulder, Col.: Westview Press, 1989.

Kim, Young. *Japanese-Soviet Relations: Interaction of Politics, Economics, and National Security*. Washington, DC: The Center for Strategic and International Studies, 1974.

Kimura Hiroshi. *Distant Neighbors (Vol. I): Japanese-Russian Relations under Brezhnev and Andropov*. Armonk, NY: M.E. Sharpe, 2000.

———. *Distant Neighbors (Vol. II): Japanese-Russian Relations under Gorbachev and Yeltsin*. Armonk, NY: M.E. Sharpe, 2000.

———. "Islands Apart." *Look Japan*, February 2001, pp. 6–15.

———. and Shaojun Li, Il-Dong Koh, "Frontiers are the Razor's Edge: Russia's borders with its eastern neighbors." In Gilbert Rozman, Mikhail Nosov, and Koji Watanabe, editors, *Russia and East Asia: The Twenty-First Century Security Environment*, Armonk, NY: East West Institute, M.E. Sharpe, 1999.

Kirkow, Peter. "Regionalism Warlordism in Russia: The Case of Primorskii Krai." *Europe-Asia Studies*, vol. 47, no. 6 (1995), pp. 923–947.

Knorr, Klaus and Sidney Verba, editors. *The International System: Theoretical Essays*. Princeton: Princeton University Press, 1961.

Kotkin, Stephen and David Wolff, editors. *Rediscovering Russia in Asia: Siberia and the Russian Far East*. Armonk, NY: M.E. Sharpe, 1995.

Kramer, Mark. "Beyond the Brezhnev Doctrine: A New Era in Soviet-East European Relations?" *International Security*, vol. 14, no. 3 (Winter 1989/90), pp. 25–67.

Krugman, Paul. *The Age of Diminished Expectations: U. S. Economic Policy in the 1990s*. Cambridge, Mass.: MIT Press, 1994.

Kuchins, Andrew. "Russia and great power security in Asia." In Gennady Chufrin, editor, *Russia and Asia: The Emerging Security Agenda*. Oxford: SIPRI/Oxford University Press, 1999.

Kutakov, Leonid. *Japanese Foreign Policy on the Eve of the Pacific War: A Soviet View*. Tallahassee, Fla.: The Diplomatic Press, 1972.

Kyogoku Junichi. *The Political Dynamics of Japan*. Tokyo: Tokyo University Press, 1993.

LaFeber, Walter. *The Clash: US-Japanese Relations Throughout History*. New York: W.W. Norton and Company, 1997.

Lane, Harold. *The Japanese Exclusion Act, 1906–1924*. Unpublished Master's Thesis, *The George Lensen Collection* (Sapporo, Japan), Haverford University, 1929.

Layne, Christopher. "Kant or Cant: The Myth of Democratic Peace." *International Security*, vol. 19, no. 2 (Fall 1994), pp. 5–49.

Leitch, Richard, Akiro Kato, and Martin Weinstein. *Japan's Role in the Post-Cold War World*. Westport, Conn.: Greenwood Press, 1995.

Lensen, G.A. *The Damned Inheritance: The Soviet Union and the Manchurian Crises, 1924–1935*. Tallahassee, Fla.: The Diplomatic Press, 1974.

———. *Japanese Recognition of the USSR: Japanese-Soviet Relations, 1921–1930*. Tallahassee, Fla.: The Diplomatic Press, 1970.

———. *The Strange Neutrality: Soviet-Japanese Relations during the Second World War, 1941–1945*. Tallahassee, Fla.: The Diplomatic Press, 1972.

Li Jingjie. "Pillars of the Sino-Russian Partnership." *Orbis*, vol. 44, no. 4 (Fall 2000), pp. 527–539.

Lincoln, Edward. *Japan's New Global Role*. Washington, DC: The Brookings Institution, 1993.

Linden, Carl. *Khrushchev and the Soviet Leadership, 1957–1964*. Baltimore: The Johns Hopkins Press, 1966.

Lobanov-Rostovsky, A. *Russia and Asia*. Ann Arbor, Mich.: The George Wahr Publishing Company, 1965.

Lowenthal, Richard. *World Communism: The Disintegration of a Secular Faith*. New York: 1964.

Lukin, Alexander. "The Image of China in Russian Border Regions." *Asian Survey*, vol. XXXVIII, no. 9 (September 1998), pp. 821–835.

Malcom, Neil, Alex Pravda, Roy Allison, and Margot Light. *Internal Factors in Russian Foreign Policy: Domestic Influences in the Post-Soviet Setting*. Oxford: Oxford University Press, 1996.

Malia, Martin. *The Soviet Tragedy: A History of Socialism in Russia, 1917–1991*. New York: The Free Press, 1994.

Mansfield, Edward and Jack Snyder. "Democratization and the Danger of War." *International Security*, vol. 20, no. 1 (Summer 1995), pp. 5–38.

May, Ernest. *"Lessons" of the Past: The Use and Misuse of History in American Foreign Policy*. Oxford: Oxford University Press, 1973.

May, Michael. *Energy and Security in East Asia*. Honolulu: Asia-Pacific Research Center, January 1998.

McDougall, Walter. *Let the Sea Make a Noise*. New York: Avon Books, 1993.

McFaul, Michael. "A Precarious Peace: Domestic Politics in the Making of Russian Foreign Policy." *International Security*, vol. 22, no. 3 (Winter 1997/98), pp. 5–35.

———. "The Far Eastern Challenge to Russian Federalism." In Sherman Garnett, editor.

Rapprochement or Rivalry?: Russia-China Relations in a Changing Asia. Washington, DC: Carnegie Endowment for International Peace, 2000, pp. 325–331.
McMahon, Robert. "The Cold War in Asia: Toward a New Synthesis?" *Diplomatic History*, vol. 12, no. 3 (Summer 1988), pp. 307–327.
Mearsheimer, John. "A Realist Reply." *International Security*, vol. 20, no. 1 (Summer 1995), pp. 82–93.
Mendelson, Sarah. *Changing Course: Idea, Politics, and the Soviet Withdrawal from Afghanistan*. Princeton: Princeton University Press, 1998.
Menon, Rajan. "Japan-Russia Relations and Northeast Asian Security." *Survival*, vol. 38, no. 2 (Summer 1996), pp. 59–78.
———. and Charles E. Ziegler. "The Balance of Power and US Foreign Policy Interests in the Russian Far East." *NBR Analysis*, vol. 11, no. 5 (December 2000).
Meyer, Peggy Falkenheim. "Is Japan's New Eurasian Diplomacy a Failure?" Unpublished paper presented to the Workshop *Russian National Security: Perceptions, Policies, and Prospects*, Army War College, Carlisle Barracks, Penn., Dec. 4–6, 2000.
———. "Sino-Japanese Relations: The Economic Security Nexus." In Tsuneo Akaha, editor, *Politics and Economics in Northeast Asia: Nationalism and Regionalism in Contention*, New York: St. Martin's Press, 1999.
Minagawa Sugo. "Political Clientelism in Primorskii Krai in the Transitional Period." *The Soviet and Post-Soviet Review*, 25, no. 2 (1998), pp. 125–148.
Minear, Richard. *Victor's Justice: The Tokyo War Crimes Trial*. Princeton: Princeton University Press, 1971.
Mingyie Yang. "Russia's Regional Security Role." in Koji Watanabe, editor, *Engaging Russia in Asia Pacific*, Tokyo: Japan Center for International Exchange, 1999, p. 49.
Mochizuki, Mike, "The Soviet/Russian Factor in Japanese Security Policy." In Tsuyoshi Hasegawa, Jonathan Haslam, and Andrew Kuchins, editors. *Russia and Japan: An Unresolved Dilemma Between Distant Neighbors*. Berkeley, Calif.: University of California Press, 1993.
———, editor. *Toward a True Alliance: Restructuring US-Japan Security Relations*. Washington, DC: The Brookings Institution, 1997.
Mochizuki Tetsuo. "Japanese Perceptions of Russian Literature in the Meiji and Taisho Eras." In J. Thomas Rimer, *A Hidden Fire: Russian and Japanese Cultural Encounters, 1868–1926*. Palo Alto, Calif.: Stanford University Press and Woodrow Wilson Center, 1995.
Moltz, James Clay. "Russian Nuclear Regionalism: Emerging Local Influences over Far Eastern Facilities." *NBR Analysis*, vol. 11, no. 4 (December 2000).
Morley, James. *The Japanese Thrust into Siberia, 1918*. New York: Columbia University Press, 1954.
———, editor. *Japan's Foreign Policy, 1868–1941*. New York: Columbia University Press, 1974.
Morse, H.B. and H.F. McNair. *Far Eastern International Relations, Vol. I*. Calif.: Riverside Press, 1931.
Mullins, Patrick. *Japanese-Soviet Relations, 1925–1940*. Unpublished Master's Thesis, *The George Lensen Collection* (Sapporo, Japan), Tallahassee, Fla.: Florida State University, 1960.
Nathan, Andrew and Robert Ross. *The Great Wall and the Empty Fortress: China's Search for Security*. New York: W.W. Norton and Company, 1997.
Neu, Charles. *The Troubled Encounter: The United States and Japan*. New York: Robert E. Krieger Publishing Company, 1975.

Neumann, William. *America Encounters Japan: From Perry to MacArthur*. Baltimore: The Johns Hopkins University Press, 1963.

Nimmo, William. *Japan and Russia: A Re-evaluation in the Post-Soviet Era*. Westport, Conn.: Greenwood Press. 1994.

Nish, Ian. *The Origins of the Russo-Japanese War*. London: Longman Press, 1985.

Nove, Alec. *An Economic History of the USSR, 1917–1991*. New York: Viking Penguin, 1972.

Ogawa Kazuo. "Economic Relations with Japan." In Rodger Swearingen, editor, *Siberia and the Soviet Far East: Strategic Dimensions in Multinational Perspective*, Palo Alto, Calif.: Hoover Institution Press, Stanford University, 1987, pp. 158–178.

Okada Kunio. "The Japanese Economic Presence in the Russian Far East." In Judith Thornton and Charles Ziegler, editors, *Russia's Far East: A Region at Risk*. Seattle, Wash.: The National Bureau of Asian Research, University of Washington Press, 2002.

Owen, John. "How Liberalism Produces Democratic Peace." *International Security*, vol. 19, no. 2 (Fall 1994), pp. 87–127.

Paik Keun Wook, editor. *Gas and Oil in Northeast Asia: Policies, Projects and Prospects*. London: The Royal Institute for International Affairs, 1995.

Panov, Alexander. "From Stalin to Brezhnev." In James Goodby, Vladimir Ivanov, and Nobuo Shimotomai, editors, *"Northern Territories" and Beyond: Russian, Japanese, and American Perspectives*. Westport, Conn.: Praeger Publishers, 1995.

Pempel, T.J. *Regime Shift: Comparative Dynamics of the Japanese Political Economy*. Ithaca, NY: Cornell University Press, 1998.

Pipes, Richard. *The Formation of the Soviet Union*. Cambridge, Mass.: Harvard University Press, 1970.

———. *Russia Under the Old Regime*. New York: Penguin Books, 1995.

Pleshakov, Constantine. "East Asian Conundrum: Geopolitics and Ideology in the Mirror of Russian Historiography." In Warren Cohen, editor, *Pacific Passage: The Study of American-East Asian Relations on the Eve of the 21st Century*. New York: Columbia University Press, 1996.

Portyakov, Vladimir. "Are the Chinese Coming? Migration Processes in Russia's Far East." *International Affairs* (Moscow), vol. 42, no. 1 (Jan.–Feb. 1996), pp. 132–140.

Prange, Gordon. *Target Tokyo: The Story of the Sorge Spy Ring*. New York: McGraw Hill Book Company, 1984.

Price, Ernest Batson. *The Russo-Japanese Treaties of 1907–1916 Concerning Manchuria and Mongolia*. Baltimore: The Johns Hopkins Press, 1933.

Pyle, Kenneth. *The Japanese Question*. Washington, DC: American Enterprise Press, 1992.

Rashid, Ahmed. "Power Play." *The Far Eastern Economic Review*, 10 Apr. 1997, pp. 22–24.

Riasonovsky, Nicholas. *A History of Russia*. Oxford: Oxford University Press, 1977.

Richter, James G. *Khrushchev's Double Bind: International Pressures and Domestic Coalition Politics*. Baltimore: Johns Hopkins University Press, 1994.

Rimer, J. Thomas, editor. *A Hidden Fire: Russian and Japanese Cultural Encounters, 1868–1926*. Palo Alto, Calif.: Stanford University Press and the Woodrow Wilson Center, 1995.

Robertson, Myles. *Soviet Policy Towards Japan: An Analysis of Trends in the 1970s and 1980s*. Cambridge: Cambridge University Press, 1988.

Rosecrance, Richard. *The Rise of the Trading State: Commerce and Conquest in the Modern World*. New York: Basic Books, 1986.

Ross, Robert. "The Geography of Peace: East Asia in the 21st Century." *International Security*, vol. 23, no. 4 (Spring 1999), pp. 81–118.

Roy, Denny. "Realism and East Asia." *The Journal of East Asian Affairs*, vol. XIV, no. 1, (Spring/Summer 2000), pp. 159–178.

Rozman, Gilbert. *Japan's Response to the Gorbachev Era, 1985–1991: A Rising Superpower Views a Declining One*. Princeton: Princeton University Press, 1992.

———. "Backdoor Japan: The Search for a Way Out via Regionalism and Decentralization." *Journal of Japanese Studies*, 25(1) (1999), pp. 3–31.

———. "A Chance for a Breakthrough in Russo-Japanese Relations: Will the Logic of Great Power Relations Prevail?" *Pacific Review*, vol. 15, no. 3 (2002), pp. 325–357.

———. "China's Quest for Great Power Identity." *Orbis*, vol. 43, no. 3 (Summer 1999), pp. 383–402.

———. "The Crisis of the Russian Far East: Who is to blame?" *Problems of Post-Communism*, vol. 44, no. 5 (September/October 1997), pp. 3–12.

———. "Decentralization in Northeast China and the Russian Far East." *The Soviet and Post-Soviet Review*, 25, no. 2 (1998), pp. 181–197.

———. "Flawed Regionalism: Reconceptualizing Northeast Asia in the 1990s." *The Pacific Review*, vol. 11, no. 1 (1998), pp. 1–27.

———. "Japan and Korea: should the US be worried about their new spat in 2001?" *The Pacific Review*, vol. 15, no. 1 (2002), pp. 1–28.

———. "A New Sino-Russian-American Triangle?" *Orbis*, vol. 44, no. 4 (Fall 2000), pp. 541–555.

———. "Russia and Japan: Mutual Misperceptions, 1992–1999." *The Carl Beck Papers*, no. 1503, University of Pittsburgh, October 2000.

———. "Russia's Resurgence in Northeast Asia: Views from the Region." *Russian Analytical Digest*, no. 25 (July 17, 2007).

———. "China, Japan, and the Post-Soviet Upheaval: Global Opportunities and Regional Risks." In Karen Dawisha, editor, *The International Dimension of Post-Communist Transitions in Russia and the New States of Eurasia*. Armonk, NY: M.E. Sharpe, 1997, pp. 146–166.

———. "Russia and the United States in the Great Power Context." In Zhang Yunling and Guo Weihong, editors, *China, US, Japan, and Russia in a Changing World*. Beijing: Social Sciences Documentation Publishing House, 2000, pp. 163–175.

———. "Russia's Calculus and Japan's Foreign Policy in Pacific Asia." In Takashi Inoguchi, editor, *Japan's Asian Policy*. New York: Palgrave, 2002.

———. "Sino-Russian Relations: Mutual Assessments and Predictions." In Sherman Garnett, editor. *Rapprochement or Rivalry?: Russia-China Relations in a Changing Asia*. Washington, DC: Carnegie Endowment for International Peace, 2000.

———. "The Strategic Quadrangle and the Northeast Asian Region." In Rouben Azizian, editor, *Strategic and Economic Dynamics of Northeast Asia: global, regional and New Zealand perspectives*. Centre for Strategic Studies, Wellington, New Zealand, 1999, pp. 8–21.

———, editor. *Japan and Russia: The Tortuous Path to Normalization, 1949–1999*. New York: St. Martin's Press, 2000.

———, Mikhail Nosov and Koji Watanabe, editors. *Russia and East Asia: The 21st Century Security Environment*. Armonk, NY: East West Institute, M.E. Sharpe, 1999.

Ruble, Blair. "Bringing America In." In J. Thomas Rimer, editor. *A Hidden Fire: Russian and Japanese Cultural Encounters, 1868–1926*. Palo Alto, Calif.: Stanford University Press and the Woodrow Wilson Center, 1995.

Ruggie, John. "The False Premise of Realism." *International Security*, vol. 20, no. 1 (Summer 1995), pp. 62–70.

Russell, Richard A. *Project Hula: Secret Soviet-American Cooperation in the War Against Japan*. Washington, DC: Naval Historical Center, The US Navy in the Modern World Series, no. 4, Department of the Navy, 1997.

Saito Motohide. *The 'Highly Crucial' Decision making Model for Postwar Japan and Prime Minister Hatoyama's Policy Toward the USSR*. Ann Arbor, Mich.: University Microfilms, 1986.

———. "Japan's 'Northward' Foreign Policy." In Gerald Curtis, editor, *Japan's Foreign Policy: After the Cold War*. Armonk, NY: M.E. Sharpe, 1993.

Sakwa, Richard. *Russian Politics and Society*. New York: Routledge, 1996.

Samuels, Richard. *'Rich Country, Strong Army': National Security and the Technological Transformation of Japan*. Ithaca, NY: Cornell University Press, 1994.

——— and Eric Heginbotham, "Mercantile Realism and Japanese Foreign Policy." *International Security*, vol. 22, no. 4 (Spring 1998), pp. 171–203.

Sang-Woo Rhee. "Russia and the new balance of power in East Asia: implications for stability on the Korean Peninsula." In Gennady Chufrin, editor, *Russia and Asia: The Emerging Security Agenda*. Oxford: SIPRI/Oxford University Press, 1999.

Sarkisov, Konstantin. "The Northern Territories Issue in the Aftermath of Yeltsin's Reelection." An unpublished paper delivered at the *International Symposium*, Nara, Japan, 1996.

Scalapino, Robert, editor. *The Foreign Policy of Modern Japan*. Berkeley: University of California Press, 1977.

Schaller, Michael. *Altered States: The United States and Japan Since the Occupation*. Oxford: Oxford University Press, 1997.

Schecter, Jerold. *Russian Negotiating Behavior*. Washington, DC: United States Institute of Peace, 1998.

Schlesinger, Jacob. *Shadow Shoguns: The Rise and Fall of Japan's Postwar Political Machine*. New York: Simon and Schuster, 1997.

Schonberger, Howard. *Aftermath of War: Americans and the Remaking of Japan, 1945–1952*. Kent, Ohio: The Kent State Press, 1989.

Seton-Watson, Hugh. *From Lenin to Khrushchev: The History of World Communism*. New York: Frederick A. Praeger, 1962.

Sevastyanov, Sergei. "Russian Reforms: Implications for Security Policy and the Status of the Military in the Russian Far East." *NBR Analysis*, vol. 11, no. 4 (December 2000).

Shimotomai Nobuo. "Japan-Soviet Relations under Perestroika: Perceptions and Interaction between Two Capitals." In Gilbert Rozman, editor, *Japan and Russia: The Tortuous Path to Normalization, 1949–1999*. New York: St. Martin's Press, 2000.

Simonia, Nodari. "Domestic Developments in Russia." In Gennady Chufrin, editor, *Russia and Asia: The Emerging Security Agenda*, Oxford: SIPRI/Oxford University Press, 1999.

———. "Russian Energy Policy in East Siberia and the Far East." Paper prepared for the James A. Baker Institute for Public Policy of Rice University, October 2004. http://www.rice.edu/energy/publications/docs/PEC_SimoniaFinal_10_2004.pdf

Slavinsky, Boris. "The Soviet-Japanese Postwar Peace Settlement: Historical Experience and Present Situation." An unpublished paper presented at the conference *Japan and Russia: Postwar Relations, Mutual Influences, and Comparisons*, Princeton University, September 1997.

Snow, Edgar. *Red Star over China*. New York: Grove Weidenfeld, 1968.

Soeya Yoshihide. "The Structure of Japan-US-China Relations and Japan's Diplomatic Strategy." *Gaikō Forum*, September 28, 1997, as quoted in FBIS-EAS-97-288.

Spence, Jonathan. *The Search for Modern China*. New York: W.W. Norton and Company, 1990.

Stephan, John. *The Kuril Islands: Russo-Japanese Frontier in the Pacific*. Oxford: Clarendon Press, 1974.

———. *The Russian Far East: A History*. Palo Alto, Calif.: Stanford University Press, 1994.

———. *Russia on the Pacific*, The Sanwa Lecture Series of the North Pacific Program at the Fletcher School of Law and Diplomacy, Winter 1989.

———. "The Political and Economic Landscape of the Russian Far East." In Tsuyoshi Hasegawa, Jonathan Haslam, and Andrew Kuchins, editors, *Russia and Japan: An Unresolved Dilemma Between Distant Neighbors*. Berkeley, Calif.: University of California Press, 1993.

Sugano Tetsuo. "Russia's Economy and Development of the Far East." In Koji Watanabe, editor, *Engaging Russian in Asia Pacific*, Tokyo: Japan Center for International Exchange, 1999.

Sugawara Ikuro. "Emerging New Players." In *The United States, Japan, and the Middle East*. Washington, DC: The Middle East Institute, 1998, pp. 21–24.

Swearingen, Rodger, editor. *Siberia and the Soviet Far East: Strategic Dimensions in Multinational Perspective*. Palo Alto, Calif.: Hoover Institution Press, Stanford University, 1987.

Takahashi Susumu. "Toward a New Era of Russian-Japanese Relations." In James Goodby, Vladimir Ivanov, and Nobuo Shimotomai, editors, *"Northern Territories" and Beyond: Russian, Japanese, and American Perspectives*. Westport, Conn.: Praeger Publishers, 1995.

Takahashi Takuma. "Economic Interdependence and Security in the East Asia-Pacific Region." In Mike Mochizuki, editor, *Toward a True Alliance: Restructuring US-Japan Security Relations*. Washington, DC: The Brookings Institution, 1997, pp. 96–133.

Tanaka Akihiko. "Domestic Politics and Foreign Policy." In Inoguchi Takashi and Purnendra Jain, editors, *Japanese Foreign Policy Today*, New York: Palgrave, 2000.

Thayer, Nathaniel. *How the Conservatives Rule Japan*. Princeton: Princeton University Press, 1973.

"The Cold War in Asia," *The Cold War International History Project Bulletin*, Woodrow Wilson International Center for Scholars, issues 6–7 (Winter 1995/1996), pp. 4–29.

Thornton, Judith and Charles Ziegler, editors. *Russia's Far East: A Region at Risk*. Seattle, Wash.: The National Bureau of Asian Research, University of Washington Press, 2002.

Tolischus, Otto. *Tokyo Record*. New York: Reynal & Hitchcock, 1943.

Tselitschev, Ivan. "Russian Economic Reforms and Japan." In James Goodby, Vladimir Ivanov, and Nobuo Shimotomai, editors, *"Northern Territories" and Beyond: Russian, Japanese, and American Perspectives*. Westport, Conn.: Praeger Publishers, 1995.

Tsuru Shigeo. *Japan's Capitalism: Creative Defeat and Beyond*. Cambridge: Cambridge University Press, 1993.

Tucker, Robert. *Stalin in Power: The Revolution from Above, 1928–41 (Vol. 2)*. New York: W.W. Norton and Company, 1990.

Tyshetskii, Igor. "The Gorbachev-Kaifu Summit: The View from Moscow." In Tsuyoshi Hasegawa, Jonathan Haslam, and Andrew Kuchins, editors, *Russia and Japan: An Unresolved Dilemma Between Distant Neighbors*, Berkeley: University of California Press, 1993.

Ulam, Adam. *Expansion and Coexistence: The History of Soviet Foreign Policy, 1917–67*. New York: Frederick Praeger Publishers, 1968.

Valencia, Mark. "Energy and Insecurity in Asia." *Survival*, no. 39 (Autumn 1997), pp. 85–106.

Valliant, Robert. "The Political Dimension." In Tsuneo Akaha, editor. *Politics and Economics in Northeast Asia: Nationalism and Regionalism in Contention*. New York: St. Martin's Press, 1999.

Verbitsky, Semyon. "Perceptions of Japan in the USSR During the Cold War and Perestroika." *The Carl Beck Papers*, no. 1503, University of Pittsburgh, October 2000.

———. "Russian Perceptions of Japan." In James Goodby, Vladimir Ivanov, and Nobuo Shimotomai, editors, *"Northern Territories" and Beyond: Russian, Japanese, and American Perspectives*, Westport, Conn.: Praeger Publishers, 1995.

Vishwanathan, Savitri. *Normalization of Japanese-Soviet Relations, 1945–1970*. Tallahassee, Fla.: The Diplomatic Press, 1973.

Vitkovskaya, Galina. "Lawlessness, Environmental Damage, and Other New Threats in the Russian Far East." In Mikhail Nosov and Koji Watanabe, editors, *Russia and East Asia: The 21st Century Security Environment*. Armonk, NY: East West Institute, M.E. Sharpe, 1999, pp. 179–199.

Waltz, Kenneth. "The Emerging Structure of International Politics." *International Security*, vol. 18, no. 2 (Fall 1993), pp. 44–79.

Watanabe Koji, editor. *Engaging Russian in Asia Pacific*. Tokyo: Japan Center for International Exchange, 1999.

Welfield, John. *An Empire in Eclipse: Japan in the Postwar American Alliance System*. Atlantic Highlands, NJ: The Athlone Press, 1988.

Wendt, Alexander. "Constructing International Politics." *International Security*, vol. 20, no. 1 (Summer 1995), pp. 71–81.

White, Gordon. *Gorbachev and After*. Cambridge: Cambridge University Press, 1992.

Whiting, Allen. *China Eyes Japan*. Berkeley, Calif.: University of California Press, 1989.

———. *Siberian Development and East Asia Threat or Promise?* Palo Alto, Calif.: Stanford University Press, 1981.

Whiting, Robert. *You Gotta Have Wa*. New York: Vintage Books, 1990.

Wich, Richard. *Sino-Soviet Crisis Politics*. Cambridge, Mass.: Harvard University Press, 1980.

Wimbush, S. Enders. "Time for a New Northeast Asian Security Order." *The Wall Street Journal* (Asia edition), October 23, 1996.

Wishnick, Elizabeth. "One Asia Policy or Two?: Moscow and the Russian Far East Debate Russia's Engagement in Asia." *NBR Analysis*, vol. 13, no. 1 (January 2002).

Wohlforth, William. "Realism and the End of the Cold War." *International Security*, vol. 19, no. 3 (Winter 1994/95), pp. 91–129.

Zagoria, Donald. *The Sino-Soviet Conflict, 1956–61*. Princeton: Princeton University Press, 1962.

Zagorsky, Aleksei. "Reconciliation in the Fifties: The Logic of Soviet Decision Making." In Gilbert Rozman, editor, *Japan and Russia: The Tortuous Path to Normalization, 1949–1999*. New York: St. Martin's Press, 2000, pp. 47–72.

Ziegler, Charles. "Russo-Japanese Relations: A New Start for the 21st Century?" *Problems of Post-Communism*, vol. 46, no. 3 (May/June 1999), pp. 15–25.

Japanese language sources (articles, books, memoirs, monographs)

Anzen Hoshō Mondai Kenkyuujo. *Kawaru Nichirō Kankei: Roshiajin kara no 88 no Shitsumon*. Tokyo: Anzen Hoshō Mondai Kenkyuujo, 1999.

Blaginsky, Sergei. "Shinrai Dekiru Paatonaa wo ikani Tuskuru-ka." *Gaikō Forum*, no. 117, (April 1998), pp. 18–25.

Boguslavsky, Sergei. "Daitōryō Sokkin Paaji no Inbō." *Aera* (September 29, 1992), pp. 16–18.

Edamura Sumio. *Teikoku Kaitai Zengo*. Tokyo: Toshishuppan, 1997.

———. "Nichirō Kōshō no 'Wakugumi' wo Tatenaose." *Chuuō Kōron* May 2002, pp. 68–73.

Enerugi Anzen Hoshō Mondai ni Kan suru Chōsa Kenkyuu. Tokyo: Sekai Heiwa Kenkyuujo, 1992.

Ferguson, Joseph. "Nichirō Kankei: Kakō, Genzai, Mirai." *Shin Bōei Ronshu*, vol. 28, no. 1 (June 2000).

Fujimura Mikio. "Hoppō Ryōdo Mondai de Nihon Shiji kara Kōtai-suru Beikoku." *Foresight Magazine*, July 1997, pp. 34–35.

Funabashi Yoichi. "Engejimento, Antei, Baransu: Ajia Taiheiyo no 21 Seiki Senryaku." *Sekai*, January 1997, pp. 90–107.

———. "Nichibeichuu Shin-Jidai ni Wa-Kan-Yo no Kyoyō to Shuuyō wo." *Foresight Magazine*, January 1997, pp. 6–10.

———, Wan Chi Su, Kokubun Ryosei. "Reisen-go Ajia ni Nichibeichuu de Kensetsu-teki Paatonaashippu wo." *Sekai*, July 1998, pp. 150–162.

Hakamada Shigeki. "Akino-shi no Nokoshita Yume." *Gaikō Forum*, no. 10 (1998), pp. 12–14.

———. "Aratana Nichirō Kankei Kochaku e no Teigen." *Foresight Magazine*, August 1997, pp. 6–9.

———. "Hashimoto-ato no Nichirō Kankei wo Yomidoku Kagi." *Foresight Magazine*, August 1998, pp. 12–15.

———. "Irukutsuku no Gensō." *Sekai Shuuhō*, May, 2001, pp. 6–9.

———. "Roshia ni Gokai wo Ataeru you na Nihon-teki 'Kikubari Gaikō' wo Aratameyo.' *Nihon no Ronten, 2001*, pp. 130–135.

———. "Roshia no Seihen to Eritsin Rainichi." *Chuuō Kōron*, April 1998, pp. 224–233.

———. "Roshia Seifu ni Tsugu." *Foresight Magazine*, March 1999, pp. 106–109.

———. "Sōgō Gensō wo Hai-shita Nichirō Kankei no Kōchiku wo." *Foresight Magazine*, April 1998, pp. 10–13.

———. " 'Tai-Ro Fushin' wo Seikaku ni Roshia ni Tsutaeyo." *Foresight Magazine*, September 2000, pp. 6–9.

Hasegawa Hiroshi. "Suzuki Muneo Nigen Gaikō no Aya." *Aera*, March 5, 2001, pp. 29–31.

Hasegawa Tsuyoshi. *Hoppō Ryōdo Mondai to Nichirō Kankei*. Tokyo: Chikuma Shobō, 2000.

Hashimoto Ryutaro. "Roshia ha Kanzen na Paatonaashippu wo." *Gaikō Forum*, no. 149 (December 2000), pp. 30–35.

Honma Hiroaki. "Hoppō Yonshima, Rankaku kara no Dappi." *Sekai*, October 2000, pp. 190–196.

Igarashi Taku. "Ajia Kinyuu Kiki ga Ukibori ni shita 'Mō hitotsu no Fuan.'" *Foresight Magazine*, February 1998, pp. 54–55.

———. "Chuutō kara Chuunanbei e Shifuto Shi-hajimeta Beikoku Sekiyuu Senryaku." *Foresight Magazine*, March 1997, pp. 42–43.

———. "Genyuu no Senryaku Bichiku ni Ugokidashita Ajia." *Foresight Magazine*, May 1997, pp. 42–43.

———. "Juuyu Ryuushutsu ga Ukibori ni shita 'Ajia Oiru Rodo' no Fuan." *Foresight Magazine*, February 1997, pp. 90–91.

Ito Shoichi. "Kiro ni tatsu Taiheiyo Paipurainu Kosō." ERINA Report, vol. 72 (Nov. 2006), pp. 23–33; vol. 73 (Jan. 2007), pp. 31–41.

Ito Tadashi. "Bei-Chuu-Nichi-Rō 'Shikyoku Kankei' no Yukue." *Foresight Magazine*, December 1997, pp. 118–121.

Japan-Russia Economic Council. *Nichirō Keizai Kōryuu Sokushin-jō no Shōgai to Teigen* (Proposals and Obstacles concerning the Promotion of Japanese-Russian Economic Exchange), July 1994.

Japonisumu no Nazo. Tokyo, Asahi Gurafu, Bessatsu Bijutsu Tokushu, 1990, no. 1.

Kato Ryozo. "Nichibei Dōmei no Kuudôka wo Osoreru." *Chuuō Kōron*, December 2000, pp. 76–84.

Kimura Hiroshi. *Borisu Eritsin: Ichi Roshia Sejika no Kidō*. Tokyo: Maruzen Library, 1997.

———. "Puchin ha Tainichi Seisaku wo Dono-yōni Kettei suru no-ka." *Gaikō Forum*, no. 149 (December 2000), pp. 12–19.

———. "Puchin no Tai-Nichi Seisaku (3)," *Journal of World Affairs* (Tokyo), April 2003.

Kishida Yoshiki. "Shushō no Gosain de Gaimushō ga Suzuki-shi ni 'Ketsubetsu Sengen'." *Sekai Shuuhō*, April 2, 2002, pp. 13–14.

Kitagawa Kazuya. "Puchin Seiken ni Ryōdo Mondai Sakiokuri no Kōjitsu." *Sekai Shuuhō*, April 2, 2002, pp. 10–12.

Kobayashi Kazuo, Sato Kazuo, Shimotomai Nobuo, Togo Kazuhiko. "Chokkan no Eritsin kara Jitsuri no Puchi he." *Gaikō Forum*, no. 149 (December 2000), pp. 20–29.

Kortunov, Andrei. "Wareware ha O-tagai wo Hitsuyō to Shiteiru no Darō-ka." *Gaikō Forum*, no. 149, December 2000, pp. 42–47.

Kovalenko, Ivan. *Tainichi Kōsakuno Kaisō*. Tokyo: Bungei Shunjuusha, 1996.

Mayama Katsuhiko. "Chuugoku kara Mita Chuurō Kankei." *Shin Bōei Ronshu*, vol. 28, no. 1 (June 2000), pp. 20–48.

Matsushima Kunihiko. "Nihon Kai Juuyu Ryuushutsu Jiko no Haigo ni Kyokutō Roshia no Keizai Hatan." *Foresight Magazine*, March 1997, pp. 36–37.

Ministry of Foreign Affairs, *Nihon Gaikō Bunsho (Documents on Japanese Foreign Policy)*, Tokyo: vol. I, 1924.

Miyamoto Nobuo. "Chuurō no Senryaku-teki Kyōchō ha Doko-made Susumu-ka." *Chuuō Kōron*, July 1996, pp. 134–139.

———. "Nichi-Bei-Chuu-Rō Shijuusō no Fukyōwaon." *Chuuō Kōron*, February 1998, pp. 138–149.

———. "Nichirō Kankei ha Naze Nichibei no Tsugi-ni Juuyō-ka." *Chuuō Kōron*, February 1999, pp. 244–251.

Mori Shinjiro. *Nyuuwa-na Manazashi no Mukō: Mori Shinjiro no Kidō*. Tokyo: Asahi Shimbun Press, 2000.

———. "Kasapi Kai no Nami ga Sawagu." *Asahi Shimbun*, February 28, 1998, p. 5.

———. "Nichirō Shuunō 'Kawana Kaidan' no Butaiura." *Sekai*, June 1998, pp. 126–132.

Murayama Yoshihisa. "2010-nen 'Shigen Yuunyuu Daikoku' Chuugoku no Kyōi." *Foresight Magazine*, November 1996, pp. 50–53.

Nagashi Takerō. "Shippai ni Owatta Tai-ro Seisaku Kudeta." *Sekai Shuuhō*, April 2, 2002, pp. 6–9.

Nakajima Yuuji. "Seiron no Takamari ga Ryōdo Mondai wo Kaiketsu-suru." *Jiyuuminshu*, no. 9 (2000), pp. 34–43.

Nakazawa Takayuki. "Ryōdo Mondai kaiketsu no Kagi to naru no ha 'Tsuyoi Roshia'."

Sekai Shuuhō, November 28, 2000, pp. 26–29.

———. "Puchin Hōnichi to Hoppō Ryōdo Mondai," *Sekai Shuuhō*, April 5, 2005, pp. 14–17.

Nishimura Yoichi. *Purometeusu no Hakaba*. Tokyo: Asahi Shimbun Press, 1997.

———. "Yurashia Gaikō Koshisuete." *Asahi Shimbun*, November, 23 1997.

———. "Yurashia Gaikō no Butaiura." *Sekai*, January 1998, pp. 138–147.

Numachi Yoshi. *Haruka-na Shiberia: Sengo 50-nen no Shōgen (Vol. I–II)*. Sapporo: Hokkaido Shimbun Press, 1995.

Ogura Yasuhiro. "Dōmei ni Kanshite no Riron-teki Kosatsu: Posuto Reisen Jidai no Nihon no Dōmei Seisaku." *Shin Bōei Ronshu*, vol. 28, no. 1 (June 2000), pp. 67–82.

Okabe Tatsumi. "Nicchuu Kankei no Kakōto Shōrai." *GaikōForum*, no. 151 (February 2001), pp. 12–20.

Okada Kunio. "Nikkei Kigyō Shinshutsu no Kōki ha?" *Gaikō Forum*, no. 149 (December 2000), pp. 56–61.

Owada Hisashi. "Kore ga Roshia shien no Ronri da." *Chuuō Kōron*, July 1993, pp. 30–38.

Panov, Aleksandr. "Nichirō Ryôkoku no Paatonaashipu Kôchiku ni Mukete." *Sekai*, December 1998, pp. 146–148.

Satō Masaru, *Kokka no Wana: Gaimushō no Rasupuchin to Yobarete*, Tokyo: Shinchosha, 2005.

Saitō Tsutomu. "Satō, Rasupuchin to Yobareta Otoko." *Bungei Shunju*, May 2002, pp. 114–121.

Sakurai Kaoru. "Enerugi Riken Sodassen ni Yusaburareru Roshia Shin-naikaku." *Foresight Magazine*, May 1998, pp. 32–33.

———. "Hashimoto 'Tai-Roshia Gaikō' ni Shizumu Shikaku." *Foresight Magazine*, November 1997, pp. 106–107.

———. "Roshia Tai-Nichi Gaikō no Shuyaku ha Dare-ka." *Foresight Magazine*, October 1997, pp. 32–33.

Sarkisov, Konstantin. "Posuto Eritsin no Hoppō Ryōdo Mondai." *Sekai Shuuhō*, January 18, 2000, pp. 18–21.

Shinoda Kenji, Mori Shinjiro, Hakamada Shigeki, Suetomi Takeo. "Nihon no Sonzaikan wo Takameru tame-ni." *Gaikō Forum*, no. 117 (April 1998), pp. 34–47.

Shiraishi Masaaki. "Gaikō Shiryo no Nichirō Hikaku." In *Gaikōshiryokanpo (Journal of the Diplomatic Record Office*. Ministry of Foreign Affairs of Japan, March 1995, pp. 64–68.

Suzuki Muneo. "Sogo Rikai no Josei ga Atarashii Nichirō Kankei wo Kizuku." *Jiyuminshu*, no. 9 (2000), pp. 26–33.

Tachiyama Yôshi. "Mega-Chuutō wo Misueru Arata-na Shisa wo." *Foresight Magazine*, September 1997, pp. 6–9.

Takubo Tadae. "Nichibei Kankei ga Shinpai da." *Seiron*, March 1997, pp. 46–59.

Tamba Minoru. *Nichirō Gaikō Hiwa*. Tokyo: Chuokoron-Shinsha, 2004.

———. "Fukugan-teki, Juusō-teki Nichirō Kankei Kochiku no Tame ni." *Gaikō Forum*, no. 149 (December 2000), pp. 36–41.

———. "Kurasunoyarusuku Kaidan kara Eritsin Daitōryō no Hōnichi he." *Gaikō Forum*, no. 117 (April 1998), pp. 12–17.

Tanabe Yasuo. "Kasupi Kai Sekiyu Shigen no Seiji-Keizai Gaku." *Gaikō Forum*, August-September 1998, pp. 64–70.

Tanaka Akihito. "97-nen, Beichuu ha Wakai suru." *Chuuō Kōron*, February 1997, pp. 52–63.

Terashima Jitsuro. "Nicchuubei Toraianguru Kuraishisu wo do Seigyo suru ka." *Chuuō Kōron*, August 1996, pp. 28–45.

Tōgō Kazuhiko. *Hoppō Ryōdo Kōshō Hiroku: Ushinawareta Gotabi no Kikai*. Tokyo: Shinchosha, 2007.

———. *Nichirō Shinjidai he no Josō: Dakai no kagi wo Motomete*. Tokyo: Saimaru, 1993.

———. Nichirō Ryokoku no Shinrai Kochiku no Tame-ni, Naze Subeki-koto." *Sekai*, October 1998, pp. 215–220.

Toshikawa Takao. "'Muneo Paji' de Miushinatta Gaikō-Gaimusho Kaikaku." *Sekai*, May 2002, pp. 78–87.

Wada Haruki. *Hoppō Ryōdo Mondai wo Kangaeru*. Tokyo: Iwanami Shoten, 1990.

———. "Nichirō Kankei 200-nen kara Mita, Shinjidai he no Henka." *Sekai*, November 1998, pp. 166–171.

———. "Sukyandaru to Gaikō." *Sekai*, May 2002, pp. 71–77.

Watanabe Koichi. "Hōkai Suru Hoppō Yon Shima." *Sekai*, January 1997, pp. 116–126.

X-shi. "Nichirō Mochikoeshi: Kunashiri, Etorofu Henkan no Dohyo Tsukuri." *Sankei Shimbun*, December 28, 2000.

———. "Nichirō Mochikoeshi: Kosho Teishi ... Ushinau mono nashi." *Sankei Shimbun*, December 29, 2000.

Yoshimura Michio. "Goto Shimpei Saigo no Hoso wo Megutte." In *Gaikōshiryokanpo*. Ministry of Foreign Affairs of Japan, March 1990, pp. 50–66.

Yoshioka Tatsuya. "Shikotan-tō Kaihatsu kore ga Chintai Jōyaku-sho da." *Aera*, September 29, 1992, pp. 13–15.

Russian language sources (articles, books, memoirs, monographs)

Abramov, Aleksei and Boris Makeev. "Kuril'skii Bar'er." *Novoe Vremya*, nos. 40–41 (October 1992), pp. 16–18, 24–25.

Akademii Nauk SSSR. *SSSR-Yaponiya: K 50-letno ustanovleniya sovietsko-yaponskikh diplomaticheskikh otnoshenii (1925–1975)*. Moscow: Nauka, 1977.

Alekseyev, Mikhail. "Ugrozhaet li Rossii Kitaiskaya Migratsiya?" *Mirovaya Ekonomika i Mezhdunarodnye Otnosheniya*, nos. 11–12 (2000), pp. 97–103, 42–50.

Aliev, Rafik. "Sovietskii Soyuz v Vostochnoi Azii: Real'nost' i Problemy." *Mirovaya Ekonomika i Mezhdunarodnye Otnosheniya*, no. 9 (1990), pp. 88–96.

Arkad'yev, A.A. "Politika Yaponii v Otnoshenii SSSR Posle Nachala Vtoroi Mirovoi Voiny." In Akademii Nauk SSSR. *SSSR-Yaponiya: K 50-letno ustanovleniya sovietsko-yaponskikh diplomaticheskikh otnoshenii (1925–1975)*. Moscow: Nauka, 1977, p. 32.

Baranets, Viktor. *Yel'tsin i ego Generaly*. Moscow: Kollektsiya 'Sovershenno Sekretno', 1998.

Bogaturov, Aleksei. *Velikie Derzhavy ha Tikhom Okeane: Istoriya i Teoriya Mezhdunarodnykh Otoshenii v Vostochnoi Azii Posle Vtoroi Mirovoi Voiny*. Moscow: RAN, Institut SShA i Kanady, 1997.

———. "Diplomatiia pered vyborom." *Novoe Vremya*, no. 32 (1989), pp. 20–23.

——— and Mikhail Nosov. "Treugol'nik bez uglov." *Novoe Vremya*, no. 18 (1989), pp. 8–9.

Bogolyubov, Gennady. "Rasshirenie NATO na Vostok i rossiisko-kitaiskie otnoshenii." *Problemy Dal'nego Vostoka*, no. 6 (1997), pp. 31–41.

Bolyatko, Anatolii. "Voenno-Tekhnicheskoe Sotrudnichestvo i Perspektivy Strategicheskogo Vsaimodeistviya Rossii i Kitaya." *Problemy Dal'nego Vostoka*, no. 3 (1997), pp. 34–38.

Bondarenko, Oleg. *Neizvestnye Kurily: Ser'eznye razmyshleniia o statuse Kuril'skikh ostrovov*. Moscow: VTI Deita Press, 1992.

Chegodar', N.I. "Russkaya i Sovietskaya Literatura v Yaponii v Posleoktyabrskii Period." In Akademii Nauk SSSR. *SSSR-Yaponiya: K 50-letno ustanovleniya sovietsko-yaponskikh diplomaticheskikh otnoshenii (1925–1975)*, Moscow: Nauka, 1977, pp. 208–224.

Cherniaev, A.S. *Shest' let s Gorbachevym*. Moscow: Kultura, 1993.

Chknaverona, Anna. *K Istorii Russko-Yaponshikh Otnoshenii*. Moscow: Sputnik, 2000.

Dergachev, Vladimir. *Geopolitika*. Kiev: VIRA, 2000.

Dolgorukov, P.D. "Torgovo-Ekonomicheskie Otnosheniya SSSR s Yaponiei." In Akademii Nauk SSSR. *SSSR-Yaponiya: K 50-letno ustanovleniya sovietsko-yaponskikh diplomaticheskikh otnoshenii (1925–1975)*, Moscow: Nauka, 1977, pp. 104–131

Dugin, Aleksandr. *Osnovy Geopolitiki*. Moscow: Arktogeya, 1997.

Fedorov, V.P. *Inostrantsy i My*. Moscow: MP Russkoe Polye, 1992.

Gadzhiev, Kamaludin. *Vvedenie v Geopolitiky*. Moscow: Logos, 2000.

Galuzin, Mikahil. "Nashi Dela s Yaponiei." *Mezhdunarodnaya Zhizn'*, March 2000, pp. 89–100.

———. "Razmerennaya Postyp' v Otnosheniyakh s Yaponiei." *Mezhdunarodnaya Zhizn'*, October 2000, pp. 73–79.

Golovnin, Vasilii. "Heokochennyi Poyedinok Bogatyrya i Samuraya." *Novoe Vremya*, no. 1–2 (1997), pp. 31–35.

———. "Russkaya Svetlana vkhodit v elitu 'Aum Sinrikyo." *Izvestia*, February 23, 2000.

Gostinyhi Dvor Mezhdunarodnoi Zhizni. "Rossiya na Vostochnom Napravlenii." *Mezhdunarodnaya Zhizn'*, October, 2000, pp. 93–109.

Ivanov, Vladimir. "Sovietskii Soyuz i Aziatsko-Tikhookeanskii Region: Evolutsiya ili Radikal'nye Peremeny?" *Mirovaya Ekonomika i Mezhdunarodnye Otnosheniya*, no. 9 (1990), pp. 97–107.

Karaganov, Sergey. "Novaya Vneshnaya Politika." *Moskovskie Novosti*, Feb. 29–March 6, 2000.

Kim Te Han. *Istoriko-politicheskie Aspekty Russko-Yaponskikh Otnoshenii*. Moscow: Tipografii Gornogo Institut, 1998.

Kistanov, Valerii. "Yaponiya i Perspektiva Rossiisko-Kitaiskogo Strategicheskogo Partnyorstva." *Problemy Dal'nego Vostoka*, no. 2 (1997), pp. 48–56.

Kitagawa, Tsuyoshi. "Strategicheskii Vybor V.V. Putina: Kitai ili Yaponia." Moscow Carnegie Center, March 15, 2005. http://www.carnegie.ru/ru/print/71530-print.htm

Kortunov, Sergey. "Rossiya isshet Soyoznikov." *Mezhdunarodnaya Zhizn'*, no. 5 (1996), pp. 17–30.

Kozyrev, Andrei. *Preobrazhenie*. Moscow: Mezhdunarodnie Otnosheniya, 1995.

Krupyanko, Mikhail. *Yaponia 90-kh: V Poiskakkh Modeli Otnoshenii s Novoi Rossiei*. Moscow: RAN, Vostochnaya Literatura, 1997.

———. *Yaponia v Sisteme Vostok-Zapad*. Moscow: Nauka, 1991.

Kulik, Boris. *Sovietsko-Kitaiskii Raskol: Prichiny i Posledstviya*. Moscow: RAN, Institut Dal'nego Vostoka, 2000.

Kunadze, G.F. "Beg v Meshkakh." *Novoe Vremya*, no. 14 (2000), pp. 26–29.

———. "Militarizm v Yaponii: Voprosy Metodologii Analiza." *Mirovaya Ekonomika i Mezhdunarodnye Otnosheniya*, no. 2 (1989), pp. 116–130.

———. "Ostrova Nevezeniya." *Novoe Vremya*, no. 14 (2001), pp. 24–25.

———. "Politika Rossii v Otnoshenii KNDR." *Mirovaya Ekonomika i Mezhdunarodnye Otnosheniya*, no. 12 (1999), pp. 36–41.

———. "Sindrom Kuz'kinoi Materi." *Novoe Vremya*, no. 37 (September 1999).

———. "V Poiskakh Novogo Myshleniya: O Politike SSSR v Otnosheniii Yaponii." *Mirovaya Ekonomika i Mezhdunarodnye Otnosheniya*, no. 8 (1990), pp. 51–67.

———. "Zagodochyi Eksprompt." *Novoe Vremya*, no. 37 (September 2000), pp. 20–22.

——— and Konstantin Sarkisov. "Razmyshliaya o sovetssko-yaponskikh otnosheniiakh." *Mirovaya Ekonomika i Mezhdunarodnye Otnosheniya*, no. 5 (1989), pp. 83–93.

Kurmazov, Alekasndr. "Rossiya i Yaponiya: Rybollovstvo v Dvustoronnikh Otnosheniyakh." *Problemy Dal'nego Vostoka*, no. 6 (1997), pp. 59–68.

Kutakov, L.N. *Rossiya i Yaponiya*. Moscow: RAN, Vostochnaya Literatura, 1988.

———. "Bor'ba SSSR za Ustanovlenie i Razvitie Dobrososedskikh Otnoshenii s Yaponiei." In Akademii Nauk SSSR. *SSSR-Yaponiya: K 50-letno ustanovleniya sovietsko-yaponskikh diplomaticheskikh otnoshenii (1925–1975)*, Moscow: Nauka, 1977, pp. 7–30.

Larin, Viktor. "Rossiya i Kitai na Poroge Tret'ego Tysyachletiya: Kto-zhe bydet otstaivat' nashi natsional'nye interesy?" In A.V. Kortunov, *Vneshaya Politika i Besopasnost' Sovremennoi Rossii (Vol. I, no. 2)*. Moscow: Moskovskii Obshestvennyi Nauchnyi Fond, 1999, pp. 159–176.

Latyshev, Igor. *Kto i kak prodaet Rossiiu: Khronika rossiisko-yaponskikh territorial'nykh torgov (1991–94 gody*). Moscow: Paleia, 1994.

———. *Yaponiya, Yapontsy, i Yaponovedy*. Moscow: Algoritm, 2001.

Loboda, Ivan. *Moskva-Pekin: Chto Dalshe?* Moscow: INFRA-M, 1995.

Lukin, Vladimir. "Strategicheskoe Partnyorstvo Rossii i Kitaya – predskazuemaya real'nost'." *Problemy Dal'nego Vostoka*, no. 3 (1997), pp. 39–44.

Makeev, Boris. "Kuril'skaya Problema: Voennyi Aspekt." *Mirovaya Ekonomika i Mezhdunarodnye Otnosheniya*, January 1993, pp. 54–59.

Markov, Andrei. *Rossiya i Yaponiya: V poiskakh soglasiya*. Moscow: RAN, Institut Dal'nego Vostoka, 1996.

Mazhorov, C.T. "Sovietsko-Yaponskie Otnosheniya v Khode Vtoroi Mirovoi Voiny (1941–1945)." In Akademii Nauk SSSR. *SSSR-Yaponiya: K 50-letno ustanovleniya sovietsko-yaponskikh diplomaticheskikh otnoshenii (1925–1975)*. Moscow: Nauka, 1977, pp. 57–63.

Moiseyev, Leonid. "Vremya Vostochnoi Politiki Kremlya." *Mezhdunarodnaya Zhinzn'*, no. 11–12 (1997), pp. 3–8.

Myasnikov, V.S. "Rossia i Kitai: Perspektivy Partnytorstva v ATR v XXI V." In A.V. Kortunov, *Vneshaya Politika i Besopasnost' Sovremennoi Rossii (Vol. I, No. 2)*. Moscow: Moskovskii Obshestvennyi Nauchnyi Fond, 1999, pp. 177–204.

Nosov, Mikhail. "Polveka Yaponskoi Vneshnei Politki." In Vadim Ramzes, editor. *Yaponiya: Polveka Obnovleniya*. Moscow: TOLK, 1995, pp. 487–567.

Panov, Aleksandr. "V Yaponii Novyi God Nastupaet so 108-m Udarom Kolokolov Khrama." *Mezhdunarodnaya Zhinzn'*, no. 1 (1998), pp. 56–66.

———. "Bor'ba Sovietskogo Soyuza za Mirnyi Demokraticheskii Put' Razvitiya Yaponii." In Akademii Nauk SSSR. *SSSR-Yaponiya: K 50-letno ustanovleniya sovietsko-yaponskikh diplomaticheskikh otnoshenii (1925–1975)*. Moscow: Nauka, 1977, pp. 65–82.

Pavlyatenko, Viktor. "Rossisko-Yaponskie Otnosheniya: Vremya Deistvovat'." *Problemy Dal'nego Vostoka*, no. 6 (1997), pp. 24–30.

——— and Aleksandr Shlyndov. "Rossisko-Yaponskie Otnosheniya: Nekotorye Itogi i Perspektivy na Starte XXI Stoletiya." *Problemy Dal'nego Vostoka*, no. 4 (2000), pp. 5–28.

Pesh, Stepan. "Zapretnyi paltus: Yaponskie brako'ery stanut rybolevami posle uregolirovaniia spora o severnykh territoriiakh." *Novoe Vremya*, no. 35 (September 1994), pp. 24–25.

Petrov, Dmitri. *Vneshnaya Politika Yaponii Posle Vtoroi Mirovoi Voiny*. Moscow: IMO, 1965.

Piadyshev, B.D. "Vzgliad iz Tokio." *Mezhdunarodnaya Zhizn'*, no. 5 (1989), pp. 26–35.

Preobrazhenskii, Konstantin. *KGB v Yaponii: Shpion, kotoryi lyubil Tokio*. Moscow: Tsenterpoligraf, 2000.

Prokhorov, Evgenii and Leonid Shevchuk. "O territorial'nykh pretenziiakh Yaponii k SSSR." *Mezhdunarodnaya Zhizn'*, no. 1 (1989), pp. 47–52.

Punzhin, S.M. "SSSR-Yaponiia: Mozhno li pri pomoshchi prava reshit' problemy 'severnykh territorii'?" *Sovestkoe gosudarstvo i pravo*, no. 7 (1991), pp. 104–119.

Rakhmanin, O.B. *K Istorii Otnoshenii Rossii-SSSR s Kitaem v XX Veke*. Moscow: RAN, Institut Dal'nego Vostoka, 2000.

Ramzes, Vadim, editor. *Yaponiya: Polveka Obnovleniya*. Moscow: TOLK, 1995.

Shmyrev, Anatolii. "Rossii nuzhno izvinit'sia pered Yaponiei." *Segodnya*, 1 September 1995.

Sidorov, Andrei. *Vneshnaya Politika Sovietskoi Rossii ha Dal'nem Vostoke (1917–1922gg.)*. Moscow: Moskovskii Gosudarstvennii Institut Mezhdunarodnykh Otnoshenii, 1997.

Slavinskii, Boris. *SSSR i Yaponiya – Ha Puti k Voine: diplomaticheskaya istoriya, 1937–45*. Moscow: ZAO Yaponiya Segodnya, 1999.

Spravochnik. *Mezhdunarodnye Issledovaniya v Rossii i SNG*. Moscow: Moskovskii Obshestvennii Nauchnii Fond, 1999.

Suzuki Muneo. "Protsess Uglubleniya Vzaimoponimaniya Sozdayot Bazu dlya Novikh Yapono-Rossiskikh Otnoshenii." *Problemy Dal'nego Vostoka*, no. 1 (2001), pp. 20–25.

Tikhvinskii, S. "Poslevoennaia normalizatsiia otnoshenii s Yaponiei." *Problemy Dal'nego Vostoka*, no. 4 (1995), pp. 108–119, and no. 5 (1995), pp. 95–105.

Titarenko, Mikhail. *Rossiya litsom k Azii*. Moscow: Respublika, 1998.

Tkachenko, Boris. "Kurilskaia problema: Istoriia i pravo." *Rossiia i ATR*, no. 3 (1995), pp. 24–29 and no. 5 (1995), pp. 10–18.

Togo Kazuhiko. *50 Let Yaponskoi Diplomatii, 1945–1995*. Moscow: Moskovskii Gosudarstvennii Institut Mezhdunarodnykh Otnoshenii, 1996.

———. *Yapono-Rossiskie Otnosheniya: Proshloe i Perspektivy*. Moscow: MFTI, 1995.

Togo Takehiro. "Yaponiya i Rossiya v XXI Veke." *Mirovaya Ekonomika i Mezhdunarodnye Otnosheniya*, May 1997, pp. 16–25.

———. "Yapono-Rossiskie Otnosheniya: Proryv v XXI Vek." *Problemy Dal'nego Vostoka*, no. 1 (1998), pp. 8–17.

Tselishev, Ivan. " 'Upravlenie po-yaponski' za predelami Yaponii." *Mirovaya Ekonomika i Mezhdunarodnye Otnosheniya*, August 1991, pp. 58–79.

———. "Yaponiya kak Politicheskii Partnyor." *Mirovaya Ekonomika i Mezhdunarodnye Otnosheniya*, June 1994, pp. 70–78.

Veki na Puti k Zaklyucheniyu Mirnogo Dogovora mezhdu Yaponiei I Rossiei: 88 Voprosov ot Grazhdan Rossii. Council on National Security Problems, Tokyo. Moscow: Maternik, 2000.

Voskresenskii, A.D. "Kitai vo Vneshnepoliticheskoi Strategii Rossii." In A.V. Kortunov, *Vneshaya Politika i Besopasnost' Sovremennoi Rossii (Vol. I, No. 2)*. Moscow: Moskovskii Obshestvennyi Nauchnyi Fond, 1999, pp. 140–158.

———. "Rossiya i Kitai: Faktory Vsaimodeistbiya." *Politiya*, no. 3(13) (Fall 1999), pp. 110–124.

Yakovlev, Aleksandr. "Rossiya i Kitai v Stroitel'stve Novogo Mirovogo Poryadka." *Problemy Dal'nego Vostoka*, no. 6 (1998), pp. 23–39.

Yel'stin, Boris. *Prezidentskii Marafon*. Moscow: AST, 2000.

Yelizar'ev, Vitalii. *Sakhalinskaya Oblast' ha Perekrestke Rossiisko-Yaponskikh Otnoshenii Kontsa XX Stoletiya*. Yuzhno-Sakhalinsk, 1999.

Yeremin, Vladimir. *Rossiya-Yaponiya: Territorial'naya Problema, Poisk Resheniya.* Moscow: Respublika, 1992.

Zakharova, Galina. *Politika Yaponii v Man'chzhurii, 1932–1945.* Moscow: Nauka, 1990.

Zilanov, Vyacheslav. "Kuril'skoe Rybolovstvo i Diplomatiya." *Nezavisimaya Gazeta*, August 31, 2000.

———, editor. *Russkie Kurily: Istoriya I Sovremennost'. Sbornik dokumentov po istorii formirovaniya russko-yaponskoi i sovietsko-yaponskoi granitsy.* Moscow: Sampo, 1995.

Newspapers and periodicals

Aera
Asahi Shimbun
Asian Survey
Bungei Shunju
Chicago Tribune
Chuuō Kōron
Comparative Connections
Current Digest of the Post-Soviet Press
Economist
Far Eastern Economic Review
Financial Times
Foreign Affairs
Foreign Broadcast Information Service (FBIS)
Foresight Magazine
Gaikō Forum
Hokkaido Shimbun
International Affairs
International Security
Izvestia
Jamestown Monitor
Japan Digest
Japan Echo
Japan Times
Journal of Asian Studies
Journal of Japanese Studies
Journal of Northeast Asian Studies
Journal of Public and International Affairs
Kommersant'
Komsomol'skaya Pravda
Los Angeles Times
Mainichi Shimbun
Mezhdunarodnaya Zhizn'
Mirovaya Ekonomika i Mezhdunarodnaya Otnosheniya (MeiMO)
Moscow Times
Moscow Tribune
Moskovskie Novosti
Moskovskii Komsomolets
National Interest

Nezavisimaya Gazeta
New York Times
Nihon Keizai Shimbun
Nikkei Shimbun
Novoe Vremya
Novoya Gazeta
Obshchaya Gazeta
Orbis
Pacific Affairs
Pacific Review
Pravda
Problems of Post-Communism
Problemy Dal'nego Vostoka
Radio Free Europe and Radio Liberty Research Report (RFE/RL)
Renmin Ribao
Rossikaya Gazeta
Russia Journal
Sankei Shimbun
Segodnya
Seiron
Sekai
Sekai Shuuhō
Shin Bōei Ronshu
Sovietskaya Rossia
Survival
Tokyo Shimbun
Versiya
Vremya Novostei
Wall Street Journal
Washington Post
Yomiuri Shimbun
Znakom'tes' – Yaponiya

Interviews

(Positions listed are those at the time of the interview)

Japan

Fuse Hiroyuke, *Yomiuri Shimbun*
Hakamada Shigeki, Aoyama Gakuin University
Hara Teruyuki, Slavic Research Center, Hokkaido University
Hasegawa Tsuyoshi, University of California at Santa Barbara
Hayashi Tadayuki, Slavic Research Center, Hokkaido University
Hirose Testuya, Ministry of Foreign Affairs
Hitachi Koji, Nissho Iwai Corporation
Ichiyanagi Mitsuhiro, Daiwa Far East and Eastern Siberia Fund
Ikeda Motohiro, *Nihon Keizai Shimbun*
Itō Koichi, Ministry of Foreign Affairs
Kawato Akio, Ministry of Foreign Affairs

Kimura Hiroshi, International Research Institute of Japanese Culture
Kodera Jirō, Ministry of Foreign Affairs
Komachi Kyōji, Ministry of Foreign Affairs
Kondo Takeshi, Itochu Corporation
Matsuda Kuninori, Ministry of Foreign Affairs
Minagawa Shugo, Slavic Research Center, Hokkaido University
Mori Shinjiro, *Asahi Shimbun*
Murakami Takashi, Slavic Research Center, Hokkaido University
Nishimura Mutsuyoshi, Ministry of Foreign Affairs
Nishimura Yoichi, *Asahi Shimbun*
Nishimura Yoshiaki, Hitotsubashi University
Nozaki Ryoichi, Itochu Corporation
Oda Takeshi, *Nihon Keizai Shimbun*
Sakanaka Tomohisa, President, Research Institute for Peace and Security
Shimotomai Nobuo, Hosei University
Shinkura Hiromasa, *Hokkaido Shimbun*
Shinoda Kenji, Ministry of Foreign Affairs
Soejima Hideki, *Asahi Shimbun*
Suetsugu Ichiro, Director, Council on National Security Problems
Sugawara Ikuro, Japan National Oil Corporation
Suzuki Osamu, Director, Japan Center Vladivostok
Tamba Minoru, Ministry of Foreign Affairs
Tōgō Kazuhiko, Ministry of Foreign Affairs
Wada Haruki, Tokyo University
Watanabe Akio, Aoyama Gakuin University
Yamaya Kenryo, *Hokkaido Shimbun*
Yokote Shinji, Keio University

Russia

Amirov, Vyacheslav, Institute of World Economy and International Relations (RAN)
Blagovolin, Sergei, Institute of World Economy and International Relations (RAN)
Bogaturov, Aleksei, Institute of the USA and Canada (RAN)
Galuzin, Mikhail, Ministry of Foreign Affairs
Kastyornov, Sergei, Ministry of Foreign Affairs
Kunadze, Georgii, Institute of World Economy and International Relations (RAN)
Latkin, Aleksandr, International Institute of Economic Studies and Prognosing, Vladivostok
Lukin, Vladimir, Deputy Speaker of the Duma of the Russian Federation
Malakhov, Vladimir, Assistant to the Deputy Governor of Primorskii Krai
Martynov, Vladen, Institute of World Economy and International Relations (RAN)
Nosov, Mikhail, Institute of the USA and Canada (RAN)
Pavlyatenko, Viktor, Institute of Far Eastern Studies (RAN)
Pushkov, Vladimir, Ministry of Foreign Affairs
Ramzes, Vadim, Institute of World Economy and International Relations (RAN)
Sarkisov, Konstantin, Hosei University (Japan)
Serov, Oleg, Ministry of Foreign Affairs
Sherstyuk, Sergei, Chief of Staff, Duma of Primorskii Krai
Simonia, Nodari, Institute of World Economy and International Relations (RAN)

Troush, Sergei, Institute of the USA and Canada (RAN)
Vishnevsky, Nikolai, Representative of the Kuril District in Yuzhno-Sakhalinsk
Yakovlev, Aleksandr, International Democracy Foundation
Yelizariev, Vitaly, Department of Foreign Economic Relations, Administration of Sakhalin Region
Zaitsev, Valerii, Institute of World Economy and International Relations (RAN)

Index